LONDON MATHEMATICAL SOCIETY LECTURE NOTE SERIES

Managing Editor: Professor M. Reid, Mathematics Institute, University of Warwick, Coventry CV4 7AL,
United Kingdom

The titles below are available from booksellers, or from Cambridge University Press at
www.cambridge.org/mathematics

London Mathematical Society Lecture Notes Series: 389

Random Fields on the Sphere

Representation, Limit Theorems and Cosmological Applications

DOMENICO MARINUCCI

University of Rome "Tor Vergata"

GIOVANNI PECCATI

University of Luxembourg

CAMBRIDGE
UNIVERSITY PRESS

CAMBRIDGE
UNIVERSITY PRESS

University Printing House, Cambridge CB2 8BS, United Kingdom

One Liberty Plaza, 20th Floor, New York, NY 10006, USA

477 Williamstown Road, Port Melbourne, VIC 3207, Australia

4843/24, 2nd Floor, Ansari Road, Daryaganj, Delhi - 110002, India

79 Anson Road, #06-04/06, Singapore 079906

Cambridge University Press is part of the University of Cambridge.

It furthers the University's mission by disseminating knowledge in the pursuit of education, learning and research at the highest international levels of excellence.

www.cambridge.org
Information on this title: www.cambridge.org/9780521175616

© D. Marinucci and G. Peccati 2011

This publication is in copyright. Subject to statutory exception and to the provisions of relevant collective licensing agreements, no reproduction of any part may take place without the written permission of Cambridge University Press.

First published 2011

A catalogue record for this publication is available from the British Library

ISBN 978-0-521-17561-6 Paperback

Cambridge University Press has no responsibility for the persistence or accuracy of URLs for external or third-party internet websites referred to in this publication, and does not guarantee that any content on such websites is, or will remain, accurate or appropriate.

A Luisa, Lorenzo e Luca
A Ieva ed Emma Elīza

Contents

Preface

Several people have worked with the authors in the last years to develop the material covered in this monograph. In particular, following the order of the book, we wish to mention Ivan Nourdin and David Nualart for recent developments on the generalized method of moments for Gaussian subordinated processes, and relationships with Stein's method; Jean-Renaud Pycke for spectral representations of isotropic random fields; Paolo Baldi for the characterizations of spherical harmonic coefficients under isotropy; Paolo Baldi, Gerard Kerkyacharian and Dominique Picard for the stochastic analysis of standard needlets, and Xiaohong Lan for the needlets bispectrum; Daryl Geller (who introduced Mexican needlets with Azita Mayeli) for the extension of the needlet paradigm to random sections of spin fiber bundles, We learned a lot from discussions with Mauro Piccioni and Igor Wigman, who have also provided very useful comments on an earlier draft, as did PhD students Mirko D'Ovidio and Claudio Durastanti.

The material of this book is strongly motivated by Cosmological applications, and it has benefited enormously from a decade-long interaction of the first author with physicists providing insights. suggestions, and applications to real data: we mention in particular (in alphabetic order) Amedeo Balbi, Paolo Cabella, Giancarlo de Gasperis, Frode Hansen, Michele Liguori, Sabino Matarrese, Paolo Natoli, Davide Pietrobon, Gianluca Polenta, Oystein Rudjord, Sandro Scodeller and Nicola Vittorio. Frode Hansen is to be thanked also for some insightful comments on the CMB description parts.

Our greatest thanks go to our families, to which the book is dedicated.

1

Introduction

1.1 Overview

The purpose of this monograph is to discuss recent developments in the analysis of isotropic spherical random fields, with a view towards applications in cosmology. We shall be concerned in particular with the interplay among three leading themes, namely:

- the connection between isotropy, representation of compact groups and spectral analysis for random fields, including the characterization of polyspectra and their statistical estimation;
- the interplay between Gaussianity, Gaussian subordination, nonlinear statistics, and recent developments in the methods of moments and diagram formulae to establish weak convergence results;
- the various facets of high-resolution asymptotics, including the high-frequency behaviour of Gaussian subordinated random fields and asymptotic statistics in the high-frequency sense.

These basic themes will be exploited in a number of different applications, some with a probabilistic flavour and others with a more statistical focus.

On the probabilistic side, we mention, for instance, a systematic study of the connections between Gaussianity, independence of Fourier coefficients, ergodicity and high-frequency asymptotics of Gaussian subordinated fields. We will also discuss at length the role of isotropy in constraining the behaviour of angular power spectra and polyspectra, thus providing a characterization of the dependence structure of random fields, as well as a sound mathematical background for establishing meaningful estimation procedures.

Among the statistical applications, we mention the estimation of angular power spectra and polyspectra, and their use to implement tests for Gaussianity, isotropy and asymmetry. A common thread of the statistical and probabilistic results will be the derivation of asymptotic results in the high-frequency sense – a concept whose rationale we shall illustrate below. In this same framework, another contribution of these lecture notes is the analysis of the stochastic properties of a new class of spherical wavelets, called needlets. For instance, we shall discuss asymptotic uncorrelation of needlets coefficients (again in the high-frequency sense) and show how this property can be exploited to derive a number of statistical procedures, related e.g. to the estimation of angular power spectra and polyspectra, as well as to the already mentioned tests of asymmetry and Gaussianity.

The book is completed by the analysis of random fields which do not take ordinary scalar values, but have a more complex geometrical structure – i.e., the so-called spin random fields. These fields can again be modeled and interpreted in terms of group representation concepts, thus maintaining a consistent connection with the leading themes of this monograph.

All our stochastic results are strongly motivated by applications, and in fact we will refer to several papers where the previous concepts have been successfully applied to the analysis of astrophysical data. Indeed, although spherical random fields may arise in a number of different circumstances, including medical imaging, atmospheric sciences, geophysics, solar physics and many others, our motivating rationale has been very much influenced by cosmological applications, in particular in connection with the analysis of the Cosmic Microwave Background (CMB) radiation data. Thus, although we believe that the results discussed here have a general mathematical interest and may find applications to different fields, we wish to provide in this Introduction an informal presentation of the foundations of CMB data analysis, which will help the reader to better understand the relevance and motivations of our work.

1.2 Cosmological motivations

Cosmology is now developing into a mature observational science, with a vast array of different experiments yielding datasets of astonishing magnitude, and nearly as great challenges for theoretical and applied statisticians. Datasets are now available on a large variety of different phenomena, but the leading part in cosmological research has been played over the past 20 years by the analysis of the Cosmic Microwave Background (CMB) radiation, an area which has already led to Nobel Prizes for Physics in 1978 and in 2006.

The nature of CMB can be loosely explained as follows (see for instance Dodelson [51] for a textbook reference and Balbi [9] for a more popular account). According to the standard cosmological model, the Universe that we currently observe has originated approximately 13.7 billion years ago in a very hot and dense state, in what is universally known as the *Big Bang*. Neglecting fundamental physics in the first fractions of seconds, we can naively imagine a fluid state where matter was completely ionized, i.e. the kinetic energy of electrons was much stronger than the attractive potential of the protons, so that no stable atomic nuclei could form. It is a consequence of quantum principles that a free electron has a much larger *cross-section* for interaction with photons than when it is bound in an atomic nucleus. Loosely speaking, it follows that the probability of interactions between photons and electrons was so high that the mean free path of the former was very short and the Universe was consequently "opaque". As the Universe expanded, the mean energy content decreased, meaning that the fluid composed of matter and radiation cooled down. The mean kinetic energy of the electrons thus decreased until it reached a critical value where it was no longer sufficient to escape the attractive electric potential of protons. Stable (and neutral) hydrogen atoms were then formed. This change of state occurred at the so-called "age of recombination", which is currently reckoned to have taken place 3.7×10^5 years after the Big Bang, i.e. when the Universe had only the 0.003% of its current age. At the age of recombination, the probability of interactions became so small that, as a first approximation, photons could start to travel freely. Neglecting second order effects, we can assume they had no further interaction up to the present epoch.

The remarkable consequence of this mechanism is that the Universe is embedded in a uniform radiation that provides pictures of its state nearly 1.37×10^{10} years ago; this is exactly the above-mentioned CMB radiation. The existence of CMB was predicted by G. Gamow in the forties; it was later discovered fortuitously by Penzias and Wilson in 1965 – for this discovery they earned the Nobel Prize for Physics in 1978. For several years, further experiments were only able to confirm the existence of the radiation, and to test its adherence to the Planckian curve of blackbody emission, as predicted by theorists. A major breakthrough occurred with NASA satellite mission *COBE*, which was launched in 1989 and publicly released the first full-sky maps of radiation in 1992; for this experiment Smoot and Mather earned the Nobel Prize for Physics in 2006 [187]. In the Figure 1.1 below we present a CMB map (the so-called ILC, Internal Linear Combination map from *WMAP* data), see Bennett et al. [21] for more details on its construction.

Full-sky maps as ILC are constructed by weighted linear interpolation of the observations across the different channels, but they are not considered fully reliable for data analysis, especially at high frequencies. Indeed, some parts of the sky are masked by the presence of foreground emission by the Milky Way and other *foreground* sources. This is a major issue for data analysis, which we shall deal with extensively in the final chapters of this monograph. In Figure 1.2, we show the map constructed from (the Q band of) *WMAP* data, where approximately 20% of the sky has been deleted to get rid of foreground emission; the missing region around the galactic plane is immediately evident.

The nature of these maps deserves further explanation. CMB is distributed in a remarkably uniform fashion over the sky, with deviations of the order of 10^{-4} with respect to the mean value (corresponding to 2.731 Kelvin). The attempts to understand this uniformity have led to very important developments in cosmology, primarily the inflationary scenario, which now dominates the theoretical landscape (see [51]). Even more important, though, are the tiny fluctuations around this mean value, which provided the seeds for stars and galaxies to form out of gravitational instability. Measuring and understanding the nature of these fluctuations has then been the core of an enormous amount of experimental and theoretical research. In particular, their stochastic properties yield a goldmine of information about a number of extremely important issues in astrophysics and cosmology, as well as many problems at the frontier of fundamental physics.

To mention just a few of these problems, we recall the issues concerning the matter content of the Universe, its global geometry, the existence and nature of (non-baryonic) *dark matter*, the existence and nature of *dark energy*, which is related to Einstein's cosmological constant, and many others. The next experimental landmark in CMB analysis followed in 2000, when two balloon-borne experiments, *BOOMERANG* and *MAXIMA* (see de Bernardis et al. [44] and Hanany et al. [91]) yielded the first high-resolution observations on small patches of the sky (less than 10° squared). These observations led to the first constraints on the global geometry of the Universe, which was found to be (very close to) Euclidean. Another major breakthrough followed with the 2003-2010 data releases from the NASA satellite experiment *WMAP*, whose observations are publicly available on the web site http://lambda.gsfc.nasa.gov/, see for instance Bennett et al. [22]. Such data releases yielded measurement of the correlation structure of the random field up to a resolution of about 0.22 degrees, i.e., approximately 30 times better than *COBE* (7-10 degrees). Finally, a major boost in data analysis is expected from the ESA satellite mission *Planck*, which was launched on May 14, 2009; data releases for the public are due in 2011-2015. *Planck* is planned to provide datasets of nearly 5×10^{10} observations, and this will allow to settle many open questions with CMB temperature data. New challenging question are expected to arise at a faster and faster pace over the next decades; for instance, *Planck* will provide high quality *polarization* data, which will set the agenda for the experiments to come. Polarization data can be viewed as spin, or tensor-valued, rather than scalar, observations.

Internal Linear Combination Map

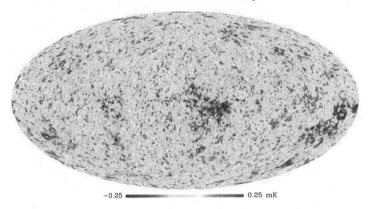

−0.25 ━━━━━━━━━━━━ 0.25 mK

Figure 1.1 The Internal Linear Combination map from NASA-*WMAP* data

As such, they require an entirely new field of statistical research, which is still in its infancy but seems very promising for future developments - in particular, as we shall discuss in the final chapter of this book, polarization data lead to the analysis of spin random fields, which in a loose sense can be viewed as random structures that at each point take as a value a random curve (e.g. an ellipse) rather than a number. Quite interestingly, this same mathematical framework covers other astrophysical applications whose analysis is growing rapidly, such as weak gravitational lensing data [28, 115].

1.3 Mathematical framework

As introduced in the previous discussion, this monograph deals with mathematical topics at the intersection of probability, mathematical statistics and harmonic analysis on groups and homogeneous spaces. In this respect, we believe that two features of our approach deserve a special mention.

On one hand, we will provide a detailed analysis of isotropic (that is, rotationally invariant) spherical random fields by using notions from group representation theory. This approach will unveil some elegant mathematical structures, based e.g. on the properties of Clebsch-Gordan intertwining matrices and Gaunt integrals, and also places our discussion into the wider framework of

Q band map

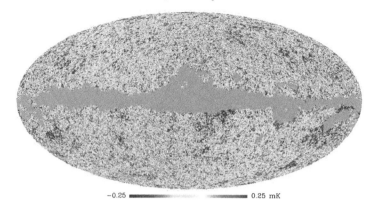

−0.25 ▬▬▬▬▬▬   ▬▬▬▬ 0.25 mK

Figure 1.2 CMB radiation from the Q band of NASA-*WMAP* data

the algebraic characterization of stochastic processes defined on homogeneous
spaces. See e.g. Diaconis [49], Gangolli [71], Guivarc'h, Keane and Roynette
[89], among others, for examples of several fascinating interactions between
probability, statistics and group theoretical concepts. Another crucial point is
that the use of isotropy is directly related to the CMB analysis. In particular,
following the standard literature, we can interpret the CMB radiation as a single realization of an isotropic, finite variance spherical random field. Note that,
since the CMB is an image of the early universe, the underlying isotropy can
be seen as a consequence of the so-called "Einstein cosmological principle",
roughly stating that, on sufficiently large distance scales, the Universe looks
identical everywhere in space (homogeneity) and appears the same in every
direction (isotropy). One should also mention that the prevailing models for
early Big Bang dynamics (the so-called *inflationary scenario*) predict the random fluctuations to be Gaussian, or polynomial (quadratic or cubic) functions
of an underlying Gaussian field (see for instance [18, 19, 20]). This point jus-
tifies the fact that, in the discussion to follow, we will very often work with
spherical random fields having a Gaussian, or Gaussian-subordinated, structure.

On the other hand, we shall systematically derive probabilistic limit theorems in the high-frequency (or high-resolution) sense. Roughly speaking, this
means that (i) we are going to decompose a given spherical random field in
terms of some deterministic basis whose elements are indexed by frequencies

(i.e., *spherical harmonics*) and, (ii) we shall study the asymptotic properties (e.g. the asymptotic Gaussianity) of some relevant statistics of the frequency components by letting the frequency diverge to infinity. We will point out that this type of limit procedures yields a number of deep and difficult mathematical challenges. It is interesting to note that these difficulties arise even in the case of elementary statistics associated with random fields with a relatively simple structure (e.g., rotationally-invariant Gaussian fields). For instance, we will see in Chapter 9 that, in order to prove high-resolution CLTs for the angular bispectra of isotropic Gaussian fields, one must perform a subtle combinatorial analysis of the moments and cumulants associated with large sums of three-products of independent and complex-valued harmonic coefficients. More generally, the high-frequency asymptotic procedures undertaken in this monograph always require us to characterize the limits of linear combinations of random summands with a very complex dependence structure, which is both determined by the isotropic assumption and by the features of the underlying homogeneous space. To some extent, this situation is similar to those encountered in the analysis of large random matrices belonging to some special ensembles (for instance, the Wigner or Ginibre families). Indeed, random matrices from these ensembles can be easily described in terms of collections of i.i.d. (independent and identically distributed) random variables, but their asymptotic spectral analysis requires us to control and assess larger and larger partial sums of highly-correlated random eigenvalues (see e.g. Guionnet [88]). Finally, we stress once again that our asymptotic results are strongly motivated by physical applications. To understand this point, we observe that cosmology is (in some sense) a science based upon a single observation (i.e. our Universe) which is observed at greater and greater degrees of resolution. As we shall discuss below, high-frequency procedures provide the proper framework to describe the environment faced by applied researchers: as experiments grow more and more sophisticated, smaller and smaller scales (and hence higher and higher harmonic components) become available for data analysis. For example, *COBE* data included observations up to multipoles (equivalent to frequencies, to be defined later) in the order of a few dozens, current data from *WMAP* have raised this limit to a few hundreds, whereas the expected bound from *Planck* is in the order of a few thousands.

1.4 Plan of the book

The plan of the book is as follows. In Chapter 2, we review some basic facts from group representation theory. Part of this material is rather technical, and well beyond the classical background of most researchers working in stochastics; however we feel it is a necessary companion for the understanding of much of the ideas to follow. In fact, group representation results play a basic role in most of the material discussed in the book, from the derivation of spectral representation theorems to the analysis of high-resolution asymptotics for Gaussian subordinated random fields, from the statistical estimation of angular power spectra and bispectra (whose definition is entirely based on group theoretic ideas) to the introduction of spin random fields. The core of this chapter is the celebrated Peter-Weyl Theorem from abstract harmonic analysis, which allows generalized Fourier expansions to take place for functions defined on arbitrary compact groups. Our discussion follows well-known textbooks such as, for instance, Faraut [63] and Vilenkin and Klimyk [197].

In Chapter 3, we specialize the previous results to representation theory for the group of rotations $SO(3)$, which will allow us to introduce Fourier analysis on the sphere as well as the most important instruments for the chapters to follow, such as Wigner's D matrices, spherical harmonics, and Clebsch-Gordan coefficients. These are the fundamental tools used throughout the book in order to understand the role of rotations and isotropy, to establish asymptotic results and to analyze nonlinear transforms and higher order statistics.

Chapter 4 is also concerned with background results, but with a much stronger probabilistic flavour: in particular, we discuss at length recent developments in the analysis of central limit theorems for Gaussian subordinated sequences. We present the classical diagram formula for the analysis of nonlinear transforms of Gaussian processes, and we discuss at length much more recent results on the characterization of asymptotic Gaussianity by the use of so-called *contractions* (see Nualart and Peccati [152]). We also make use of some recent developments by Nourdin and Peccati [148] connecting Malliavin's calculus to the so-called Stein's method for probabilistic approximations. These techniques greatly simplify the derivation of Central Limit Theorems results and their use is widespread throughout the book. In this same chapter, the graphical method for dealing with convolutions of Clebsch-Gordan coefficients is also recalled (see Varshalovich, Moskalev and Khersonskii [195] for a rather exhaustive account of available results).

In Chapter 5 we introduce spectral representation results, again in connection with the abstract harmonic analysis setting of the previous chapter. These representation results can also be derived via more probabilistic techniques, and we discuss the interaction between the different approaches. Note that the group theoretic point of view is necessary in order to make some of the further developments more transparent.

Chapter 6 is more directly concerned with isotropic spherical random fields and their characterization. We start by discussing results from Baldi and Marinucci [10], which prove that the random harmonic coefficients of a spherical random field are independent if and only if the field is Gaussian. We then characterize the higher order moments of these coefficients in the general (not necessarily Gaussian) isotropic case by means of Clebsch-Gordan coefficients, which allow us to evaluate multiple integrals of spherical harmonics in a neat and sharp way. We can then introduce the definition and main properties of angular polyspectra, a fundamental tool in the statistical analysis of the chapters to follow (see Marinucci and Peccati [136]).

In Chapter 7 we present our first results on high-frequency asymptotics. In particular, the question we try to address is the following: given a nonlinear transform of a Gaussian field, what can be said about the limiting distribution of its Fourier components? This question was investigated by the authors in [134, 135] and is again strongly motivated by physical applications. Indeed, as discussed in the book, a major theme of cosmological research is related to the investigation of non-Gaussian features in CMB radiation. These non-Gaussian features typically take the form of nonlinear transformation of a subordinating Gaussian field, (see [18, 19, 20, 67, 68, 117, 184]); the exact form of the nonlinear terms depend on different scenarios for the Big Bang dynamics, i.e. different versions of the celebrated inflationary model first introduced by Guth in 1981 (see again [51]). As mentioned above, statistical inference is restricted to a single realization of the Universe, observed at higher and higher frequencies as the experiments get more sophisticated. The issue is then, whether the components at these frequencies allow for the consistent estimation of non-Gaussian components. We provide some answers to this question, in terms of conditions on the decay of the angular power spectra and random walks on the representations of the group of rotations $SO(3)$.

In Chapter 8 we dwell more directly into mathematical statistics issues. We start by reviewing very briefly some background issues on the construction of CMB maps and the analysis of instrumental noise. We then focus more directly on angular power spectra estimation, bias testing and bias correction, and we prove consistency and weak convergence under Gaussianity assumptions, in the idealistic circumstances of fully observed CMB maps. We then consider angular power spectra estimation under non-Gaussianity. Our main results (which follow [137]) can be connected in a surprising way to those of Chapter 7. In particular, the possibility of consistently estimating (as always, in the high-frequency sense) the angular power spectrum of the random fields turns out to be closely related to the discussion in Chapter 7 on asymptotic Gaussianity. Loosely speaking, we show in fact that under broad conditions, ergodicity of the empirical spectral measure (i.e., high-frequency consistency of the angular power spectrum) and asymptotic Gaussianity are equivalent. These results suggest a more general connection between ergodicity and Gaussianity, in a high-resolution sense.

Statistical analysis is further developed in Chapter 9, where the results on the characterization of higher moments under isotropy are exploited to introduce the (sample) bispectrum and its asymptotic behaviour. We review recent results on CLT and FCLT for these statistics (compare [131], [132]), with a heavy use of the diagram formula machinery which was introduced in previous chapters. The definition and asymptotic analysis of the bispectrum is entirely based on the properties of Clebsch-Gordan (or Wigner's) coefficients, so that the group-theory point of view proves once again to be mandatory in this field. We discuss some remarkable statistical features such as consistency, i.e. the divergence to infinity of bispectrum-based statistics under non-Gaussianity, even in the presence of a single realization of the CMB field.

In Chapter 10 we start to discuss needlets and their stochastic properties. Needlets are a new form of spherical wavelets which were introduced into the mathematical literature by Narcowich, Petrushev and Ward [143, 144]; the derivation of their stochastic properties is first due to Baldi, Kerkyacharian, Marinucci and Picard [13, 14] (see also [123, 124, 140]), while the first applications to CMB data are due to [162, 139], with many further developments provided for instance by [46, 65, 163, 164, 165, 175, 176], among others. We also discuss the wider class of Mexican needlets, introduced by Geller and Mayeli in [76, 77, 78], and applied to CMB in [179]. Needlets enjoy a number of important analytical properties that we shall discuss in some detail; from the perspective of this book, however, the key feature that makes them valuable for stochastic analysis is the asymptotic uncorrelation of needlets coefficients, in the high-resolution sense. In a loose sense, the uncorrelation property is stating that in the Gaussian case we can derive a growing array of asymptotically i.i.d. coefficients (after normalization) out of a single spherical random field. This of course, opens the way to the implementation of several statistical procedures.

In particular, in Chapter 11 we are concerned with issues such as angular power spectrum estimation, estimation of the bispectrum, testing for asymmetries and directional dependence, and many others, all exploiting the uncorrelation property and the other features of needlets, i.e. their localization in real and harmonic space. The localization properties allow us also to deal most naturally with one of the main features of practical data analysis, i.e. the presence of unobserved regions due to the masking effect of the Milky Way and other foreground sources. We discuss also geometric techniques for the detection of asymmetries in CMB data, a theme which has drawn an enormous amount of attention in the physical literature. Needlets can serve for several other purposes in the broader area of astrophysical data analysis, for instance in [15] they are applied to minimax density estimation for directional data, as those arising from the Cosmic Rays experiments. Since this monograph is devoted to random fields, we do not dwell into these issues here.

Finally, in Chapter 12 we present some preliminary results on so-called spin random fields. Loosely speaking, the latter can be envisaged as random fields such that at each location $x \in S^2$, a random ellipse instead than a random number is observed. This is again strongly motivated by physics, in particular by the so-called polarization component of CMB radiation. From the mathematical point of view, we shall model polarization as random sections of a fiber bundles on the sphere; spectral representations turn out to be again possible by a natural application of the Peter-Weyl Theorem entailing a generalization of the spectral representation for the scalar case (for physical discussions, see [33, 108, 181]). It turns out to be possible to define needlets in this broader framework as well, and to establish again their uncorrelation properties, so that much of the previous statistical procedures can be extended almost unaltered, as done by Geller and Marinucci [74, 75], Geller et al. [73, 79]. As recalled earlier, we note that spin needlet techniques are likely to be applied in several other fields, for instance in the analysis of gravitational lensing data, but we leave these as open issues for further research.

2

Background Results in Representation Theory

2.1 Introduction

A basic result of standard harmonic analysis on $\mathbb{R}$ is that every 2π-periodic function can be expanded into an orthonormal basis of trigonometric polynomials of the type $\{\exp(ikx) : k = 0, \pm 1, \pm 2, ...\}$. This fundamental fact is exploited in several branches of applied mathematics: in particular – as we shall mention more extensively in the subsequent chapters – it is a main staple in the analysis of stationary stochastic processes. A proof can be provided by elementary techniques, requiring little more than standard calculus – see for instance Rudin [174, p. 88]. It is also well-known (see e.g. [63]) that one can deduce a similar result for fields defined on the sphere S^2, where the appropriate orthonormal basis is given by the class of *spherical harmonics.*

The aim of this introductory chapter is to put these two results into the broader framework of abstract harmonic analysis on compact groups.

At first sight, this choice may seem unduly complicated, but it is indeed necessary for the purposes we are aiming at. On one hand, our standpoint allows us to develop quite straightforwardly Fourier analysis on the sphere. On the other hand, group representation ideas are necessary to understand most of the material presented in the subsequent parts of the book, namely: the characterizations of isotropy, the explicit computation of higher order moments and angular polyspectra, the derivation of high-frequency asymptotic properties, and the discussion of spin ideas at the very end of the book. The material to follow, hence, cannot be dispensed with: for this reason, we decided to put it in the very first chapter, rather than confining it to an appendix.

2.2 Preliminary remarks

Fix $T > 0$, and consider a T-periodic complex-valued function $f : \mathbb{R} \to \mathbb{C}$; suppose further that f is locally square-integrable with respect to the Lebesgue measure. Then, a classic result of harmonic analysis states that, on any finite interval, the function f is the L^2 limit of sums of complex exponentials of the type $x \mapsto \exp\left(i\frac{2n\pi}{T}x\right)$, $n \in \mathbb{Z}$. This means that there exist complex numbers $\{c_n : n \in \mathbb{Z}\} \in \ell^2(\mathbb{Z})$ such that, for every $t \in \mathbb{R}$ and $s > 0$,

$$\lim_{N \to \infty} \int_t^{t+s} \left[f(x) - \sum_{n=-N}^{N} c_n \exp\left(i\frac{2n\pi}{T}x\right) \right]^2 \lambda(dx) = 0, \qquad (2.1)$$

where $\lambda(\cdot)$ stands for the Lebesgue measure. This classic approximation result admits a well-known group theoretical interpretation. Indeed, by periodicity, there is no loss of generality in regarding f as a complex-valued mapping defined on the interval $\mathbb{T} = [0, T)$, endowed with the commutative group operation $x \circ y = (x + y)\mathrm{mod}\,(T)$. Note that $\mathbb{T} \simeq \mathbb{R}/(T\mathbb{Z})$, where the symbol "$\simeq$" stands for a group isomorphism (see below for definitions and discussions). The following two facts are noteworthy: **(i)** the normalized Lebesgue measure $\lambda_T(dx) := T^{-1}\lambda(dx)$ is the unique normalized *Haar* measure on $\mathbb{T}$, that is, λ_T is the unique probability measure on $\mathbb{T}$ with the property that, for every bounded f,

$$\int_{\mathbb{T}} f(x)\lambda_T(dx) = \int_{\mathbb{T}} f(x \circ y)\lambda_T(dx), \quad \text{for every } y \in \mathbb{T}, \qquad (2.2)$$

(in other words, λ_T is the unique translation-invariant probability measure on $\mathbb{T}$), and **(ii)** the collection of the mappings $x \mapsto \exp\left(i\frac{2n\pi}{T}x\right)$, $n \in \mathbb{Z}$, coincides with the *dual* of $\mathbb{T}$, that is, with the class of all continuous homomorphisms (known as *characters*) from $\mathbb{T}$ into the unit circle S^1. As a consequence, relation (2.1) can be rephrased as follows: *the characters of $\mathbb{T}$ form an orthonormal basis of the separable complex Hilbert space $L^2(\mathbb{T}, d\lambda_T)$.* A phenomenon analogous to (2.1) holds for much more general Abelian groups: indeed, the fact that characters are a basis of the Haar L^2 space is a feature shared by all locally compact Abelian groups. This result is the starting point of the so-called Pontryagin theory of harmonic analysis on commutative groups (see e.g. the classic Rudin's monograph [173]).

Pontryagin theory can be successfully applied to study random processes whose laws are invariant with respect to the action of some locally compact Abelian group – for instance processes defined on the circle, or on the two-dimensional torus, whose laws are rotationally invariant (see also [134], and Chapter 7 below). However, as anticipated in the Introduction, this monograph is mainly concerned with spectral decompositions of random processes defined on the sphere

$$S^2 = \left\{(x_1, x_2, x_3) \in \mathbb{R}^3 : x_1^2 + x_2^2 + x_3^2 = 1\right\},$$

and whose laws are *isotropic*, in the sense that they are invariant with respect to the action on S^2 of the group of rotations $SO(3)$. Observe that $SO(3)$ is a compact and non-Abelian Lie group. It follows that the Pontryagin approach can no longer be applied in this framework, and we are naturally led to use tools from the more general spectral theory on groups and homogeneous spaces, that is, the so-called *representation theory of compact groups*, revolving around a fundamental representation result, known as the *Peter-Weyl Theorem*.

Remark 2.1 In short (more rigorous definitions and discussions are provided below) let us recall that a n-dimensional *matrix representation* of a general (compact) group G is an application $\pi : G \to \Pi_n$ from the group to an adequate space of $n \times n$ matrices, preserving the group multiplicative structure, i.e.: for all $g_1, g_2 \in G$, $\pi(g_1)\pi(g_2) = \pi(g_1 g_2)$, where $\pi(g_1), \pi(g_2), \pi(g_1 g_2) \in \Pi_n$. As we shall prove below, for compact groups a crucial role is played by *unitary irreducible representations*. A representation π is unitary if $\pi(g)$ is a unitary matrix for every g. A representation is irreducible if one cannot decompose the image of π into two $\pi(G)$-invariant non-trivial subspaces. In the case of commutative groups, it can be proved that the set of characters (that is, continuous homomorphism from the group to S^1) and the class of unitary irreducible representations coincide (in particular, every irreducible representation of a locally compact commutative group is one-dimensional).

Roughly speaking, the Peter-Weyl Theorem states that, given a topological compact group G,

- the complete set of unitary irreducible representations of G is a countable family of $d_\ell \times d_\ell$ matrix-valued functions $\pi^\ell(.)$, $\ell = 1, 2, ...$, with d_ℓ denoting the dimension of π_ℓ;
- the matrix coefficients $\left\{\sqrt{d_\ell}\pi_{uv}^\ell(.)\right\}$ form an orthonormal basis of the space $L^2(G)$ of square-integrable functions with respect to the Haar measure.

It follows that the Peter-Weyl Theorem can be viewed as an extension of (2.1) to a non-commutative framework, with matrix coefficients of irreducible representations generalizing complex exponentials (which, as we said, can be viewed as elements of one-dimensional matrices). Much of the difficulty, richness and interest of studying the harmonic analysis for random and deterministic functions on the sphere is related to the non-commutative nature of the group $SO(3)$. The Peter-Weyl Theorem will hence play a crucial role throughout the book; it is the theoretical cornerstone around which we will define such fundamental objects as *Wigner matrices, spherical harmonics* and *Clebsch-Gordan coefficients*, and it will be the main ingredient to establish the spectral representation for isotropic spherical random fields in the chapters to follow.

In the following sections, we shall provide a succinct overview of the results from representation theory that are necessary for our analysis. We start by discussing some basic background material and definitions in Section 2.3; Section 2.4 presents the general theory of the representations of compact groups, while Section 2.5 focuses on the aforementioned Peter-Weyl Theorem. Note that our presentation is by no means exhaustive. Excellent general references on group representations are e.g. Bump [32], Diaconis [49], Faraut [63], Duistermaat and Kolk [57], Miller [142], Serre [183] and Simon [180].

2.3 Groups: basic definitions

2.3.1 First definitions and examples

We start with the following definition.

Definition 2.2 A **group** G is a set of elements, denoted as $g_1, g_2, ...$, with a binary operation $\circ$ such that the following properties are verified:

- $(g_1 \circ g_2) \in G$ for all $g_1, g_2 \in G$;
- there exists an identity element $e \in G$ such that $g \circ e = g = e \circ g$ for all $g \in G$;
- for all $g \in G$, there exists an inverse element $g^{-1} \in G$ such that $g \circ g^{-1} = g^{-1} \circ g = e$;
- for all $(g_1, g_2, g_3) \in G$, the equality $(g_1 \circ g_2) \circ g_3 = g_1 \circ (g_2 \circ g_3)$ is satisfied.

Note that a group G need not be countable (uncountable groups are indeed the main object of this book).

In words, a group is a (countable or uncountable) set with an associative binary operation which admits an identity element and an inverse. We note that the inverse and the identity elements are always unique. If the binary operation is commutative, i.e. if $g_1 \circ g_2 = g_2 \circ g_1$ for all $g_1, g_2 \in G$, then the group is said to be *Abelian* (or *commutative*). A *subgroup* of a group G is a subset which is itself a group under the same binary operation. The empty sets and G are called *improper* subgroups, while all the others are called *proper*. In the sequel, to simplify the notation, we shall write $g_1 g_2$, rather than $g_1 \circ g_2$, when there is no risk of ambiguity.

Definition 2.3 Consider two groups G and H. A (group) **homomorphism** from G to H is a mapping $\phi : G \to H$ such that, for every $g_1, g_2 \in G$, $\phi(g_1)\phi(g_2) = \phi(g_1 g_2)$. The **kernel** of a homomorphism $\phi : G \to H$ is the collection of all those $g \in G$ such that $\phi(g) = e_H$, where e_H is the identity element of H. A homomorphism $\phi : G \to H$ is called a (group) **isomorphism** if ϕ is one-to-one, and ϕ^{-1} is also a homomorphism (from H to G). If there exists an isomorphism between G and H, we say that G and H are **isomorphic**, and we write $G \simeq H$.

Example 2.4 Let $0 < T_1 < T_2$. For $i = 1, 2$, endow the interval $[0, T_i)$ with the commutative group operation $x \circ y = (x + y) \mathrm{mod}(T_i)$. Then, the mapping $\phi : [0, T_1) \to [0, T_2)$ given by $\phi(x) = x T_2 / T_1$ is a group isomorphism.

Example 2.5 The following are classic examples of groups.

(1) The real line $\mathbb{R}$, with the addition $+$ as group operation. Here, the inverse of x is of course $-x$ and the identity element is 0. This is an infinite Abelian group. Proper subgroups are e.g. the integers and the even integers.
(2) The strictly positive real numbers $\mathbb{R}^+$ with multiplication $\times$ as group operation. The inverse of x is $1/x$, and the identity element is 1. This is another infinite Abelian group. Examples of proper subgroups are the integers and the positive rational numbers.
(3) The set of non-singular $n \times n$ matrices with real coefficients is called the **real general linear group**, and it is denoted by $\mathbf{GL}(n, \mathbb{R})$. The binary operation is ordinary matrix multiplication and the identity element is the identity matrix of order n.
(4) Among the proper subgroups of $\mathbf{GL}(n, \mathbb{R})$, of crucial importance are: (4a) the **special linear group**, i.e. the subgroup of the elements of $\mathbf{GL}(n, \mathbb{R})$ with determinant equal to 1, denoted by

$$SL(n, \mathbb{R}) = \{A : A \in \mathbf{GL}(n, \mathbb{R}) \text{ and } |A| = 1\},$$

(4b) the **orthogonal group**

$$O(n) = \{A : A \in \mathbf{GL}(n, \mathbb{R}) \text{ and } A'A = I_n\},$$

and (4c) the **special orthogonal group**

$$SO(n) = \{A : A \in \mathbf{GL}(n, \mathbb{R}), A'A = I_n \text{ and } |A| = 1\} .$$

The group $SO(n)$ (in particular, $SO(3)$) plays a fundamental role in this book: indeed it admits a geometrical representation as the set of vector rotations in $\mathbb{R}^3$.

(5) The **complex general linear group** $\mathbf{GL}(n, \mathbb{C})$ is the set of all nonsingular $n \times n$ matrices with complex coefficients, again with matrix multiplication as group operation. Very important subgroups of $\mathbf{GL}(n, \mathbb{C})$ are: (5a) the **unitary group**

$$U(n) = \{A : A \in \mathbf{GL}(n, \mathbb{C}) \text{ and } A^*A = AA^* = I_n\},$$

(where A^* indicates the complex conjugate of A) and (5b) the **special unitary group**

$$SU(n) = \{A : A \in \mathbf{GL}(n, \mathbb{C}), A^*A = AA^* = I_n \text{ and } |A| = 1\}.$$

(6) For $n = 1$, $U(1)$ is just the set of complex numbers such that $|z| = 1$ and is usually labelled the **circle group**.

(7) The **cyclic group** Z_n is the set of integers $\{1, 2, ..., n\}$ with addition mod(n) as group operation. The group is isomorphic to the set of the complex numbers

$$C_n = \left\{\exp\left(i2\pi\frac{k}{n}\right) : k = 1, 2, ..., n\right\},$$

where the isomorphism is given by $k \mapsto \exp(i2\pi\frac{k}{n})$ and the group operation on C_n is given by pointwise multiplication.

(8) For $n \geq 1$, the **symmetric group** (usually labelled S_n) is the set of all permutations of a class of n objects (or, equivalently, the collection of all bijections of a class of n objects onto itself).

(9) More generally, given a set X, the class Perm(X) of all bijections of X onto itself. In this case, the group operation is given by function composition.

2.3.2 Cosets and quotients

Definition 2.6 Let G be a group and H be a subgroup. A **left coset** (resp. **right coset**) of H is any set with the form

$$gH = \{gh : h \in H\} \quad (\text{resp. } Hg = \{hg : h \in H\})$$

where $g \in G$. A subgroup H of G is said to be **normal** if $gH = Hg$ (or, equivalently, $gHg^{-1} = H$) for every $g \in G$.

We recall that the collection of all the distinct left cosets (resp. right cosets) of a subgroup H constitutes a partition of G. The class of distinct left cosets will be denoted by G/H. It is well-known that, whenever the subgroup H of

G is normal, then the class G/H of the cosets of H (in this case, right and left cosets are the same) constitutes a group, with respect to the binary operation

$$(gH) \circ (kH) = (g \circ k)H, \quad g, k \in G.$$

In this case, the group G/H is customarily called the *quotient group* of H.

Example 2.7   Let $G = \mathbb{R}$ (with group operation $+$), fix $a > 0$ and let

$$H = a\mathbb{Z} = \{..., -2a, -a, 0, a, 2a, ...\}.$$

We have that $a\mathbb{Z}$ is a normal subgroup of $\mathbb{R}$ (indeed, G is Abelian) and the coset $x(a\mathbb{Z})$ is the class of all those real y such that $y - x$ is an integer multiple of $+a$ or $-a$. Also, the quotient $\mathbb{R}/(a\mathbb{Z})$ is in this case isomorphic to the interval $[0, a)$, endowed with the group operation $x \circ y = (x + y)\mathrm{mod}(a)$.

Further examples are discussed later in the book.

2.3.3 Actions

Let X be any set. A (left) *action A* of a group G on X is a mapping

$$A : G \times X \to X : (g, x) \mapsto A(g, x),$$

verifying the two properties: (i) $A(g, A(h, x)) = A(gh, x)$, for every $g, h \in G$, and (ii) $A(e, x) = x$, where e is the identity element of G. When there is no risk of confusion, we write $g \cdot x$ instead of $A(g, x)$. A group action is said to be *transitive* if, for every $x, y \in X$, there exists $g \in G$ such that $y = g \cdot x$.

A set X admitting an action (resp. a transitive action) of a group G is called a *G-set* (resp. a *G-homogeneous space*).

Example 2.8   Every group G acts transitively on itself, the transitive action being given by $A(g, h) = g \circ h$, where $\circ$ is the group operation of G.

Example 2.9   Let H be a subgroup of G, and let $X = G/H$ be the collection of all left cosets of H. Then, G acts transitively on X. The corresponding action A_H is given by

$$A_H(g, kH) = gkH.$$

For every fixed g, the mapping $A_H(g, \cdot) : G/H \to G/H$ is a bijection, and the mapping $g \mapsto A_H(g, \cdot)$ is a group homomorphism from G to $\mathrm{Perm}(G/H)$.

Example 2.10   Let $X = \mathbb{R}^m$, $m \geq 2$, and take $G = S_m$ (the symmetric group of order m). Then, an action of S_m on X is given by

$$A(\sigma, (x_1, ..., x_m)) = (x_{\sigma(1)}, ..., x_{\sigma(m)}).$$

Plainly, the action A is not transitive.

Example 2.11   Group representations (the topic of the next Section 2.4) are special instances of group actions.

To conclude, we recall a well-known fact showing that every homogeneous space can be identified with a class of cosets. Let X be a G-homogeneous space, with action $A(g, x) = g \cdot x$. For every fixed $x \in X$, we write

$$H_x = \{g \in G : g \cdot x = x\}, \tag{2.3}$$

for the *isotropy subgroup* of x, that is, H_x is the subgroup of G whose action fixes x. Then, the following statement holds:

Proposition 2.12   *For every fixed x, the mapping $\rho_x : G/H_x \to X$ given by*

$$\rho_x(aH_x) = a \cdot x$$

is a bijection intertwining A and A_{H_x} (see Example 2.9), that is

$$A(g, \cdot) \circ \rho_x = \rho_x \circ A_{H_x}(g, \cdot).$$

The content of the previous proposition is customarily written $X \simeq G/H_x$, for every $x \in X$.

2.4 Representations of compact groups

2.4.1 Basic definitions

Throughout the book, we shall often deal with integrals of functions defined over some group. In order for these objects to be properly defined, the group must be endowed with a *topology* (i.e., a family of open sets), thus allowing us to extend to these framework the measure-theoretic notions of continuity, measurability, Lebesgue-Stieltjes integration, and so on. We shall now recall some basic definitions, and refer again to textbooks such as ([63, 197]) for more details and discussion.

A *Hausdorff space* (also called a T_2 space) is a topological space in which distinct points have distinct open neighborhoods. A Hausdorff space is *locally compact* if each of its points has a *compact neighbourhood*, that is, a neighbourhood that can be covered by a finite number of open sets. A *topological group* is a pair $(G, \mathbb{G})$, where G is a group and $\mathbb{G}$ is a topology such that the following three conditions are satisfied: (i) G is a Hausdorff topological space, (ii) the multiplication $G \times G \mapsto G : (g, h) \mapsto gh$ is continuous, and (iii) the inversion $G \mapsto G : g \mapsto g^{-1}$ is continuous. In what follows, when no further specification is given, G will always denote a topological group (the topology $\mathbb{G}$ being implicitly defined) which is also compact, and such that $\mathbb{G}$ has a countable basis (this means that there exists a countable family $\{A_i : i \in \mathbb{N}\} \subset \mathbb{G}$ verifying the property that every $A \in \mathbb{G}$ can be written as $A = \cup \{A_i : A_i \subset A\}$). For such a G, we will denote by $C(G)$ the class of continuous, complex-valued functions on G. The symbol $\mathcal{G}$ denotes the Borel σ-field generated by $\mathbb{G}$.

An immediate consequence (see e.g. [57, Section 10.3]) of the structure imposed on G, is that G always carries a (unique) positive Borel measure, noted dg and known as the (normalized) *Haar measure*, such that $\int_G dg = 1$, and $\forall f \in C(G)$ and $\forall h \in G$

$$\int_G f(g)\,dg = \int_G f\left(g^{-1}\right)dg \tag{2.4}$$

$$\int_G f(hg)\,dg = \int_G f(gh)\,dg = \int_G f(g)\,dg \quad \text{(left and right invariance)}. \tag{2.5}$$

The Haar measure can clearly be viewed as an extension to topological groups of the uniform (or Lebesgue) measure. We shall denote by $L^2(G,\mathcal{G},dg) = L^2(G)$ the Hilbert space of complex-valued functions on G that are square-integrable with respect to dg, endowed with the usual inner product $\langle f_1, f_2 \rangle_G = \int_G f_1(g)\,\overline{f_2(g)}\,dg$. We write $\|\cdot\|_G$ to indicate the norm associated with $\langle \cdot, \cdot \rangle_G$, and we observe that $L^2(G)$ is the completion of $C(G)$ with respect to $\|\cdot\|_G$.

Example 2.13 Any finite group is a topological compact group (just take $\mathbb{G}$ to be the discrete topology, where each element of the group is itself an open set), and the associated normalized Haar measure is given by the normalized counting measure

$$dg = \frac{1}{|G|} \sum_{h \in G} \delta_h(dg)\,,$$

where $\delta_h(\cdot)$ stands for the Dirac mass concentrated at h, and $|G|$ is the cardinality of G.

Example 2.14 Let $G = \mathbb{T} = [0,1)$, endowed with the commutative group operation $xy = (x+y)\mathrm{mod}(1)$. Then, G is a connected compact Abelian group (to prove compactness, use the fact that G is isomorphic to the circle S^1) and we have

$$dg = \lambda(dg)\,,$$

where λ is the restriction of the Lebesgue measure to $\mathbb{T}$.

Example 2.15 Let $G = G_1 \times \cdots \times G_d = \mathbb{T}_1 \times \cdots \times \mathbb{T}_d$, where $\mathbb{T}_i = [0, 1)$, endowed with the following commutative group operation: for $x = (x_1, ..., x_d) \in G$ and $y = (y_1, ..., y_d) \in G$,

$$xy = (x_1 y_1, ..., x_d y_d) = ((x_1 + y_1)\mathrm{mod}\,(1), ..., (x_d + y_d)\mathrm{mod}\,(1)).$$

Then, being a product of topological compact groups, G is itself a compact Abelian group and we have that

$$dg = \lambda\,(dg_1) \cdots \lambda\,(dg_d)\,,$$

where $g = (g_1, ..., g_d)$.

Example 2.16 We anticipate some content of the forthcoming Chapter 3. The group $SU\,(2)$ is the group of all matrices of the type

$$g = \begin{pmatrix} \alpha & \beta \\ -\overline{\beta} & \overline{\alpha} \end{pmatrix}, \text{ with } \alpha, \beta \in \mathbb{C} \text{ and } |\alpha|^2 + |\beta|^2 = 1, \tag{2.6}$$

where the bar indicates complex conjugation. Being homeomorphic to the unit sphere in $\mathbb{C}^2$, $SU\,(2)$ is a connected non-commutative compact group. The Haar measure of $SU\,(2)$ can be characterized as follows. Write A_0 for the subset of $SU(2)$ composed of those g as in (2.6) such that $|\alpha|, |\beta| > 0$. Then, the mapping

$$\Phi : [0, 2\pi) \times (0, \pi) \times [-2\pi, 2\pi) \to A_0 : (\varphi, \vartheta, \psi) \mapsto \Phi(\varphi, \vartheta, \psi),$$

such that $\Phi(\psi, \vartheta, \varphi) = \begin{pmatrix} \alpha & \beta \\ -\overline{\beta} & \overline{\alpha} \end{pmatrix}$ is defined by $\alpha = \cos\frac{\vartheta}{2}e^{i\frac{\varphi+\psi}{2}}$ and $\beta = \sin\frac{\vartheta}{2}e^{i\frac{\varphi-\psi}{2}}$, is one-to-one. Then, for every bounded f on $SU\,(2)$,

$$\int_{SU(2)} f\,(g)\,dg = \int_{A_0} f\,(g)\,dg \tag{2.7}$$

$$= \frac{1}{16\pi^2} \int_{-2\pi}^{2\pi} \int_0^\pi \int_0^{2\pi} f\,(\Phi\,(\varphi, \vartheta, \psi))\sin\vartheta d\varphi d\vartheta d\psi.$$

The groups we deal with in this book will be infinite-dimensional and typically have also the structure of smooth real manifolds, i.e. they can be parametrized by means of local smooth maps. This leads to the important notion of a *Lie group*.

Definition 2.17 A **real Lie group** is a group which is also a finite-dimensional smooth real manifold, and in which the group operations of multiplication and inversion are smooth.

Example 2.18 The groups **GL**$(n, \mathbb{R})$, $SO(n)$ and $O(n)$ are real Lie groups. Other examples of real Lie groups are given by the groups $U(n)$, $SU(n)$ defined above.

In words, the requirement that a Lie group G be a smooth manifold means that for all $g_0 \in G$ we can choose a set of local coordinates A such that the group takes locally the form $A(g)$ for g in a neigborhood of g_0; also, the components of A are smooth, and so are also those of the applications A^{-1} and AB^{-1}, for any other choice of local coordinates B. For Lie groups, the notion of continuous functions is especially simple: indeed, if G is a Lie group, a function $f : G \rightarrow \mathbb{C}$ is continuous at $g \in G$ if and only if it is a continuous function in a local coordinate system at g. Every Lie group is a topological group, while the reverse need not be the case (just consider any finite or discrete topological group).

2.4.2 Group representations and Schur Lemma

A fundamental tool in the analysis of compact groups is the idea of "representation". Loosely speaking, the intuition behind this concept is that the only crucial feature in the structure of a given group is its multiplication table $g_i g_j = g_k$, i.e. the rule according to which two elements in the group are associated to a third. Once the multiplication table of a group G is known, the theory of group representations allows to study G by first translating its structure into the "universal language" of linear applications on vector spaces. The translation is realized in terms of some homomorphism. As we shall see below, the most natural choice for compact groups is to build homomorphisms with some adequate spaces of matrices with complex-valued entries.

More rigorously, let V be a normed vector space over $\mathbb{C}$. A *representation* of G in V is a homomorphism π, from G into **GL** (V) (the set of isomorphisms, i.e. invertible linear applications, of V onto itself), such that the mapping $G \times V \rightarrow V : (g, v) \mapsto \pi(g)(v)$ is continuous. The homomorphism property is equivalent to saying that, for every $g, g_1, g_2 \in G$,

$$\pi\left(g^{-1}\right) = \pi(g)^{-1} \quad \text{and} \quad \pi(g_1)\pi(g_2) = \pi(g_1 g_2) \ .$$

A trivial remark is that the class of the representations of a given group is nonempty: indeed, we can always take $V = \mathbb{C}$ and associate with every $g \in G$ the identity application $\pi(g) z = z$, thus obtaining the so-called *trivial representation* of G. Depending on the notational convenience (and on the context) we shall write (π, V) (instead of π) in order to stress that the underlying vector space is V. The *dimension* of a representation (π, V), equivalently denoted by $\dim \pi$ or d_π, is defined to be the dimension of V.

Given a representation (π, V), we say that a subset $A \subseteq V$ is π-*invariant* if, for every $v \in A$ and for every $g \in G$, one has that $\pi(g)v \in A$. A representation π of G in V is *irreducible* if the only closed π-invariant subspaces of V are $\{0\}$ and V. Note that, if V is finite-dimensional, then (π, V) is irreducible if and only if the only π-invariant subspaces of V are $\{0\}$ and V (the closedness requirement is immaterial): in particular, any one-dimensional representation is irreducible.

Two representations (π, V) and (π', V') are said to be *equivalent*, written $\pi \sim \pi'$, if there exists an isomorphism $A : V \to V'$ (called an *intertwining operator*) such that, for every $g \in G$,

$$A\pi(g) = \pi'(g)A. \tag{2.8}$$

The terminology is well chosen: indeed, the relation "$\sim$" defines an equivalence relation over the class of all representations of G. Given a representation π, we denote by $[\pi]$ the equivalence class associated with $\sim$ and π.

Remark 2.19 (Matrix representations.) If $V = \mathbb{C}^d$, $d < \infty$, then a representation (π, V) of G is just a homomorphism between G and $\mathbf{GL}(d, \mathbb{C})$. Recall that the (general linear) group $\mathbf{GL}(d, \mathbb{C})$ has already been defined as the group of all $d \times d$ invertible matrices with complex entries. In this case, one says that π is a *matrix representation*.

Note that, if (π, V) is such that $\dim V = d < \infty$, then (π, V) is equivalent to a matrix representation. To see this, let $(v_1, ..., v_d)$ and $(w_1, ..., w_d)$ be, respectively, a basis of V and a basis of $\mathbb{C}^d$, and define the isomorphism $A : V \to \mathbb{C}^d$ via the relation

$$A(\lambda_1 v_1 + \cdots + \lambda_d v_d) = \lambda_1 w_1 + \cdots + \lambda_d w_d$$

(in particular, $A(v_i) = w_i$). Now, for every $g \in G$, let $\{\pi_{ij}(g) : 1 \le i, j \le d\}$ be the matrix of $\pi(g)$ with respect to the basis $(v_1, ..., v_d)$, that is $\pi(g)v_j = \sum_i \pi_{ij}(g)v_i$. Then, it is easily seen that the two representations (π, V) and $(\widetilde{\pi}, \mathbb{C}^d)$, where $\widetilde{\pi}(g) = \{\pi_{ij}(g)\}$, are equivalent (the intertwining operator being the just defined isomorphism A).

Example 2.20 The matrix groups $\mathbf{GL}(n, \mathbb{C})$ and $SL(n, \mathbb{C})$ are n-dimensional matrix representations of themselves.

Example 2.21 For the torus $\mathbb{T} = [0, 1)$, endowed with the commutative group operation $xy = (x+y)\bmod(1)$ (see Example 2.14), a family of one-dimensional (and therefore irreducible) matrix representations $\{\exp(i2n\pi\cdot), \mathbb{C}\}$, $n \in \mathbb{Z}$, is provided by the mappings of the type

$$x \mapsto \exp(i2n\pi x).$$

Example 2.22 Consider the square with centre at the origin and vertices $A(-1, 1)$, $B(1, 1)$, $C(1, -1)$, $D(-1, -1)$. There are exactly eight Euclidean transformations that map the square into itself: these transformations can be either written in the form

$$\begin{bmatrix} A & B & C & D \\ A & B & C & D \end{bmatrix}, \dots, \begin{bmatrix} A & B & C & D \\ A & D & C & B \end{bmatrix},$$

where each element of the first row is mapped into the corresponding element of the second row (for instance, the first bracket represents the identity and the last is a reflection along the diagonal AC), or graphically represented as in Fig. 2.1.

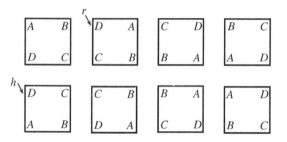

Figure 2.1 The group D_4

We verify immediately that the above eight transformations are indeed a group. This group is isomorphic to a subgroup of S_4 (the set of all permutations of $\{1, 2, 3, 4\}$) and it is customarily called the *dihedral group* of order four, written D_4. Now define (see again Fig. 2.1)

$$r := \begin{bmatrix} A & B & C & D \\ B & C & D & A \end{bmatrix}, \quad h := \begin{bmatrix} A & B & C & D \\ D & C & B & A \end{bmatrix}.$$

Routine computations show that D_4 has the form

$$D_4 = \left\{ e, r, r^2, r^3, h, hr, hr^2, hr^3 \right\}.$$

A matrix representation $\{T, \mathbb{C}^2\}$ is then obtained by taking as a vector space $\mathbb{C}^2$ with orthonormal basis $e_1 = (1, 0)'$, $e_2 = (0, 1)'$, and then by setting

$$T(r) = \begin{pmatrix} 0 & 1 \\ -1 & 0 \end{pmatrix}, \quad T(h) = \begin{pmatrix} 1 & 0 \\ 0 & -1 \end{pmatrix},$$

and

$$T(e) = I_2 = \begin{pmatrix} 1 & 0 \\ 0 & 1 \end{pmatrix}, \quad T(r^2) = \begin{pmatrix} -1 & 0 \\ 0 & -1 \end{pmatrix} = T^2(r),$$

$$T(r^3) = \begin{pmatrix} 0 & -1 \\ 1 & 0 \end{pmatrix} = T^3(r), \quad T(hr) = \begin{pmatrix} 0 & 1 \\ 1 & 0 \end{pmatrix} = T(h)T(r),$$

$$T(hr^2) = \begin{pmatrix} -1 & 0 \\ 0 & 1 \end{pmatrix} = T(h)T^2(r), \quad T(hr^3) = \begin{pmatrix} 0 & -1 \\ -1 & 0 \end{pmatrix} = T(h)T^3(r).$$

We will see below that this representation is irreducible.

Example 2.23 Let T be an n-dimensional matrix representation. Any change of basis $S \in \mathbf{GL}(n, \mathbb{C})$ yields an equivalent representation $T'(g) = ST(g)S^{-1}$. In some sense, then, equivalent matrix representations can be viewed as being associated to the same operator representation, with a different choice of basis.

Now suppose that the representation (π, V) is such that V is a (complex) Hilbert space with respect to some scalar product $\langle \cdot, \cdot \rangle_V$, and write $\|\cdot\|_V = \langle \cdot, \cdot \rangle_V^{1/2}$. We say that π is *unitary* if, for every $g \in G$ and every $u, v \in V$, we have that $\langle u, v \rangle_V = \langle \pi(g)u, \pi(g)v \rangle_V$, which is equivalent to saying that, for every $g \in G$ and every $v \in V$, $\|\pi(g)v\|_V = \|v\|_V$.

Remark 2.24 (1) Let $(\pi, \mathbb{C}^d)$ be a matrix representation. Then, π is unitary with respect to the usual inner product $\langle \cdot, \cdot \rangle_{\mathbb{C}^d}$ if and only if the following holds: for every orthonormal basis $(e_1, ..., e_d)$ of $\mathbb{C}^d$ and every $g \in G$,

$$\langle \pi(g)e_j, \pi(g)e_i \rangle_{\mathbb{C}^d} = \delta_{ij}, \quad 1 \le i, j \le d,$$

where δ_{ij} is the Kronecker symbol. It follows that π is unitary if and only if every matrix of the type $\pi_{ij}(g) = \langle \pi(g)e_j, e_i \rangle_{\mathbb{C}^d}$ is unitary.

(2) Suppose that $(\pi, \mathbb{C}^d)$ and $(\pi', \mathbb{C}^d)$ are two unitary matrix representations, and fix a basis $(e_1, ..., e_d)$ of $\mathbb{C}^d$. Then, π and π' are equivalent if and only if there exists a $d \times d$ unitary matrix A such that, for every $g \in G$, the two matrices

$$\pi_{ij}(g) = \langle \pi(g)e_j, e_i \rangle_{\mathbb{C}^d}, \quad 1 \le i, j \le d,$$
$$\pi'_{ij}(g) = \langle \pi'(g)e_j, e_i \rangle_{\mathbb{C}^d}, \quad 1 \le i, j \le d,$$

are such that

$$A\pi(g) = \pi'(g)A, \quad \text{or (equivalently)} \quad \pi(g) = A^*\pi'(g)A,$$

where $*$ indicates a conjugate transpose. In this case, we say that π and π' are *unitary equivalent*.

(3) The discussion at Point (2) is indeed a specific case of a more general result: *unitary representations are equivalent if and only if they are unitarily equivalent*, in the sense that their intertwining operator is a unitary operator.

The collection of all equivalence classes of irreducible unitary representations is denoted by $\hat{G}$, and is called the *dual* of G. Further characterizations of $\hat{G}$ are provided in the forthcoming discussion. The following result is known as the "averaging trick": it says that, given a representation π on a Hilbert space V, it is always possible to modify $\langle \cdot, \cdot \rangle_V$ in order to obtain a unitary representation.

Proposition 2.25   *Let (π, V) be a representation of the compact group G.*

(1) If V is an Hilbert space, then there exists a scalar product on V such that π is unitary.

(2) If V is finite-dimensional, then there exists a scalar product on V such that π is unitary.

Proof   Let $\langle \cdot, \cdot \rangle_V$ be a scalar product on V, and denote by dg the Haar measure of G. Then, the application

$$(u, v) \longmapsto \int_G \langle \pi(g)v, \pi(g)u \rangle_V \, dg,$$

defines a scalar product over V such that π is unitary. This proves Point 1. The proof of Point 2 follows from the fact that it is always possible to endow a finite-dimensional vector space with a scalar product.   □

Example 2.26   Let G be a compact group and let $L^2(G)$ be the complex Hilbert space of square-integrable functions with respect to the normalized Haar measure dg. Then, a representation of G over $L^2(G)$, known as the *right regular representation*, is given by

$$R(g)f(h) = f(hg), \quad h, g \in G, f \in L^2(G). \tag{2.9}$$

Note that, due to (2.5), the representation R is unitary. Also, for $h, g_1, g_2 \in G$, $f \in L^2(G)$

$$R(g_2)R(g_1)f(h) = R(g_2)f(hg_1)$$
$$= f(hg_2g_1)$$
$$= R(g_2g_1)f(h),$$

which shows the representation is properly defined. Analogously the (unitary) *left regular representation* of G over $L^2(G)$, denoted by L, is given by

$$L(g)f(h) = f\left(g^{-1}h\right), \quad h,g \in G, f \in L^2(G) . \tag{2.10}$$

In general, a crucial (and easily checked) feature of a unitary representation (π, V) is that any π-invariant subspace $V_0 \subset V$ is necessarily such that $V_0^{\perp}$ (that is, the orthogonal of V_0 with respect to $\langle \cdot, \cdot \rangle_V$) is also π-invariant. This elementary fact yields the following decomposition result for finite-dimensional representations.

Proposition 2.27 *Let (π, V) be a finite-dimensional representation of G. Then, there exist spaces $V_1,...,V_k$ ($k < \infty$) such that (i) each V_i is π-invariant, and (ii) we have*

$$V = V_1 \oplus \cdots \oplus V_k, \tag{2.11}$$

that is, V is the direct sum of the V_i's (meaning that every $v \in V$ admits a unique representation of the type $v = \sum_i v_i, v_i \in V_i$).

Remark 2.28 Of course, if (π, V) is irreducible, we can take $k = 2$, $V_1 = V$ and $V_2 = \{0\}$.

Proof of Proposition 2.27 If V is irreducible, there is nothing to prove. If not, endow V with a scalar product making π unitary. If there exists a π-invariant proper subspace $V_1 \subset V$, then the space $V_1^{\perp}$ (the orthogonal with respect to the scalar product) is also π-invariant, and we must check the reducibility of $V_1^{\perp}$. If $V_1^{\perp}$ is irreducible, then the proof is concluded. If not, write $V_1^{\perp} = V_2 + V_2^{\perp}$, where V_2 is irreducible and π-invariant and $V_2^{\perp}$ is the orthogonal of V_2 in $V_1^{\perp}$. Now observe that $V_2^{\perp}$ is also π-invariant, so that the same procedure can be iterated until finding an irreducible orthogonal $V_k^{\perp}$. □

The previous result suggests that irreducible representations are the basic building blocks for all finite-dimensional representations. It is easily seen that the decomposition (2.11) is not unique: to see this, take for instance the case where dim $V > 1$ and $\pi(g)$ equals the identity operator for every g.

The following result is known as *Schur Lemma*. It will be used in several occasions throughout this monograph: for instance, a surprisingly powerful application is detailed in Chapter 6, where we obtain an exhaustive characterization of harmonic coefficients associated with isotropic fields on the sphere.

Theorem 2.29 (Schur Lemma)   (1) *Let (π, V) and (π', V') be two irreducible finite-dimensional representations of a topological compact group G. Let $A : V \to V'$ be a linear mapping such that, for every $g \in G$,*

$$A\pi(g) = \pi'(g)A. \tag{2.12}$$

Then, either $A = 0$, or A is an isomorphism from V to V'.

(2) *Let V be a finite-dimensional vector space over $\mathbb{C}$, let (π, V) be an irreducible representation of the topological compact group G, and let $A : V \to V$ be a linear mapping such that, for every $g \in G$,*

$$A\pi(g) = \pi(g)A. \tag{2.13}$$

Then, there exists $\lambda \in \mathbb{C}$ such that

$$A = \lambda I,$$

where I stands for the identity operator.

(3) *Let V be a finite-dimensional vector space over $\mathbb{C}$, and let (π, V) be a representation of the topological compact group G. Assume that the only linear mappings $A : V \to V$ such that*

$$A\pi(g) = \pi(g)A, \quad \text{for every } g \in G, \tag{2.14}$$

are the multiples of the identity operator. Then, (π, V) is irreducible.

Proof   (1) Relation (2.12) implies that the two spaces

$$\ker(A) = \{v \in V : Av = 0\} \quad \text{and} \quad \text{Im}(A) = \{v' \in V' : \exists v \in V : Av = v'\}$$

are, respectively, π-invariant and π'-invariant. Since π and π' are both irreducible, this implies that either $\ker(A) = V$ and $\text{Im}(A) = \{0\}$, or $\ker(A) = \{0\}$ and $\text{Im}(A) = V'$. In the first case we have that $A = 0$, and in the second case we have that A is an isomorphism. This concludes the proof.

(2) Since A is a linear operator over a finite-dimensional complex vector space, A admits an eigenvalue $\lambda \in \mathbb{C}$. In particular, $A - \lambda I$ is not an isomorphism from V onto itself. On the other hand, relation (2.13) implies that, for every $g \in G$,

$$(A - \lambda I)\pi(g) = \pi(g)(A - \lambda I).$$

It follows from Point 1 in the statement that $A - \lambda I = 0$.

(3) Assume by contradiction that (π, V) is reducible. Then, there exist two proper π-invariant subspaces, say V_1 and V_2, such that $V = V_1 \oplus V_2$. Writing P_1 for the projection operator onto V_1, it is immediately seen that $P_1\pi(g) = \pi(g)P_1$ for every $g \in G$. Since P_1 is not a multiple of the identity, we deduce that (π, V) must be irreducible.   □

Schur Lemma is very useful for the characterization of irreducible representations. For instance, an interesting consequence of Theorem 2.29 is the following.

Proposition 2.30 *Let G be a topological compact commutative group. Then, every irreducible representation of G has dimension one.*

Proof Let (π, V) be an irreducible representation of G, and assume by contradiction that $\dim V > 1$. Fix $h \in G$. Then, for every $g \in G$ and since G is commutative, it follows that

$$\pi(h)\pi(g) = \pi(hg) = \pi(gh) = \pi(g)\pi(h) .$$

This last relation, combined with the second part of Schur Lemma, implies that for every $h \in G$ there exists $\lambda_h \in \mathbb{C}$ such that

$$\pi(h) = \lambda_h I .$$

Hence every one-dimensional subspace $V_0 \subset V$ is π-invariant, thus yielding a contradiction with the initial assumptions on V. It follows that $\dim V = 1$. □

Example 2.31 Let us consider again the group D_4, to check whether the representation $\{T, \mathbb{C}^2\}$ defined in Example 2.22 is irreducible. To do so, consider $T(.)$ to be defined over the complex field, and take a generic complex matrix of the type

$$A = \begin{pmatrix} a & b \\ c & d \end{pmatrix} .$$

We have

$$AT(g) = T(g)A, \text{ for all } g \in D_4$$

if and only if

$$AT(r) = T(r)A, \text{ and } AT(h) = T(h)A .$$

It is easy to check that these conditions imply $a = d$, $b = -c$ and $b = c = 0$, respectively. Thus

$$A = \begin{pmatrix} a & 0 \\ 0 & a \end{pmatrix},$$

and Part (3) of Schur Lemma implies that the representation is irreducible.

2.4.3 Direct sum and tensor product representations

Let (π_i, V_i), $i = 1, ..., k$, be representations of the group G. Then, one can obtain further representations of G in at least two ways: either by taking a *direct sum*, or a *tensor product*. We will now discuss the two procedures.

Direct sums. The *direct sum representation* of G, built from (π_i, V_i), $i = 1, ..., k$, is

$$(\pi_1 \oplus \cdots \oplus \pi_k, V_1 \oplus \cdots \oplus V_k),$$

where $V_1 \oplus \cdots \oplus V_k$ is the direct sum of the spaces V_i and the operator $\pi_1 \oplus \cdots \oplus \pi_k$ is defined by the relation: for every $v = \lambda_1 v_1 + \cdots + \lambda_k v_k \in V_1 \oplus \cdots \oplus V_k$ (with $\lambda_i \in \mathbb{C}$ and $v_i \in V_i$),

$$(\pi_1 \oplus \cdots \oplus \pi_k)(g) v = \lambda_1 \pi_1(g) v_1 + \cdots + \lambda_k \pi_k(g) v_k.$$

Note that $\dim (\pi_1 \oplus \cdots \oplus \pi_k) = \dim \pi_1 + \cdots + \dim \pi_k$. For instance, Proposition 2.27 can be rephrased as follows: every reducible representation (π, V) is the direct sum of k representations of the type (π_i, V_i), where each $V_i \subset V$ is π-invariant and π_i is the restriction of π on V_i. Note that the direct sum of unitary representations is unitary.

Now suppose that each (π_i, V_i), $i = 1, ..., k$, is a matrix representation, with $V_i = \mathbb{C}^{d_i}$ for some $d_i < \infty$. Then $(\pi_1 \oplus \cdots \oplus \pi_k, V_1 \oplus \cdots \oplus V_k)$ is equivalent to the d-dimensional matrix representation (π, V), obtained by taking $d = d_1 + \cdots + d_k$, $V = \mathbb{C}^d$ and $\pi(g)$ equal to the $d \times d$ diagonal block-matrix

$$\pi_1(g) \oplus \cdots \oplus \pi_d(g) = \begin{pmatrix} \pi_1(g) & 0 & 0 & 0 \\ 0 & \pi_2(g) & 0 & 0 \\ 0 & 0 & \ddots & 0 \\ 0 & 0 & 0 & \pi_d(g) \end{pmatrix}.$$

Tensor products. The *tensor product representation* of G, built from (π_i, V_i), $i = 1, ..., k$, is

$$(\pi_1 \otimes \cdots \otimes \pi_k, V_1 \otimes \cdots \otimes V_k),$$

where $V_1 \otimes \cdots \otimes V_k$ is the tensor product of the vector spaces V_i and $\pi_1 \otimes \cdots \otimes \pi_k$ is defined as follows: for every

$$v = v_1 \otimes \cdots \otimes v_k \in V_1 \otimes \cdots \otimes V_k$$

(with $v_i \in V_i$) we have that

$$(\pi_1 \otimes \cdots \otimes \pi_k)(g) v = \pi_1(g) v_1 \otimes \cdots \otimes \pi_k(g) v_k.$$

It is easily seen that $\dim (\pi_1 \otimes \cdots \otimes \pi_k) = \dim \pi_1 \times \cdots \times \dim \pi_k$. If each V_i is a Hilbert space and π_i is unitary, then the tensor product representation is also unitary.

Now suppose that each (π_i, V_i), $i = 1, ..., k$, is a matrix representation, with $V_i = \mathbb{C}^{d_i}$ for some $d_i < \infty$. Then $(\pi_1 \otimes \cdots \otimes \pi_k, V_1 \otimes \cdots \otimes V_k)$ is equivalent to the d-dimensional matrix representation (π, V), obtained by taking $d = d_1 \times \cdots \times d_k$, $V = \mathbb{C}^d$ and $\pi(g)$ equal to the $d \times d$ matrix

$$\pi_1(g) \otimes \cdots \otimes \pi_d(g)$$

given by the *Kronecker product* of the matrices $\pi_i(g)$, $i = 1, ..., k$ (see e.g. [128] or, to a lesser extent, Section 6.4 below). A crucial point, that we will explore in detail in the subsequent discussion, is that the direct product of irreducible representations is, in general, reducible.

2.4.4 Orthogonality relations

In this section, we follow quite closely [63, Ch. VI]. Let (π, V) be a unitary representation of the topological compact group G (in particular, V is a separable complex Hilbert space endowed with a scalar product $\langle \cdot, \cdot \rangle_V$). One of the key objects of this section is the family of operators $K_v : V \to V$, $v \in V$, defined as follows: let $\{e_i : i \geq 1\}$ be an orthonormal basis of V, then, for every $w \in V$

$$K_v w = \sum_{i=1}^{\infty} \left[\int_G \langle w, \pi(g) v \rangle_V \overline{\langle e_i, \pi(g) v \rangle_V} dg \right] e_i := \sum_{i=1}^{\infty} \alpha(v, w, i) e_i. \quad (2.15)$$

Note that $\{\alpha(v, w, i) : i \geq 1\} \in \ell^2$, indeed (by several applications of Cauchy-Schwartz and by the Parseval relation)

$$\sum_{i=1}^{\infty} |\alpha(v, w, i)|^2 \leq \left[\int_G |\langle w, \pi(g) v \rangle_V|^2 dg \right] \times \sum_{i=1}^{\infty} \left[\int_G |\langle e_i, \pi(g) v \rangle_V|^2 dg \right]$$

$$= \left[\int_G |\langle w, \pi(g) v \rangle_V|^2 dg \right] \times \int_G \|\pi(g) v\|_V^2 dg$$

$$= \left[\int_G |\langle w, \pi(g) v \rangle_V|^2 dg \right] \times \|v\|_V^2 \leq \|w\|_V^2 \times \|v\|_V^4 < \infty.$$

The following properties can be verified: (a) for $v = 0$, $K_v = 0$, (b) for every $v \neq 0$, $K_v \neq 0$ and K_v is bounded, compact and self-adjoint, and (c) for every $g \in G$ and $v \in V$, the following intertwining relation holds:

$$K_v \pi(g) = \pi(g) K_v. \quad (2.16)$$

Also, we have the following representation:

$$K_v w = \int_G \langle w, \pi(g) v \rangle_V \left[\sum_{i=1}^{\infty} \langle \pi(g) v, e_i \rangle_V e_i \right] dg = \int_G [\langle w, \pi(g) v \rangle_V \pi(g) v] dg .$$

The following result ensures that unitary representations of compact groups are necessarily finite-dimensional.

Proposition 2.32 *Let (π, V) be unitary representation of the topological compact group G. Then, V contains a π-invariant subspace of finite dimension. It follows that every irreducible representation of G is finite-dimensional.*

Proof Fix $v \neq 0$. Being non-zero, bounded, compact and self-adjoint, the operator K_v admits an eigenvalue which is different from zero. The associated eigenspace is finite-dimensional and (in view of relation (2.16)) π-invariant. This proves that V contains a π-invariant finite-dimensional subspace. $\square$

As anticipated by the title of this section, we now focus on some orthogonality (and isometric) properties of the mappings $g \mapsto \langle \pi(g) v, w \rangle_V$, where (π, V) is a unitary irreducible representation of the topological compact group G. These relations (which are sometimes referred as *Schur's orthogonality relations*) announce the Peter-Weyl theory developed in Section 2.5, and are one of the recurrent themes of this monograph; they can be viewed as extensions to compact groups of well-known results from standard Fourier analysis, such as the Plancherel identity (see [190]). Note that the mapping $g \mapsto \langle \pi(g) v, w \rangle_V$ is an element of $L^2(G)$: indeed, by Cauchy-Schwartz and by the fact that π is unitary,

$$\int_G |\langle \pi(g) v, w \rangle_V|^2 dg \leq \|v\|_V^2 \times \|w\|_V^2 < \infty.$$

Theorem 2.33 *Let (π, V) be a unitary irreducible representation of the compact group G, where V is a Hilbert space with dimension $d_\pi < \infty$ and scalar product $\langle \cdot, \cdot \rangle_V$.*

(1) For every $v, v', w, w' \in V$,

$$\int_G \langle \pi(g) v, w \rangle_V \overline{\langle \pi(g) v', w' \rangle_V} dg = \frac{1}{d_\pi} \langle v, v' \rangle_V \overline{\langle w, w' \rangle_V} . \qquad (2.17)$$

(2) Let (π', V') be another irreducible unitary representation such that π and π' are not equivalent. Then, for every $v, w \in V$ and every $x, y \in V'$

$$\int_G \langle \pi(g) v, w \rangle_V \overline{\langle \pi'(g) x, y \rangle_{V'}} dg = 0 . \qquad (2.18)$$

Proof (1) It is sufficient to prove that, for every $v, w \in V$

$$\int_G |\langle \pi(g)v, w \rangle_V|^2 \, dg = \frac{1}{d_\pi} \|v\|_V^2 \|w\|_V^2 \ .$$

Start by observing that, since (2.16) is in order, according to Theorem 2.29 one has that, for every $v \in V$,

$$K_v = \lambda(v) I \ ,$$

for some $\lambda(v) \in \mathbb{C}$. Now select $v, w \in V$ and observe that

$$\int_G |\langle \pi(g)v, w \rangle_V|^2 \, dg = \langle K_v w, w \rangle_V = \lambda(v) \|w\|_V^2 \ ,$$

and also, by using the fact that π is unitary as well as (2.4),

$$\int_G |\langle \pi(g)v, w \rangle_V|^2 \, dg = \int_G \left| \left\langle v, \pi\left(g^{-1}\right) w \right\rangle_V \right|^2 \, dg = \int_G |\langle \pi(g)w, v \rangle_V|^2 \, dg$$

$$= \langle K_w v, v \rangle_V = \lambda(w) \|v\|_V^2 = \lambda(v) \|w\|_V^2 \ .$$

It follows that $\lambda(v) = \lambda_0 \|v\|_V^2$ for some positive constant λ_0. One now easily verifies that $\lambda_0 = d_\pi^{-1}$, and consequently

$$\int_G |\langle \pi(g)v, w \rangle_V|^2 \, dg = \langle K_v w, w \rangle_V = \frac{1}{d_\pi} \|v\|_V^2 \|w\|_V^2 \ .$$

(2) Fix $v, w \in V$ and $x, y \in V'$ as in the statement, and consider the operators $A, B : V \to V'$

$$Av = \langle v, w \rangle_V \, y \quad \text{and} \quad Bv = \int_G [\pi'\left(g^{-1}\right) A\pi(g)v] \, dg \ .$$

It is now straightforward to check that, for every $h \in G$ and thanks to the "translation invariance" property (2.5) of the Haar measure dg,

$$B\pi(h) = \pi'(h) B \ .$$

As a consequence, since π and π' are not equivalent and by virtue of the Schur Lemma, one has that $B = 0$, entailing in turn that

$$\langle Bv, x \rangle_{V'} = 0 \ .$$

The conclusion is deduced by observing that

$$\langle Bv, x \rangle_{V'} = \int_G \langle \pi(g)v, w \rangle_V \, \overline{\langle \pi'(g)x, y \rangle_{V'}} \, dg \ .$$

$\square$

Example 2.34 Consider two representations of the torus $\mathbb{T} = [0, 1)$,

$$\{\exp(i2m\pi\cdot), \mathbb{C}\} \quad \text{and} \quad \{\exp(i2n\pi\cdot), \mathbb{C}\},$$

and denote them, respectively, by π and π'. The two representations are equivalent if and only if $m = n$; for any $z_1, z_2, z_3, z_4 \in \mathbb{C}$, we have trivially

$$\int_{\mathbb{T}} \langle \pi(g) v, w \rangle_V \overline{\langle \pi'(g) v', w' \rangle_V} dg$$

$$= \int_0^1 \langle \exp(i2m\pi x) z_1, z_2 \rangle_{\mathbb{C}} \overline{\langle \exp(i2n\pi x) z_3, z_4 \rangle_{\mathbb{C}}} dx$$

$$= \int_0^1 \exp(i2(m-n)\pi x) dx \langle z_1, z_2 \rangle_{\mathbb{C}} \overline{\langle z_3, z_4 \rangle_{\mathbb{C}}}$$

$$= \delta_m^n \langle z_1, z_2 \rangle_{\mathbb{C}} \overline{\langle z_3, z_4 \rangle_{\mathbb{C}}} .$$

Remark 2.35 (1) (Orthogonality relationships.) Let (π, V) be a unitary representation of the compact group G and let $(v_1, ..., v_{d_\pi})$ be an orthonormal basis of V. We define $\{\pi_{ij}(g) : 1 \le i, j \le d_\pi\}$ to be the matrix elements of the application $\pi(g) : V \to V$ associated with the basis $(v_1, ..., v_{d_\pi})$, that is,

$$\pi(g) v_j = \sum_{i=1}^{d_\pi} \pi_{ij}(g) v_i, \quad \pi_{ij}(g) = \langle \pi(g) v_j, v_i \rangle_V . \tag{2.19}$$

Then, relation (2.17) implies that, for every $i, j, k, l \in \{1, ..., d_\pi\}$,

$$\langle \pi_{ij}, \pi_{kl} \rangle_{L^2(G)} = \int_G \langle \pi(g) v_j, v_i \rangle_V \overline{\langle \pi(g) v_l, v_k \rangle_V} dg = \frac{1}{d_\pi} \delta_{il} \delta_{ik} , \tag{2.20}$$

where δ_{ab} is the Kronecker symbol, that is: $\delta_{ab} = 1$ if $a = b$ and $\delta_{ab} = 0$ if $a \ne b$. More explicitly, the elements of the matrix representations $\{\pi_{ij}\}$ are mutually orthogonal.

(2) (The spaces M_π.) Let $[\pi] \in \hat{G}$, i.e. $[\pi]$ is an equivalence class of unitary representations of G. We denote by M_π the *space of matrix coefficients* associated with $[\pi]$, that is, M_π is the closed subspace of $L^2(G)$ generated by the mappings of the type $g \mapsto \langle \pi(g) v, w \rangle_V$, where (π, V) is a representative element of $[\pi]$ and $v, w \in V$. Note that the definition of of M_π is well given, since it does not depend on the choice of the representative element of $[\pi]$. Now select a basis $(v_1, ..., v_{d_\pi})$ of V, and define the mapping $g \mapsto \pi_{ij}(g)$, $1 \le i, j \le d_\pi$, according to (2.19). Then, relation (2.20) implies that the set

$$\sqrt{d_\pi} \times \pi_{ij}, \quad 1 \le i, j \le d_\pi ,$$

is an orthonormal basis of M_π, and therefore that $\dim M_\pi = d_\pi^2$.

(3) (Orthogonality of the spaces M_π.) Relation (2.18) implies that, if $[\pi] \neq [\sigma]$, then M_π and M_σ are two orthogonal subspaces of $L^2(G)$. Since $L^2(G)$ is a complex *separable* Hilbert space, it admits a countable basis and we hence deduce that the dual $\hat{G}$ of a topological compact group G is *at most countable*.

Example 2.36 Write $T_{ab}(g)$, $a, b = 1, 2$, for the matrix elements of the representation for the group D_4 (see again Example 2.22). Using (2.13), it is easy to check that

$$\langle T_{ab}, T_{cd} \rangle_{L^2(D_4)} = \frac{1}{8} \sum_{g \in D_4} T_{ab}(g) T_{cd}(g) = \frac{1}{2} \delta_a^c \delta_b^d , \ a, b, c, d = 1, 2 .$$

Another irreducible one-dimensional unitary representation is provided by the trivial representation $(\mathbf{1}, \mathbb{C})$, $\mathbf{1}(g) \equiv 1$ for all $g \in D_4$; it is clearly not equivalent to $(T, \mathbb{C}^2)$, and indeed

$$\frac{1}{8} \sum_{g \in D_4} T_{ab}(g) \mathbf{1}(g) = \frac{1}{8} \sum_{g \in D_4} T_{ab}(g) = 0 .$$

2.4.5 Characters

With every finite-dimensional representation (π, V) of G we associate the mapping

$$\chi_\pi : G \to \mathbb{C} : g \mapsto \mathbf{Trace} \ \pi(g) , \tag{2.21}$$

called the *character* of π. In the following discussion, we demonstrate that the application χ_π completely characterizes a representation π. We start by discussing a few easy examples.

Example 2.37 When G is a commutative locally compact group, one customarily calls 'character' every continuous mapping $\chi : G \to S^1 = \{z \in \mathbb{C} : |z| = 1\}$ verifying the homomorphic relation

$$\chi(g)\chi(h) = \chi(gh), \quad \text{for every } g, h \in G.$$

Indeed, Proposition 2.30 implies that every *irreducible and unitary* representation of a compact topological commutative group G can be identified with a character of G. For instance, if $G = \mathbb{T} = [0, 1)$, endowed with the commutative group operation $xy = (x + y) \mathrm{mod}(1)$, then one has that the characters of G are given by the mappings of the type

$$x \mapsto \exp(i2n\pi x), \quad n \in \mathbb{Z};$$

note that, for $n = 0$, one recovers the trivial representation.

Example 2.38   Another remarkable example of a commutative compact group is the so-called *Cantor group* $\{-1, 1\}^{\mathbb{N}}$. It consists of all infinite sequences

$$\omega = (\omega_1, \omega_2, ...) , \quad \omega_i \in \{-1, 1\}$$

endowed with the following commutative group operation: for every $\omega, \omega' \in \{-1, 1\}^{\mathbb{N}}$

$$\omega\omega' = (\omega_1\omega_1', \omega_2\omega_2', ...) .$$

To see that $\{-1, 1\}^{\mathbb{N}}$ is compact, we can use e.g. the Tychonoff theorem. One can directly verify that the unique normalized Haar measure associated with $\{-1, 1\}^{\mathbb{N}}$ is $dg = \left\{ \frac{1}{2}\delta_{-1}(dg) + \frac{1}{2}\delta_1(dg) \right\}^{\mathbb{N}}$, where δ_a stands for the Dirac mass at a. The characters of $\{-1, 1\}^{\mathbb{N}}$ are given by the identity mapping $\omega \mapsto 1$, and all mappings of the type

$$\omega \longmapsto \omega_{i_1} \times \omega_{i_2} \times \cdots \times \omega_{i_k} ,$$

where $\{i_1, ..., i_k\}$ is some finite set of indices. See Blei [26, Ch. VII] for an exhaustive discussion of the Cantor group.

Example 2.39   For the representation T of the symmetry group D_4 introduced in Example 2.22, we have

$$\chi_T(e) = 2 , \chi_T(r^2) = -2 ,$$

$$\chi_T(r) = \chi_T(h) = \chi_T(r^3) = \chi_T(hr) = \chi_T(hr^2) = \chi_T(hr^3) = 0 .$$

Let us now start with some straightforward properties (the proof is elementary and left to the reader).

Proposition 2.40   *Let χ_π be the character of a representation (π, V), with dimension $d_\pi < \infty$, of the compact group G. Then, the following holds.*

(1) If e is the unity of G, then $\chi_\pi(e) = d_\pi$.
(2) For every $g \in G$, $\chi\left(g^{-1}\right) = \overline{\chi(g)}$.
(3) For every $g, h \in G$, $\chi\left(ghg^{-1}\right) = \chi(h)$.
(4) If (ρ, U) is another representation of G such that $d_\rho < \infty$, then the character of the direct sum of π and ρ is $\chi_{\pi \oplus \rho} = \chi_\pi + \chi_\rho$.
(5) If (ρ, U) is a representation of G such that $d_\rho < \infty$, then the character of the tensor product of π and ρ is $\chi_{\pi \otimes \rho} = \chi_\pi \times \chi_\rho$.

Property (3) in the above statement means that χ_π is a *class function*, in the sense that χ_π is constant over conjugacy classes. We will see in the next section that characters are indeed an orthonormal basis of class functions. This result will partially based on the following theorem, which is an immediate consequence of the orthogonality and isometric relations stated in Theorem 2.33.

Theorem 2.41 *Let χ_π be the character of a unitary representation (π, V) of the compact group G, with finite dimension d_π. Then, the following holds.*

(1) The representation (π, V) is irreducible if and only if

$$\langle \chi_\pi, \chi_\pi \rangle_{L^2(G)} = \int_G |\chi_\pi(g)|^2 \, dg = 1.$$

(2) If (π, V) is irreducible and if (ρ, U) is another irreducible (and hence finite-dimensional) representation of G, then ρ and π are not equivalent if and only if

$$\langle \chi_\pi, \chi_\rho \rangle_{L^2(G)} = \int_G \chi_\pi(g) \chi_\rho\left(g^{-1}\right) dg = 0. \tag{2.22}$$

Example 2.42 Consider again the Abelian group $\mathbb{T}$ of the Example (2.14). Here, the unitary representation $\{x \mapsto \exp(i2n\pi x), n \in \mathbb{Z}\}$ and the characters coincide, and we have trivially

$$\langle \chi_n, \chi_n \rangle_{L^2(\mathbb{T})} = \int_0^1 |\exp(i2n\pi x)|^2 \, dx = 1$$

and

$$\langle \chi_m, \chi_n \rangle_{L^2(\mathbb{T})} = \int_0^1 \exp(i2(m-n)\pi x) \, dx = 0 \text{ for } m \neq n,$$

as expected from Theorem 2.41.

Example 2.43 Focussing again on the group D_4 of Examples (2.22-2.39) and using (2.13), we have

$$\langle \chi_T, \chi_T \rangle_{L^2(D_4)} = \frac{1}{8} \sum_{g \in D_4} |\chi_\pi(g)|^2 = 1.$$

Consider the trivial representation $(\rho, U) = (\mathbf{1}, \mathbb{C})$, where $\mathbf{1}(g) \equiv 1$ for all $g \in D_4$. It is obvious that ρ and T are not equivalent, and indeed

$$\langle \chi_T, \chi_\rho \rangle_{L^2(D_4)} = \frac{1}{8} \sum_{g \in D_4} \chi_T(g) = 0.$$

Finally, the next result proves that the character (2.21) completely determines a given finite-dimensional representation (π, V).

Theorem 2.44 *Let (π, V) and (ρ, U) be, respectively, a finite-dimensional unitary representation and a unitary irreducible representation of the compact group G. Let χ_π and χ_ρ be the characters of π and ρ.*

(1) Suppose that there exist π-invariant subspaces $V_1, ..., V_k$ such that

$$V = V_1 \oplus \cdots \oplus V_k. \tag{2.23}$$

Then, the number of spaces V_j in the decomposition (2.23) such that $\left(\pi, V_j\right) \in [\rho]$ is given by

$$m_\rho = \left\langle \chi_\pi, \chi_\rho \right\rangle_{L^2(G)} = \int_G \chi_\pi(g) \chi_\rho\left(g^{-1}\right) dg. \tag{2.24}$$

(2) For every $\rho \in \hat{G}$, the number m_ρ at the previous point is independent of the choice of the decomposition (2.23) of V into π-invariant subspaces.

(3) Suppose (π', V') is another finite-dimensional unitary representation of G, such that $\chi_\pi = \chi_{\pi'}$. Then, π and π' are equivalent.

(4) The representation (π, V) is irreducible if and only if

$$\|\chi_\pi\|_{L^2(G)}^2 = \int_G \chi_\pi(g) \chi_\pi\left(g^{-1}\right) dg = 1.$$

Proof (1) By invariance, one has that, for every $j = 1, ..., k$, $\left(\pi, V_j\right) \in \left[\rho_j\right]$ for some irreducible representation ρ_j. It follows from Proposition 2.40 that

$$\chi_\pi = \chi_{\rho_1} \oplus \cdots \oplus \chi_{\rho_k},$$

and the result is a consequence of (2.22).

(2) The quantity (2.24) does not depend on the chosen decomposition of V. Point (3) and Point (4) are now easy consequences of the previous discussion. $\square$

Remark 2.45 (1) Theorem 2.44 implies that characters of equivalent representations are the same, so that character χ_π characterizes the equivalence class $[\pi]$.

(2) We deduce from the previous discussion that, for every finite-dimensional unitary representation (π, V), there exists a unique set $[\rho_1], ..., [\rho_h] \in \hat{G}$, as well as integers $m_{\rho_1}, ..., m_{\rho_h} \geq 1$ such that

$$\chi_\pi = m_{\rho_1} \chi_{\rho_1} + \cdots + m_{\rho_h} \chi_{\rho_h},$$

and therefore that (π, V) is equivalent to a direct sum of the type

$$\left[\pi^{(1,1)} \oplus \cdots \oplus \pi^{\left(1, m_{\rho_1}\right)}\right] \oplus \cdots \oplus \left[\pi^{(h,1)} \oplus \cdots \oplus \pi^{\left(h, m_{\rho_h}\right)}\right]$$

$$= \bigoplus_{j=1}^{h} \left[\pi^{(j,1)} \oplus \cdots \oplus \pi^{\left(j, m_{\rho_j}\right)}\right],$$

where, for $j = 1, ..., h$, one has that $\pi^{(j,l)} \in \left[\rho_j\right]$ for every $l = 1, ..., m_{\rho_j}$.

2.5 The Peter-Weyl Theorem

The representation theorems contained in this section will be used as fundamental tools for the rest of the book. The following discussion is again close to [63, Section VI.4].

Let (π, V) be an irreducible unitary representation of the compact group G. According to Proposition 2.32, V has finite dimension, say $d_\pi < \infty$. Now select an orthonormal basis $(e_1, ..., e_{d_\pi})$ of V, and write

$$\pi_{ij}(g) = \left\langle \pi(g) e_j, e_i \right\rangle_V, \quad g \in G, \quad 1 \le i, j \le d_\pi. \tag{2.25}$$

According to the discussion of Section 2.4.4, the functions $\sqrt{d_\pi} \pi_{ij}$ represent an orthonormal basis of the space of matrix coefficients M_π, associated with the equivalence class $[\pi]$. The point we want to make is that the direct sum of this spaces covers $L^2(G)$, that is, the space of square integrable functions on the group. We will hence establish that for any compact group, the elements of its irreducible matrix representations provide an orthonormal basis for the space of square integrable functions. This will be the basis for all spectral representations results in the chapters to follow, and for the derivation of stochastic properties for isotropic random fields.

More precisely, fix $k = 1, ..., d_\pi$, and denote by $M_\pi^{(k)}$ the d_π-dimensional subspace of M_π spanned by the mappings $g \mapsto \pi_{kj}(g)$, $j = 1, ..., d_\pi$, corresponding to the kth line of the matrix (2.25). We recall that the right and left regular representations (denoted R and L) of G on $L^2(G)$ are defined, respectively, according to (2.9) and (2.10). The following result states that each space $M_\pi^{(k)}$ is both R- and L-invariant, and that the associated representation is equivalent to π.

Proposition 2.46 *For every $k = 1, ..., d_\pi$, the space $M_\pi^{(k)}$ is R-invariant and L-invariant, and the two representations $(R, M_\pi^{(k)})$ and $(L, M_\pi^{(k)})$ are both equivalent to π. In particular, (R, M_π) and (L, M_π) are both equivalent to the direct* *sum*

$$\pi \oplus \cdots \oplus \pi,$$

whose order is d_π.

Proof We provide the proof only for the right representation R (the case of L is dealt with analogously). To see that the space $M_\pi^{(k)}$ is R-invariant, just use the fact that π is a representation of G to deduce that

$$R\pi_{kj}(h) = \pi_{kj}(hg) = \sum_{l=1}^{d_\pi} \pi_{kl}(h)\pi_{lj}(g),$$

yielding that $R\pi_{kj} \in M_\pi^{(k)}$. To see that $(R, M_\pi^{(k)}) \in [\pi]$ we have to show the existence of an intertwining isomorphism A between (π, V) and $(R, M_\pi^{(k)})$. This is given by the application

$$A : \sum_{j=1}^{d_\pi} c_j e_j \mapsto \sum_{j=1}^{d_\pi} c_j \pi_{kj},$$

where $(e_1, ..., e_{d_\pi})$ is the orthonormal basis of V appearing in (2.25). $\square$

We are now ready to state and prove the announced Peter-Weyl Theorem.

Theorem 2.47 (Peter-Weyl) *Let G be a topological compact group, and let $\hat{G}$ be its dual. For every $[\pi] \in \hat{G}$, denote by M_π the space of matrix coefficients associated with M_π.*

(1) The following orthogonal decomposition holds:

$$L^2(G) = \bigoplus_{\pi \in \hat{G}} M_\pi,$$

where $\oplus$ indicates a possibly infinite orthogonal sum.

(2) For every $f \in L^2(G)$ and every $[\pi] \in M_\pi$, the projection of f on the space M_π is given by the mapping

$$h \mapsto d_\pi \times \int_G f(g) \chi_\pi\left(g^{-1}h\right) dg, \tag{2.26}$$

where χ_π is the character of π.

Proof (1) We must show that the orthogonal (in $L^2(G)$) of the space $\bigoplus M_\pi$, say H, is such that $H = \{0\}$. To do this, we can argue by contradiction, and suppose that $H \neq 0$. By using the fact that $\bigoplus M_\pi$ is R-invariant (due to Proposition 2.46) as well as the invariance property (2.5) of the Haar measure, we see that H is indeed R-invariant. Since H is also closed, it must contain a nontrivial R-invariant subspace Y of finite dimension (due to Proposition 2.32), such that $(R, Y) \in [\sigma]$ for some $[\sigma] \in \hat{G}$. Now select a basis $(y_1, ..., y_{d_\sigma})$ of Y, and observe that, by definition, for every $j = 1, ..., d_\sigma$ the function

$$h \mapsto F_j(h) = \int_G R(h) f(g) \overline{y_j(g)} dg = \int_G f(gh) \overline{y_j(g)} dg = \langle R(h) f, y_j \rangle$$

is an element of M_σ. By an appropriate use of the Fubini Theorem, one can now show that F_j is orthogonal (in $L^2(G)$) to every mapping of the type $g \mapsto \langle \sigma(g) v, w \rangle_W$, where $(\sigma, W) \in [\sigma]$, and therefore that $\|F_j\|_{L^2(G)} = 0$. It follows that there exists a set U of Haar measure equal to 1 such that, for every $h \in U$,

$$\int_G f(gh) \overline{y_j(g)} dg = 0, \quad \text{for every } j = 1, ..., d_\sigma,$$

implying that, for every $h \in U$, the mapping $g \mapsto f(gh)$ is equal to zero. By right-invariance (2.5), this means that $f = 0$, thus yielding a contradiction. The proof is concluded.

(2) Select a representative element (π, V) for every equivalence class $[\pi] \in \hat{G}$, as well as an orthonormal basis $(e_1, ..., e_{d_\pi})$ of V. Then, the mappings

$$g \mapsto \sqrt{d_\pi} \pi_{ij}(g) = \sqrt{d_\pi} \langle \pi(g) e_j, e_i \rangle_V, \quad g \in G, \quad 1 \le i, j \le d_\pi,$$

constitute an orthonormal basis of M_π. It follows that the projection of f on M_π is given by the application

$$h \longmapsto d_\pi \sum_{i,j=1}^{d_\pi} \int_G f(g) \overline{\pi_{ij}(g)} dg \times \pi_{ij}(h)$$

$$= d_\pi \sum_{i,j=1}^{d_\pi} \int_G f(g) \pi_{ji}\left(g^{-1}\right) dg \times \pi_{ij}(h)$$

$$= d_\pi \sum_{j=1}^{d_\pi} \int_G f(g) \left[\sum_{i=1}^{d_\pi} \pi_{ji}\left(g^{-1}\right) \pi_{ij}(h) \right] dg$$

$$= d_\pi \sum_{j=1}^{d_\pi} \int_G f(g) \pi_{jj}\left(g^{-1}h\right) dg = d_\pi \int_G f(g) \chi_\pi\left(g^{-1}h\right) dg.$$

$\square$

As noted earlier, the Peter-Weyl Theorem can be alternatively stated as follows: let $\{\pi^\ell : \ell = 1, 2, ...\}$ be any enumeration of the dual of G, and $d_\ell = d_{\pi^\ell}$; for every $f \in L^2(G)$, the following spectral representation holds

$$f(g) = \sum_\ell \sum_{i,j=1}^{d_\ell} b_{ij}^\ell \sqrt{d_\ell} \pi_{ij}^\ell(g),$$

where the expansion holds in the L^2 sense and

$$b_{ij}^{\ell} := \left\langle f, \sqrt{d_{\ell}} \pi_{ij}^{\ell} \right\rangle_V = \sqrt{d_{\ell}} \int_G f(g) \overline{\pi_{ij}^{\ell}(g)} dg , \ \pi^{\ell} \in \widehat{G} .$$

We have also the *Plancherel identity*

$$\int_G f(g) h(g) dg = \sum_{\ell} \left\langle f, \sqrt{d_{\ell}} \pi_{ij}^{\ell} \right\rangle_V \left\langle h, \sqrt{d_{\ell}} \pi_{ij}^{\ell} \right\rangle_V .$$

In particular, the application $f \mapsto \left\langle f, \sqrt{d_{\ell}} \pi_{ij}^{\ell} \right\rangle_V$ is an isometry from the space of square-integrable functions $L^2(G)$ into the space of square summable sequences $\left\{ \sqrt{d_{\ell}} b_{ij}^{\ell} \right\}$, i.e.

$$\int_G f^2(g) dg = \sum_{\ell} \sum_{ij} d_{\ell} \left\{ b_{ij}^{\ell} \right\}^2 ,$$

a fact will have a great importance in the spectral theory of spherical random fields developed in the following chapters.

Remark 2.48   (1) Let G be a compact commutative group. Then, in this case the Peter-Weyl Theorem reduces to the statement that the characters, that is, the continuous homomorphisms from G to $S^1 = \{z \in \mathbb{C} : |z| = 1\}$ are an orthonormal basis of $L^2(G)$. As already pointed out, a particular case of this phenomenon is the orthogonal decomposition (2.1).

(2) Suppose that G is a *finite* commutative group. Then, the Peter-Weyl Theorem implies that $|\widehat{G}| = |G|$. To show this, it is sufficient to observe that $\left\{ \chi_{\pi} : [\pi] \in \widehat{G} \right\}$ and $\left\{ \mathbf{1}_{\{g\}} : g \in G \right\}$, where $\mathbf{1}_{\{g\}}(h) = 1$ if $h = g$ and $= 0$ otherwise, are both orthogonal basis of $L^2(G)$.

(3) When G is not commutative, one still has that characters are an orthonormal basis of the linear subspace of $L^2(G)$ composed of *conjugacy-invariant (or class, or central) functions*. We recall that a function f is conjugacy-invariant if

$$f\left(ghg^{-1} \right) = f(h) \quad \text{for every } h, g \in G .$$

In particular, one can use this fact (along with the discussion at Point 2 above) in order to prove that, if G is finite, then $|G| \geq |\widehat{G}|$ and that the equality holds if and only if G is commutative. See e.g. [63, Section VI.5] for further details on this point.

Remark 2.49   The spaces M_{π} are sometimes called the *π-isotypical spaces* of $L^2(G)$ (associated with the right and left representations)

An interesting consequence of the Peter-Weyl Theorem is the following characterization of finite groups (containing Point (2) of Remark 2.48 as a special case).

Corollary 2.50 *Let G be a finite group. Then,*

$$|G| = \sum_{[\pi] \in \hat{G}} \dim M_\pi = \sum_{[\pi] \in \hat{G}} d_\pi^2$$

Proof Just observe that, for a finite group G, the dimension of $L^2(G)$ is equal to $|G|$.   □

Example 2.51 As already recalled, an example of a finite commutative group is given by the cyclic group of length n, which is defined as the set of integers $Z_n = \{1, ..., n\}$, endowed with the group operation $xy = (x + y) \bmod(n)$. It is easily seen that the characters of Z_n are the complex exponentials

$$\left\{ \exp\left(i\frac{2\pi j}{n}t\right) : t \in C_n \right\}, \quad j = 1, ..., n . \tag{2.27}$$

The Peter-Weyl Theorem therefore entails the well-known result that every function f on Z_n can be written as

$$f(t) = \sum_{j=1}^{n} a_j \exp\left(i\frac{2\pi j}{n}t\right), \quad t \in Z_n , \tag{2.28}$$

where $a_j = \frac{1}{n}\sum_{u=1}^{n} f(u) \exp\left(-i\frac{2\pi j}{n}u\right)$. Plainly, this result can be restated by saying that every n-periodic function on the integers can be represented as a linear combination of the discrete complex exponentials $t \mapsto \exp\left(i\frac{2\pi j}{n}t\right)$. In this sense, the Peter-Weyl Theorem provides a strict generalization of usual Fourier analysis results on the cyclic group.

Example 2.52 We want to use Corollary 2.50 in order to completely characterize the dual of the dihedral group D_4, as introduced in Example 2.22. We already know that the two-dimensional representation $(T, \mathbb{C}^2)$ introduced therein is irreducible (see Example 2.31). Since $|D_4| = 8$ and $8 = 2^2 + 1 + 1 + 1 + 1$, it follows from Corollary 2.50 that it suffices to find four one-dimensional non-equivalent representations of D_4, say T_1, T_2, T_3 and T_4. These four representations are obtained by attaching to r and h the values 1 and -1, in all possible configurations, namely: $T_1(r) = T_1(h) = 1$ (the trivial representation); $T_2(r) = 1, T_2(h) = -1$; $T_3(r) = -1, T_3(h) = 1$; $T_4(r) = -1, T_4(h) = -1$. In particular,

$$\hat{D}_4 = \{[T]; [T_i] : i = 1, 2, 3, 4\}.$$

By using Theorem 2.47 one can also prove the following converse of Proposition 2.30.

Corollary 2.53 *Let G be a compact group, and suppose that every irreducible representation of G is one-dimensional. Then, G is commutative.*

Proof According to the Peter-Weyl Theorem, there exists a sequence

$$\{\chi_n : n \geq 1\}$$

of orthonormal homomorphisms from G into S^1 such that, for every $f \in L^2(G)$,

$$f = \sum_n c_n \chi_n, \quad \text{for some } \{c_n\} \in \ell^2,$$

with convergence in $L^2(G)$. It follows that, for every $f \in L^2(G)$ and every $h \in G$,

$$\int_G f(gh)\,dg = \sum_n c_n \int_G \chi_n(g)\chi_n(h)\,dg = \sum_n c_n \int_G \chi_n(h)\chi_n(g)\,dg$$
$$= \int_G f(hg)\,dg.$$

This yields that, for every $h \in G$ and dg-almost every g, $gh = hg$. The desired conclusion follows by using the continuity of the applications $g \mapsto gh$ and $g \mapsto hg$. □

To conclude this section, we present the following consequence of the Stone-Weierstrass Theorem (see [173]). The proof is omitted (see e.g. [57]).

Theorem 2.54 *The spaces M_π, $[\pi] \in \hat{G}$, are dense in $C(G)$, the class of continuous functions on G, endowed with the supremum norm.*

3

Representations of $SO(3)$ and Harmonic Analysis on S^2

3.1 Introduction

In this chapter, we shall specialize the results of Chapter 2 to the compact group which is central for our analysis, namely the "special group of rotations" $SO(3)$. The latter can be realized as the space of 3×3 real matrices A such that $A'A = I_3$ (where I_3 is the three-dimensional identity matrix) and $\det(A) = 1$. In particular, we shall carry out an explicit construction for a complete set of irreducible representations of $SO(3)$. To do so, we shall first establish a more general result, namely, we will provide (following a classical argument) a complete family of irreducible representations for the group $SU(2)$; we will then recall a well-known relationship between $SO(3)$ and $SU(2)$ (i.e. that the latter "covers" the former twice, i.e. $SO(3) \simeq SU(2)/\{I_2, -I_2\}$, where the 2×2 identity matrix I_2 is the identity element of $SU(2)$) and hence show that the representations of $SO(3)$ are a subset of the representations of $SU(2)$. We will then develop Fourier analysis on the sphere, largely by means of the Peter-Weyl Theorem discussed in the previous chapter. In particular, we shall prove that functions on the sphere can be identified with a subset of those on the group $SO(3)$, so that their spectral representation will require only a subset of the matrix coefficients in the representation of the latter (more formally, we shall identify the sphere S^2 as the quotient space $SO(3)/SO(2)$). We will also introduce many fundamental tools that are needed for this monograph, such as *Wigner's matrices, spherical harmonics* and *Clebsch-Gordan coefficients*. In particular, we will discuss how to exploit group representation techniques to deal with multiple integral of spherical harmonics, a topic which will have great relevance in the chapters to follow – especially in connection with non-linear statistics of Gaussian fields. The reader is referred to Varshalovich, Moskalev and Khersonskii[195] , Vilenkin and Klimyk [197], Sternberg [191] and Miller [142] for useful references and further discussions on the topics presented in this chapter.

3.2 Euler angles

3.2.1 Euler angles for $SU(2)$

As mentioned above, the purpose of this chapter is to analytically derive a complete family of irreducible representations for the group of rotations $SO(3)$, and to exploit these representations in order to develop some harmonic analysis procedures on the sphere. To do this, it is convenient to start with the more general group $SU(2)$. First, recall first that every $g \in SU(2)$ can be written as

$$g = \begin{pmatrix} \alpha & \beta \\ -\bar{\beta} & \bar{\alpha} \end{pmatrix}, \quad \alpha, \beta \in \mathbb{C}, \, \alpha\bar{\alpha} + \beta\bar{\beta} = 1 .$$

Complex numbers α, β such that $|\alpha|^2 + |\beta|^2 = 1$ can be given in terms of three real parameters, for instance $|\alpha|$, $\arg \alpha$ and $\arg \beta$. It is more convenient, though, to use the *Euler angles* φ, ϑ, ψ which are defined implicitly by the formulae

$$|\alpha| = \cos\frac{\vartheta}{2} , \quad \arg \alpha = \frac{\varphi + \psi}{2} , \quad \arg \beta = \frac{\varphi - \psi}{2} , \tag{3.1}$$

where

$$0 \leq \varphi < 2\pi , \quad 0 \leq \vartheta \leq \pi , \quad \text{and} \, -2\pi \leq \psi < 2\pi .$$

It can be checked that the correspondence between α, β and φ, ϑ, ψ is one-to-one, whenever one is restricted to complex numbers $\alpha, \beta \neq 0$ (or, equivalently when one considers angles $(\varphi, \vartheta, \psi)$ such that $\vartheta \neq 0, \pi$); plainly, when $\beta = 0$ (resp. $\alpha = 0$), only the sum $\varphi + \psi$ (resp. the difference $\varphi - \psi$) is determined. We can hence write any element of $SU(2)$ in the form

$$\begin{pmatrix} \alpha & \beta \\ -\bar{\beta} & \bar{\alpha} \end{pmatrix} = \begin{pmatrix} \cos\frac{\vartheta}{2} \exp\left\{i\frac{\varphi+\psi}{2}\right\} & \sin\frac{\vartheta}{2} \exp\left\{i\frac{\varphi-\psi}{2}\right\} \\ -\sin\frac{\vartheta}{2} \exp\left\{-i\frac{\varphi-\psi}{2}\right\} & \cos\frac{\vartheta}{2} \exp\left\{-i\frac{\varphi+\psi}{2}\right\} \end{pmatrix} \tag{3.2}$$

$$= \begin{pmatrix} \exp\left\{i\frac{\varphi}{2}\right\} & 0 \\ 0 & \exp\left\{-i\frac{\varphi}{2}\right\} \end{pmatrix} \begin{pmatrix} \cos\frac{\vartheta}{2} & \sin\frac{\vartheta}{2} \\ -\sin\frac{\vartheta}{2} & \cos\frac{\vartheta}{2} \end{pmatrix} \begin{pmatrix} \exp\left\{i\frac{\psi}{2}\right\} & 0 \\ 0 & \exp\left\{-i\frac{\psi}{2}\right\} \end{pmatrix} .$$

Formula (3.2) is the key to study the relation linking $SU(2)$ and $SO(3)$. For every $x = (x_1, x_2, x_3) \in \mathbb{R}^3$, consider the complex 2×2 matrix (which is Hermitian and has trace zero)

$$h_x := \begin{pmatrix} x_3 & x_1 + ix_2 \\ x_1 - ix_2 & -x_3 \end{pmatrix} ,$$

and consider the transformation

$$T(g)h_x := gh_xg^* \, .$$

It is simple to see that the matrix $T(g)h_x$ is Hermitian with trace equal to zero, hence there exist unique real numbers y_1, y_2, y_3 such that

$$T(g)h_x = \begin{pmatrix} y_3 & y_1 + iy_2 \\ y_1 - iy_2 & -y_3 \end{pmatrix} =: h_y.$$

It follows that we can regard each $T(g)$ as a linear mapping $T(g) : \mathbb{R}^3 \to \mathbb{R}^3$, written $T(g)x = y$. It is immediately verified that

$$x_1^2 + x_2^2 + x_3^2 = -\det h_x = -\det h_y = y_1^2 + y_2^2 + y_3^2 , \quad \det T(g) = 1 ,$$

so that $T(g) \in SO(3)$. In other words, $T(g)$ maps $SU(2)$ to $SO(3)$; it can be checked that the mapping is actually a homomorphism onto $SO(3)$, with kernel $\{I_2, -I_2\}$, so that $SU(2)$ and $SO(3)$ are locally isomorphic. More precisely, as claimed in the introduction to this chapter,

$$SO(3) \simeq SU(2)/\{I_2, -I_2\} ,$$

that is, $SO(3)$ is isomorphic to the quotient group $SU(2)/\{I_2, -I_2\}$, where I_2 is the identity of $SU(2)$. Since the kernel $\{I_2, -I_2\}$ contains exactly two elements, one customarily says that "$SU(2)$ covers $SO(3)$ twice". Using the elementary relations $e^{i\frac{\psi}{2}} = -e^{i\frac{\psi-2\pi}{2}}$ and $e^{-i\frac{\psi}{2}} = -e^{-i\frac{\psi-2\pi}{2}}$, we see that, as far as the parametrization of $SO(3)$ through T is concerned, the range of the Euler angles can be restricted to $0 \le \varphi < 2\pi$, $0 \le \vartheta \le \pi$, $0 \le \psi < 2\pi$. This remark leads to a more geometric interpretation, which we discuss in the next section.

3.2.2 Euler angles for $SO(3)$

Let us introduce $\{e_1, e_2, e_3\}$ as the standard, right-handed basis for $\mathbb{R}^3$, i.e.

$$e_1 = (1, 0, 0) \, , \, e_2 = (0, 1, 0) \, , \, e_3 = (0, 0, 1).$$

Note that, for $g \in SU(2)$ corresponding to the angles $\{\varphi, 0, 0\}$, we have

$$T(g)h_x = gh_xg^*$$

$$= \begin{pmatrix} \exp\{i\frac{\varphi}{2}\} & 0 \\ 0 & \exp\{-i\frac{\varphi}{2}\} \end{pmatrix} \begin{pmatrix} x_3 & x_1 + ix_2 \\ x_1 - ix_2 & -x_3 \end{pmatrix} \begin{pmatrix} \exp\{-i\frac{\varphi}{2}\} & 0 \\ 0 & \exp\{i\frac{\varphi}{2}\} \end{pmatrix}$$

$$= \begin{pmatrix} x_3 & \exp\{i\varphi\}\{x_1 + ix_2\} \\ \exp\{-i\varphi\}\{x_1 - ix_2\} & -x_3 \end{pmatrix}.$$

It is clear that such a g corresponds to an element $T(g) \in SO(3)$ given by a rotation by an angle φ around the $\mathbf{e}_3$ axis; we shall label it $R_{\mathbf{e}_3}(\varphi)$. A similar computation shows that the action

$$T \begin{pmatrix} \cos\frac{\vartheta}{2} & \sin\frac{\vartheta}{2} \\ -\sin\frac{\vartheta}{2} & \cos\frac{\vartheta}{2} \end{pmatrix} h_x$$

can be realized in $\mathbb{R}^3$ as a rotation by an angle ϑ around the $\mathbf{e}_2$ axis. We now use this decomposition to provide a geometric interpretation of (3.2) (see e.g. [191], page 21-22, for a classic discussion of this point). To do this, we introduce three one-dimensional subgroups $R_{\mathbf{e}_1}(.), R_{\mathbf{e}_2}(.), R_{\mathbf{e}_3}(.)$ of $SO(3)$, defined by the following parametrized matrices:

$$R_{\mathbf{e}_1}(\alpha) = \begin{pmatrix} 1 & 0 & 0 \\ 0 & \cos\alpha & -\sin\alpha \\ 0 & \sin\alpha & \cos\alpha \end{pmatrix}, R_{\mathbf{e}_2}(\alpha) = \begin{pmatrix} \cos\alpha & 0 & \sin\alpha \\ 0 & 1 & 0 \\ -\sin\alpha & 0 & \cos\alpha \end{pmatrix},$$

$$R_{\mathbf{e}_3}(\alpha) = \begin{pmatrix} \cos\alpha & -\sin\alpha & 0 \\ \sin\alpha & \cos\alpha & 0 \\ 0 & 0 & 1 \end{pmatrix}, \quad \alpha \in \mathbb{R}.$$

The following proposition yields a complete description of the elements of $SO(3)$ in terms of three elementary rotations. Two proofs, one analytic and one more geometric, are provided.

Proposition 3.1 *Each rotation R in $\mathbb{R}^3$ can be realized sequentially as*

$$R = R_{\mathbf{e}_3}(\varphi)R_{\mathbf{e}_2}(\vartheta)R_{\mathbf{e}_3}(\psi), \, 0 \le \varphi < 2\pi, \, 0 \le \vartheta \le \pi, \, 0 \le \psi < 2\pi. \quad (3.3)$$

Representation (3.3) is unique whenever $\vartheta \ne 0, \pi$. If $\vartheta = 0$, then only the sum $\varphi + \psi$ is uniquely defined. If $\vartheta = \pi$, then only the difference $\varphi - \psi$ is uniquely defined.

Proof (Geometric proof.) It is straightforward that for any rotation $R \in SO(3)$, we have $R\mathbf{e}_3 = \widetilde{R}\mathbf{e}_3$, where $\widetilde{R} = R_{\mathbf{e}_3}(\varphi)R_{\mathbf{e}_2}(\vartheta)$, some ϑ, φ (to check the latter claim, it suffices to note that for each vector $\widetilde{R}\mathbf{e}_3 \in S^2$, there exist $\vartheta \in [0, \pi]$, $\varphi \in [0, 2\pi)$, such that $\widetilde{R}\mathbf{e}_3 = (\cos\varphi\sin\vartheta, \sin\varphi\sin\vartheta, \cos\vartheta)$). It follows that R must necessarily have the form $R = R_{\widetilde{R}\mathbf{e}_3}\widetilde{R}$, where $R_{\widetilde{R}\mathbf{e}_3}$ belongs to the isotropy group of the vector $\widetilde{R}\mathbf{e}_3$, that is to the group of rotations that leave this vector unchanged and hence take the form $R_{\widetilde{R}\mathbf{e}_3} = \widetilde{R}R_{\mathbf{e}_3}(\psi)\widetilde{R}^{-1}$, some $\psi \in [0, 2\pi)$. Thus

$$R = R_{\widetilde{R}\mathbf{e}_3}\widetilde{R} = \widetilde{R}R_{\mathbf{e}_3}(\psi)\widetilde{R}^{-1}\widetilde{R} = \widetilde{R}R_{\mathbf{e}_3}(\psi) = R_{\mathbf{e}_3}(\varphi)R_{\mathbf{e}_2}(\vartheta)R_{\mathbf{e}_3}(\psi),$$

as claimed.

(Analytic proof.) Using (3.2) and the discussion at the beginning of this section, we deduce that every $g \in SU(2)$ can be written as $g = g_1 g_2 g_3$, where

$$T(g_1) = R_{\mathbf{e}_3}(\varphi), \; T(g_2) = R_{\mathbf{e}_2}(\vartheta), \; \text{and} \; T(g_3) = R_{\mathbf{e}_3}(\psi),$$

for some angles such that

$$0 \le \varphi < 2\pi, \; 0 \le \vartheta \le \pi, \; 0 \le \psi < 2\pi$$

(recall that, when mapping $SU(2)$ through T, we can restrict the angle ψ to the interval $[0, 2\pi)$). To conclude, just use the fact that T is an homomorphism whose image coincides with $SO(3)$. $\square$

We have hence shown that every $R \in SO(3)$ can be expressed in coordinates $\{\varphi, \vartheta, \psi\}$ $(0 \le \varphi < 2\pi, 0 \le \vartheta \le \pi, 0 \le \psi < 2\pi)$ as

$$R = R(\varphi, \vartheta, \psi) = R_{\mathbf{e}_3}(\varphi) R_{\mathbf{e}_2}(\vartheta) R_{\mathbf{e}_3}(\psi) =$$

$$\begin{pmatrix} \cos\varphi\cos\psi - \sin\varphi\sin\psi\cos\vartheta & -\cos\varphi\sin\psi - \sin\varphi\cos\psi\cos\vartheta & \sin\varphi\sin\vartheta \\ \sin\varphi\cos\psi + \cos\varphi\sin\psi\cos\vartheta & -\sin\varphi\sin\psi + \cos\varphi\cos\psi\cos\vartheta & -\cos\varphi\sin\vartheta \\ \sin\psi\sin\vartheta & \cos\psi\cos\vartheta & \cos\vartheta \end{pmatrix}.$$

$$(3.4)$$

We refer to [195], Chapter 1 for further discussions of rotations in $\mathbb{R}^3$. In words, (3.3) is stating that each rotation can be realized by rotating first by an angle ψ around the axis $\mathbf{e}_3$, then rotating around the $\mathbf{e}_2$ axis by and angle ϑ, finally rotating by an angle φ around the $\mathbf{e}_3$ axis; this realization is unique except for $\vartheta = 0$, where only the sum $\varphi + \psi$ is determined, and for $\vartheta = \pi$, where only the difference $\varphi - \psi$ is determined. As mentioned before, the three angles φ, ϑ, ψ are known as the *Euler angles* of $SO(3)$. Note that $R_{\mathbf{e}_1}$ rotates $\mathbf{e}_2$ towards $\mathbf{e}_3$, $R_{\mathbf{e}_2}$ rotates $\mathbf{e}_3$ towards $\mathbf{e}_1$, and $R_{\mathbf{e}_3}$ rotates $\mathbf{e}_1$ towards $\mathbf{e}_2$. It is clear that the last two rotations identify one point on the sphere, so (3.3) could be also interpreted as moving the North Pole to a new orientation in S^2 after rotating by ψ the tangent plane at the original location. This decomposition of rotations in $SO(3)$ is somewhat canonical, and it is known as the "z-y-z convention".

Remark 3.2 The three steps could be alternatively visualized as i) rotation by an angle φ around $\mathbf{e}_3$, ii) rotation by an angle ϑ around the new axis $\mathbf{e}_2' = R_{\mathbf{e}_3}(\varphi)\mathbf{e}_2$, and iii) rotation by an angle ψ around this new axis $\mathbf{e}_3' = R_{\mathbf{e}_3}(\varphi)R_{\mathbf{e}_2}(\vartheta)\mathbf{e}_3$, see [195, pp. 21-22]. Fig. 3.1 provides an illustration of this point.

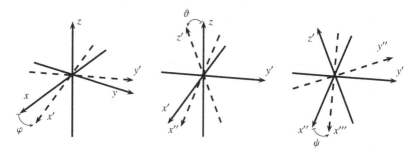

Figure 3.1 Performing a rotation $R = (\varphi, \vartheta, \psi) \in SO(3)$: (a) rotate around the z axis by an angle φ, (b) rotate around the new y' axis by an angle ϑ, and (c) rotate around the new z' axis by an angle ψ. The resulting Cartesian system is given by the three new axis x''', y'' and z'.

Remark 3.3 The above discussion implies that $SO(3)$ acts transitively on the sphere S^2, that is, S^2 is a homogeneous space. Since for every $x \in S^2$ the isotropy subgroup (see (2.3))

$$H_x = \{g \in SO(3) : g \cdot x = x\}$$

is trivially isomorphic to $SO(2)$, we deduce from Proposition 2.12 the well-known fact that $S^2 \simeq SO(3)/SO(2)$.

Remark 3.4 Of course, other parametrizations of $SU(2)$ and $SO(3)$ are possible. For instance, in [197, Ch. 6], the Euler angles $\hat{\varphi}, \vartheta, \hat{\psi}$ of $SU(2)$ are defined through the relations

$$|\alpha| = \cos\frac{\hat{\vartheta}}{2} \, , \quad \arg\alpha = \frac{\hat{\varphi} + \hat{\psi}}{2} \, , \quad \arg\beta = \frac{\hat{\varphi} - \hat{\psi} + \pi}{2},$$

or, equivalently,

$$\hat{\vartheta} = \vartheta, \, \hat{\varphi} = \varphi + \frac{\pi}{2}, \, \hat{\psi} = \psi - \frac{\pi}{2}.$$

This choice leads to a different definition of the Euler angles of $SO(3)$, which is usually labeled as the "z-x-z convention". Further details on the effects of this convention are given in the subsequent sections.

The invariant (Haar) measure for the group $SO(3)$ is obtained as follows (more details can be found e.g. in [197], pp. 274-275). Let $R(\varphi, \vartheta, \psi)$ be the rotation associated with the Euler angles φ, ϑ, ψ via Proposition 3.1. Then, for every bounded function f on $SO(3)$,

$$\int_{SO(3)} f(g)dg = \frac{1}{8\pi^2} \int_0^{2\pi} \int_0^\pi \int_0^{2\pi} f(R(\varphi, \vartheta, \psi)) \sin\vartheta \, d\varphi \, d\vartheta \, d\psi \qquad (3.5)$$

Note that the set of points of the type $(\varphi, 0, \psi)$ and (φ, π, ψ) are not charged by the product Lebesgue measure. The constant of proportionality is justified by the relations

$$\int_0^{2\pi} \int_0^{\pi} \int_0^{2\pi} \sin \vartheta d\varphi d\vartheta d\psi = 4\pi^2 \int_0^{\pi} \sin \vartheta d\vartheta = 8\pi^2.$$

3.3 Wigner's D matrices

In this section, we shall show that a complete set of irreducible matrix representations for $SO(3)$ is provided by the so-called *Wigner's D matrices* $\{D^l(g) : l = 0, 1, ...\}$, having dimension $(2l + 1) \times (2l + 1)$ for every $l = 0, 1, 2, ...$. In the sequel, we will (slightly abusively) use the same notation for these matrices as a function of the elements of $SU(2)$ in their different parametrizations, i.e. we shall write both $D^l(g)$, $g \in SU(2)$ and $D^l(\varphi, \vartheta, \psi) = D^l(\Phi(\varphi, \vartheta, \psi)) = \{D^l_{mn}(\varphi, \vartheta, \psi)\}_{m,n=-l,...,l}$ where $(\varphi, \vartheta, \psi)$ are the Euler angles associated with g, and the symbol Φ denotes the mapping $\Phi(\varphi, \vartheta, \psi) \mapsto g$, that is: $\Phi(\varphi, \vartheta, \psi)$ is the element of $SU(2)$ associated with the Euler angles $(\varphi, \vartheta, \psi)$ via (3.1), and m, n are, respectively, rows and column indexes.

Every Wigner matrix operates irreducibly and equivalently on $(2l+1)$ vector spaces (whose direct sum gives one of the "isotypical spaces" of matrix coefficients M_π, as appearing in the Peter-Weyl Theorem, see Remark 2.49), each of them spanned by a different column, say n, of the matrix representation itself. As we shall discuss extensively later in this book, the elements of column n can be related to the so-called *spin n* spherical harmonics, which enjoy a great importance in particle physics and in harmonic expansions for tensor-valued random fields (note that, due to the specific structure of Wigner matrices, one could replace "columns" by "rows" in the previous discussion).

In the sequel, we follow closely [122], see also [197], Chapter 6 (but one has to be careful to the different parametrization of the Euler angles). The idea is to build a family of representations for $SU(2)$, and then exploit the results in the previous section to deal with $SO(3)$.

3.3.1 A family of unitary representations of $SU(2)$

Consider the semi-integers $l \in \{0, \frac{1}{2}, 1, \frac{3}{2}, ...\}$, and let $\mathcal{H}_l$ be the space of homogeneous polynomials of degree $2l$ in two complex variables z_1, z_2. The monomials $z_1^{l-m}z_2^{l+m}$, $m = -l, -l + 1, ..., l$ form a basis of $\mathcal{H}_l$ (which has dimension

$2l + 1$). It is useful to normalize monomials as follows:

$$\psi_m^l(z_1, z_2) := \left(\begin{array}{c} 2l \\ l-m \end{array} \right)^{\frac{1}{2}} z_1^{l-m} z_2^{l+m}, \quad m = -l, -l+1, \ldots, l .$$

Note that, in the above notation, every $l - m$ is an integer. The basis $\{\psi_m^l\}$ is orthonormal with respect to the inner product on $\mathcal{H}_l$ defined by

$$\langle p, q \rangle_{\mathcal{H}_l} := \frac{1}{(2l)!} p \left(\frac{\partial}{\partial z_1}, \frac{\partial}{\partial z_2} \right) \overline{q(z_1, z_2)}, \quad p, q \in \mathcal{H}_l.$$

For $z = (z_1, z_2)$ and $g \in SU(2)$, we define

$$(D^l(g)f)(z) := f(g'z) ,$$

or, more explicitly,

$$\text{for } g = \left(\begin{array}{cc} \alpha & \beta \\ -\bar{\beta} & \bar{\alpha} \end{array} \right), \quad \alpha\bar{\alpha} + \beta\bar{\beta} = 1 ,$$

$$(D^l(g)f)(z) = \left(D^l \left(\begin{array}{cc} \alpha & \beta \\ -\bar{\beta} & \bar{\alpha} \end{array} \right) f \right)(z) = f(\alpha z_1 - \bar{\beta} z_2, \beta z_1 + \bar{\alpha} z_2) .$$

It is easy to see that each pair $(\mathcal{H}_l, D^l)$ defines a representation of $SU(2)$. To see this, first observe that, for $\alpha = 1$ and $\beta = 0$ (i.e. for $g = I_2$, the 2×2 identity matrix) we have $(D^l(g)f)(z) = (D^l(I_2)f)(z) = f(z)$, for every $f \in \mathcal{H}_l$. Also, for general $g_1, g_2 \in SU(2)$ and $f \in \mathcal{H}_l$,

$$(D^l(g_1 g_2)f)(z) = f((g_1 g_2)'z) = f(g_2' g_1' z) = (D^l(g_2)f)(g_1' z)$$
$$= (D^l(g_1)(D^l(g_2)f)(z) = (D^l(g_1)D^l(g_2)f)(z) .$$

We can now determine the elements of the matrix representation, with respect to the basis $\{\psi_m^l\}$ introduced above, by means of the identity

$$D^l(g)\psi_n^l = \sum_{m=-l}^l D_{mn}^l(g)\psi_m^l , \quad g \in SU(2), \quad n = -l, \ldots, l,$$

which we rewrite as

$$\left(\begin{array}{c} 2l \\ l-n \end{array} \right)^{\frac{1}{2}} (\alpha z_1 - \bar{\beta} z_2)^{l-n} (\beta z_1 + \bar{\alpha} z_2)^{l+n} = \sum_{m=-l}^l \left(\begin{array}{c} 2l \\ l-m \end{array} \right)^{\frac{1}{2}} D_{mn}^l \left(\begin{array}{cc} \alpha & \beta \\ -\bar{\beta} & \bar{\alpha} \end{array} \right) z_1^{l-m} z_2^{l+m} .$$

$$(3.6)$$

Remark 3.5 (Remark on notation) When l is not an integer, the entries of the matrix D^l are labeled by letters m, n ranging in the set $\{-l, -l+1, \ldots, l-1, l\}$. For the rest of this book, one must also read accordingly sums of the type $\sum_{k=-l}^{l}$. For instance

$$\sum_{k=-5/2}^{5/2} f(k) = f(-5/2) + f(-3/2) + f(-1/2) + f(1/2) + f(3/2) + f(5/2).$$

Fixing $n = l$ we deduce that

$$D_{ml}^l \begin{pmatrix} \alpha & \beta \\ -\bar\beta & \bar\alpha \end{pmatrix} = \begin{pmatrix} 2l \\ l - m \end{pmatrix}^{\frac{1}{2}} \beta^{l-m} \bar\alpha^{l+m},$$

where we have used the elementary relation

$$(a + b)^{2l} = \sum_{m=-l}^{l} \begin{pmatrix} 2l \\ l - m \end{pmatrix} a^{l-m} b^{l+m}.$$

Also, multiplying both sides of (3.6) by

$$\begin{pmatrix} 2l \\ l - n \end{pmatrix}^{\frac{1}{2}} w_1^{l-n} w_2^{l+n},$$

and summing over n, we obtain

$$(\alpha z_1 w_1 - \bar\beta z_2 w_1 + \beta z_1 w_2 + \bar\alpha z_2 w_2)^{2l}$$

$$= \sum_{m,n=-l}^{l} \begin{pmatrix} 2l \\ l - m \end{pmatrix}^{\frac{1}{2}} \begin{pmatrix} 2l \\ l - n \end{pmatrix}^{\frac{1}{2}} D_{mn}^l \begin{pmatrix} \alpha & \beta \\ -\bar\beta & \bar\alpha \end{pmatrix} z_1^{l-m} z_2^{l+m} w_1^{l-n} w_2^{l+n}. \qquad (3.7)$$

The $(2l + 1) \times (2l + 1)$ matrices $\{D^l\}_{m,n=-l,\ldots,l}$ are called *Wigner's D matrices*. Some further manipulations on (3.6)-(3.7) yields the symmetry relationships

$$D_{mn}^l \begin{pmatrix} \alpha & -\bar\beta \\ \beta & \bar\alpha \end{pmatrix} = D_{nm}^l \begin{pmatrix} \alpha & \beta \\ -\bar\beta & \bar\alpha \end{pmatrix}, \qquad (3.8)$$

$$D_{mn}^l \begin{pmatrix} \alpha & -\bar\beta \\ \beta & \bar\alpha \end{pmatrix} = D_{-m,-n}^l \begin{pmatrix} \bar\alpha & \beta \\ -\bar\beta & \alpha \end{pmatrix}, \qquad (3.9)$$

$$D_{mn}^l \begin{pmatrix} \alpha & -\bar\beta \\ \beta & \bar\alpha \end{pmatrix} = D_{-n,-m}^l \begin{pmatrix} \bar\alpha & -\bar\beta \\ \beta & \alpha \end{pmatrix}. \qquad (3.10)$$

We are now in the position to establish the following characterization

Proposition 3.6 *For every nonnegative semi-integer l, the pair $(\mathcal{H}_l, D^l)$ is a unitary representation of SU(2).*

Proof We have trivially

$$
\begin{pmatrix} \alpha & \beta \\ -\overline{\beta} & \overline{\alpha} \end{pmatrix}^{-1} = \begin{pmatrix} \alpha & \beta \\ -\overline{\beta} & \overline{\alpha} \end{pmatrix}^{*} = \begin{pmatrix} \overline{\alpha} & -\beta \\ \overline{\beta} & \alpha \end{pmatrix},
$$

therefore to prove that $D^l(g)^* = D^l(g^{-1})$ for all $g \in SU(2)$ we just need to show that, for all α, β,

$$
\overline{D^l_{mn}}\begin{pmatrix} \alpha & \beta \\ -\overline{\beta} & \overline{\alpha} \end{pmatrix} = D^l_{nm}\begin{pmatrix} \overline{\alpha} & -\beta \\ \overline{\beta} & \alpha \end{pmatrix}.
$$

By (3.6), D^l_{nm} is a polynomial in $\alpha, \overline{\alpha}, \beta, \overline{\beta}$ with real coefficients, and therefore

$$
\overline{D^l_{mn}}\begin{pmatrix} \alpha & \beta \\ -\overline{\beta} & \overline{\alpha} \end{pmatrix} = D^l_{mn}\begin{pmatrix} \overline{\alpha} & \overline{\beta} \\ -\beta & \alpha \end{pmatrix} = D^l_{nm}\begin{pmatrix} \overline{\alpha} & -\beta \\ \overline{\beta} & \alpha \end{pmatrix},
$$

where the last identity follows from (3.8). $\qquad\square$

We conclude this section by providing an analytic expression for the coefficients D^l_{mn}.

Proposition 3.7 *For all $g = \begin{pmatrix} \alpha & \beta \\ -\overline{\beta} & \overline{\alpha} \end{pmatrix} \in SU(2)$, we have*

$$
D^l_{mn}\begin{pmatrix} \alpha & \beta \\ -\overline{\beta} & \overline{\alpha} \end{pmatrix} = \tag{3.11}
$$

$$
\frac{\begin{pmatrix} 2l \\ l-n \end{pmatrix}^{\frac{1}{2}}}{\begin{pmatrix} 2l \\ l-m \end{pmatrix}^{\frac{1}{2}}} \sum_{j=0\vee(-m-n)}^{(l-m)\wedge(l-n)} \begin{pmatrix} l-n \\ j \end{pmatrix}\begin{pmatrix} l+n \\ l-m-j \end{pmatrix} \alpha^j (-\overline{\beta})^{l-n-j}\beta^{l-m-j}\overline{\alpha}^{n+m+j}.
$$

Proof Using the binomial formula, we can rewrite the left-hand side of (3.6) as

$$
\begin{pmatrix} 2l \\ l-n \end{pmatrix}^{\frac{1}{2}} (\alpha z_1 - \overline{\beta}z_2)^{l-n}(\beta z_1 + \overline{\alpha}z_2)^{l+n}
$$

$$
= \begin{pmatrix} 2l \\ l-n \end{pmatrix}^{\frac{1}{2}} \sum_{j=0}^{l-n}\sum_{k=0}^{l+n} \begin{pmatrix} l-n \\ j \end{pmatrix}\begin{pmatrix} l+n \\ k \end{pmatrix} \alpha^j \beta^k (-\overline{\beta})^{l-n-j}\overline{\alpha}^{l+n-k}z_1^{j+k}z_2^{2l-j-k}.
$$

Now consider the change of variables $(j,k) \mapsto (j,m)$, where $j+k = l-m$, with inverse map $(j,m) \mapsto (j, l-m-j)$ and domains of summations

$$
\{0 \le j \le l-n, \ 0 \le k \le l+n\} \iff
$$
$$
\{-l \le m \le l, \ 0 \le j \le l-n, \ -m-n \le j \le l-m\}.
$$

The previous sum then becomes

$$\binom{2l}{l-n}\sum_{m=-l}^{l}\sum_{j=0\vee(-m-n)}^{(l-m)\wedge(l-n)}\binom{l-n}{j}\binom{l+n}{l-m-j}\alpha^{j}(-\bar{\beta})^{l-n-j}\beta^{l-m-j}\bar{\alpha}^{n+m+j}z_1^{l-m}z_2^{l+m}.$$

(3.12)

The result is established by equating the coefficients in (3.12) and (3.6). □

3.3.2 Expressions in terms of Euler angles and irreducibility

It is convenient to derive an expression for the elements of Wigner's D matrices in terms of Euler angles. As mentioned before, with a slight abuse of notation, we shall write $D_{mn}^{l}(\varphi,\vartheta,\psi)$ instead of $D_{mn}^{l}(\Phi(\varphi,\vartheta,\psi))$ for the elements of D^{l} as a function of the Euler angles (φ,ϑ,ψ) (see (3.1)). These elements are known as *Wigner's D functions*. Writing as before $\alpha = \cos\frac{\vartheta}{2}\exp\left\{i\left(\frac{\varphi+\psi}{2}\right)\right\}$, $\beta = \sin\frac{\vartheta}{2}\exp\left\{i\left(\frac{\varphi-\psi}{2}\right)\right\}$, substituting into (3.11) and performing some manipulations, we obtain (see also Koornwinder [122], Vilenkin and Klimyk [197])

$$D_{mn}^{l}(\varphi,\vartheta,\psi) = e^{-im\varphi}d_{mn}^{l}(\vartheta)e^{-in\psi}, \quad m,n = -l,...,l,$$

(3.13)

where $d_{mn}^{l}(.)$ is a real function given by

$$d_{mn}^{l}(\vartheta) = (-1)^{l-n}\left[(l+m)!(l-m)!(l+n)!(l-n)!\right]^{1/2}$$

(3.14)

$$\times\sum_{k}(-1)^{k}\frac{\left(\cos\frac{\vartheta}{2}\right)^{m+n+2k}\left(\sin\frac{\vartheta}{2}\right)^{2l-m-n-2k}}{k!(l-m-k)!(l-n-k)!(m+n+k)!},$$

(3.15)

and the sum runs over all integers k such that the factorials are nonnegative; see also [195, Chapter 4] for a huge collection of alternative expressions. In particular, the following set of symmetry properties hold:

$$d_{mn}^{l}(\vartheta) = (-1)^{m-n}d_{-m-n}^{l}(\vartheta) = (-1)^{m-n}d_{nm}^{l}(\vartheta) = d_{-n-m}^{l}(\vartheta),\quad (3.16)$$

$$d_{mn}^{l}(-\vartheta) = (-1)^{m-n}d_{mn}^{l}(\vartheta) = d_{nm}^{l}(\vartheta),\quad (3.17)$$

$$d_{mn}^{l}(\pi-\vartheta) = (-1)^{l-n}d_{-mn}^{l}(\vartheta) = (-1)^{l+m}d_{m-n}^{l}(\vartheta).\quad (3.18)$$

Remark 3.8 The matrix D^{l} can be expressed in terms of the Euler angles $\hat{\varphi},\hat{\vartheta},\hat{\psi}$ introduced in Remark 3.4. This can be done by using the straightforward relation

$$D_{mn}^{l}(\varphi,\vartheta,\psi) = i^{m-n}D_{mn}^{l}(\hat{\varphi},\hat{\vartheta},\hat{\psi}).$$

See [197, Chapter 6].

In view of (3.16)-(3.18), it is sufficient to look for an analytic expression of $d_{mn}^{l}(\vartheta)$ when $m-n, m+n \geq 0$; all the other cases can then be recovered by (3.16)-(3.17). For this purpose, fix nonnegative reals $a, b > -1$, and let $\{P_n^{(a,b)}(.) : n = 0, 1, 2, ...\}$ denote the sequence of Jacobi polynomials with parameters a, b (see the Appendix for more details), that is, polynomials on the interval $[-1, 1]$ that are orthogonal with respect to the measure

$$w_{ab}(x)dx = (1-x)^a(1+x)^b dx .$$

More precisely, Jacobi polynomials are completely determined by the following relation, valid for every $n_1, n_2 \geq 0$,

$$\int_{-1}^{1} P_{n_1}^{(a,b)}(x)P_{n_2}^{(a,b)}(x)w_{ab}(x)dx = \frac{2^{a+b+1}}{2n_1+a+b+1} \frac{\Gamma(n_1+a+1)\Gamma(n_1+b+1)}{\Gamma(n_1+a+b+1)n_1!}\delta_{n_1}^{n_2} . \tag{3.19}$$

Now, using the trigonometric identities $(0 \leq \vartheta \leq \pi)$

$$\sin\frac{\vartheta}{2} = \sqrt{\frac{1-\cos\vartheta}{2}} , \quad \cos\frac{\vartheta}{2} = \sqrt{\frac{1+\cos\vartheta}{2}}$$

and some manipulations, (3.15) yields

$$d_{mn}^l(\vartheta) = 2^{-m}\left[\frac{(l-m)!(l+m)!}{(l-n)!(l+n)!}\right]^{\frac{1}{2}}(1-\cos\vartheta)^{\frac{m+n}{2}}(1+\cos\vartheta)^{\frac{m-n}{2}}P_{l-m}^{(m-n,m+n)}(\cos\vartheta) \tag{3.20}$$

$$= 2^{-m}\left[\frac{(l-m)!(l+m)!}{(l-n)!(l+n)!}\right]^{\frac{1}{2}}\{w_{m-n,m+n}(\cos\vartheta)\}^{\frac{1}{2}}P_{l-m}^{(m-n,m+n)}(\cos\vartheta) \tag{3.21}$$

(compare for instance with [195, formula 4.3.4.13], and also with [197, p. 285-288], where analogous expressions are deduced in terms of the Euler angles of Remark 3.4). The relationship between Wigner's D functions and Jacobi polynomials will be exploited in the Appendix to verify the orthonormality properties of the elements of $\{D^l\}$.

Remark 3.9   From (3.13), (3.20-3.21), we have that

$$|D_{mn}^l(\varphi, \vartheta, \psi)|^2 = 2^{-2m}\left[\frac{(l-m)!(l+m)!}{(l-n)!(l+n)!}\right]\{w_{m-n,m+n}(\cos\vartheta)\}\left\{P_{l-m}^{(m-n,m+n)}(\cos\vartheta)\right\}^2 ,$$

which is a polynomial of degree $2l$ in $\cos\vartheta = x_3$ (the choice of the symbol "x_3" will become clear in Section 3.4).

Example 3.10   For $l = 0$, we have trivially $d_{00}^0(\vartheta) = D_{00}^0(\varphi, \vartheta, \psi) \equiv 1$. For $l = 1$, $\{d_{mn}^1(\vartheta)\}$ is provided by:

$m\backslash n$	-1	0	1
-1	$\frac{1+\cos\vartheta}{2}$	$\frac{\sin\vartheta}{\sqrt{2}}$	$\frac{1-\cos\vartheta}{2}$
0	$-\frac{\sin\vartheta}{\sqrt{2}}$	$\cos\vartheta$	$\frac{\sin\vartheta}{\sqrt{2}}$
1	$\frac{1-\cos\vartheta}{2}$	$-\frac{\sin\vartheta}{\sqrt{2}}$	$\frac{1+\cos\vartheta}{2}$

Note that

$$\left|D^1_{00}\right|^2 = \cos^2\vartheta = x_3^2,$$

$$\left|D^1_{11}\right|^2 = \left|D^1_{-1,-1}\right|^2 = \frac{1}{4}\{1+\cos\vartheta\}^2 = \frac{1}{4}\{1+x_3\}^2,$$

$$\left|D^1_{1,-1}\right|^2 = \left|D^1_{-1,1}\right|^2 = \frac{1}{4}\{1-\cos\vartheta\}^2 = \frac{1}{4}\{1-x_3\}^2,$$

$$\left|D^1_{10}\right|^2 = \left|D^1_{0,1}\right|^2 = \left|D^1_{-1,0}\right|^2 = \left|D^1_{0,-1}\right|^2 = \frac{1}{4}\sin^2\vartheta = \frac{1}{4}\{1-x_3^2\}.$$

Example 3.11 For $l = 2$, $\left\{d^2_{mn}(\vartheta)\right\}$ is provided by:

$m\backslash n$	-2	-1	0	1	2
-2	$\frac{(1+\cos\vartheta)^2}{4}$	$\frac{\sin\vartheta(1+\cos\vartheta)}{2}$	$\frac{1}{2}\sqrt{\frac{3}{2}}\sin^2\vartheta$	$\frac{\sin\vartheta(1-\cos\vartheta)}{2}$	$\frac{(1-\cos\vartheta)^2}{4}$
-1	$\frac{-\sin\vartheta(1+\cos\vartheta)}{2}$	$\frac{2\cos^2\vartheta+\cos\vartheta-1}{2}$	$\sqrt{\frac{3}{2}}\sin\vartheta\cos\vartheta$	$\frac{-2\cos^2\vartheta-\cos\vartheta-1}{2}$	$\frac{\sin\vartheta(1-\cos\vartheta)}{2}$
0	$\frac{1}{2}\sqrt{\frac{3}{2}}\sin^2\vartheta$	$-\sqrt{\frac{3}{2}}\sin\vartheta\cos\vartheta$	$\frac{3\cos^2\vartheta-1}{2}$	$\sqrt{\frac{3}{2}}\sin\vartheta\cos\vartheta$	$\frac{1}{2}\sqrt{\frac{3}{2}}\sin^2\vartheta$
1	$\frac{-\sin\vartheta(1-\cos\vartheta)}{2}$	$\frac{-2\cos^2\vartheta-\cos\vartheta-1}{2}$	$-\sqrt{\frac{3}{2}}\sin\vartheta\cos\vartheta$	$\frac{2\cos^2\vartheta+\cos\vartheta-1}{2}$	$\frac{\sin\vartheta(1+\cos\vartheta)}{2}$
2	$\frac{(1-\cos\vartheta)^2}{4}$	$-\frac{\sin\vartheta(1-\cos\vartheta)}{2}$	$\frac{1}{2}\sqrt{\frac{3}{2}}\sin^2\vartheta$	$-\frac{\sin\vartheta(1+\cos\vartheta)}{2}$	$\frac{(1+\cos\vartheta)^2}{4}$

Once again, it is immediate to check that $\left|D^2_{mn}\right|^2$ is a polynomial of order 4 in the variable $\cos\vartheta = x_3$.

We are now going to show that the representations of $SU(2)$ given by the family $\left\{(\mathcal{H}_l, D^l) : l = 0, \frac{1}{2}, 1, ...\right\}$ are (i) irreducible, and (ii) such that their equivalence classes are exhaustive of the dual $\widehat{SU(2)}$ (see Section 2.4.2 for a formal definition of the dual of a topological compact group). Let us first note that the character (see Section 2.4.5) χ_l of Wigner's representation $(\mathcal{H}_l, D^l)$, $l = 0, \frac{1}{2}, ...$ can be expressed in terms of the Euler's angles as

$$\chi_l(\varphi, \vartheta, \psi) = \sum_{m=-l}^{l} D^l_{mm}(\varphi, \vartheta, \psi) = \sum_{m=-l}^{l} \exp\{-im(\psi + \varphi)\} d^l_{mm}(\vartheta),$$

For instance,

$$\chi_1(\varphi, \vartheta, \psi) = \cos\vartheta + \frac{1+\cos\vartheta}{2}\left[\exp\{-i(\psi + \varphi)\} + \exp\{i(\psi + \varphi)\}\right]$$

$$= \cos\vartheta\left[1 + \cos(\psi + \varphi)\right] + \cos(\psi + \varphi).$$

An alternative expression for the character of D^l can be deduced as follows. We recall the character is a *central*, or conjugacy invariant, function (this means that it is invariant over conjugacy classes $g = g_1 u g_1^{-1}$). It is easily checked that the elements of $SU(2)$ can be diagonalized as $g = Q \Lambda Q^{-1}$, where $Q \in SU(2)$ and $\Lambda = \text{diag} \left\{ \exp(i\frac{t}{2}), \exp(-i\frac{t}{2}) \right\}$ is the matrix having the two eigenvalues of g (written

$$\exp \left(i\frac{t}{2} \right) \quad \text{and} \quad \exp \left(-i\frac{t}{2} \right)$$

for some $0 \leq t \leq 2\pi$) on the main diagonal, and equal to zero elsewhere. Clearly, $D^l(\Lambda)$ is given by the diagonal matrix

$$D^l(\Lambda) = \text{diag} \left\{ \exp(ilt), ..., \exp(-ilt) \right\} ,$$

(just use (3.13)), so that the corresponding character is given by

$$\chi_l(g) = \chi_l(Q \Lambda Q^{-1}) = \chi_l(\Lambda) = \sum_{k=-l}^{l} \exp \left\{ -ikt \right\}$$

$$= \frac{\exp(-ilt) - \exp(i(l+1)t)}{1 - \exp\{it\}} = \frac{\sin \left\{ \frac{2l+1}{2} t \right\}}{\sin \left\{ \frac{t}{2} \right\}} .$$

Note that

$$\frac{\sin \left\{ \frac{2l+1}{2} t \right\}}{\sin \left\{ \frac{t}{2} \right\}} = U_{2l} \left(\cos \frac{t}{2} \right) , \tag{3.22}$$

where $U_{2l}(.)$ denotes the *Chebyshev polynomial of the second kind* of order $2l$; see the Appendix for more details.

Remark 3.12 The collection of the Chebyshev polynomials of the second kind $\{U_n : n \geq 0\}$ is completely characterized by the relation

$$\frac{2}{\pi} \int_{-1}^{1} U_m(x) U_n(x) (1 - x^2)^{\frac{1}{2}} dx = \delta_m^n,$$

that is, the family $\{U_n : n \geq 0\}$ is orthonormal with respect to the measure

$$\frac{2}{\pi} w_{\frac{1}{2} \frac{1}{2}}(x) dx = \frac{2}{\pi} (1 - x^2)^{1/2} dx.$$

Using (3.19) one can therefore rewrite U_n in terms of the Jacobi polynomials of index $(\frac{1}{2}, \frac{1}{2})$, that is

$$U_n(x) = \frac{P_n^{(\frac{1}{2}, \frac{1}{2})}(x)}{2} \times \frac{2 \cdot 4 \cdots (2n+2)}{1 \cdot 3 \cdots (2n+1)},$$

see [193, formula (4.17)].

A quick analysis of the characteristic equation of the matrix g (see [197] p. 359) reveals that the relationship between t and the Euler angles $(\varphi, \vartheta, \psi)$ of g is given by

$$\cos \frac{t}{2} = \cos \frac{\vartheta}{2} \cos \frac{\varphi + \psi}{2} . \tag{3.23}$$

Remark 3.13 Suppose that the function f on $SU(2)$ is central, that is, $f(g) = f(hgh^{-1})$, for every $g, h \in SU(2)$. Then, arguing as above yields that $f(g) = f(t)$, that is: with a slight abuse of notation, f depends on g only through the parameter $t \in [0, 2\pi]$ determining the two eigenvalues e^{it} and e^{-it} of g. Using (3.23) together with the characterization of the Haar measure of $SU(2)$ given in (2.7), it is therefore possible to show that, for every integrable central f,

$$\int_{SU(2)} f(g)dg = \frac{1}{\pi} \int_0^{2\pi} f(t) \sin^2 \frac{t}{2} dt \tag{3.24}$$

(see [197, p. 262]).

We shall now exploit the properties of characters, as discussed in Section 2.4.5, in order to establish crucial properties of the Wigner's representations $\{D^l\}$. More precisely, one has the following

Theorem 3.14 *(1) The representations of $SU(2)$ given by $(\mathcal{H}_l, D^l)$, $l = 0, \frac{1}{2}, ...,$ are irreducible.*

(2) Every irreducible unitary representation of $SU(2)$ is equivalent to one of the representations D^l, $l = 0, \frac{1}{2}, 1, ...,$ that is

$$\widehat{SU(2)} = \left\{ [D^l] : l = 0, \frac{1}{2}, 1, ... \right\}.$$

Proof (1) Fix, $l, k = 0, \frac{1}{2}, ...$ Using (3.24), (3.22) and the fact that characters are central functions, it follows that

$$\int_{SU(2)} \chi_l(g)\overline{\chi_k(g)}dg = \frac{1}{\pi} \int_0^{2\pi} U_{2l}\left(\cos \frac{t}{2}\right) U_{2k}\left(\cos \frac{t}{2}\right) \sin^2 \frac{t}{2} dt$$

$$= \frac{2}{\pi} \int_{-1}^1 U_{2l}(x)U_{2k}(x)(1 - x^2)^{1/2}dx = \delta_k^l.$$

By virtue of Theorem 2.41, this proves that the representations $(D^l, \mathcal{H}_l)$, $l = 0, \frac{1}{2}, ...$ are not equivalent and irreducible.

(2) Owing to Remark 2.48-(3), it is sufficient to show that the collection $\{\chi_l : l \geq 0\}$ is an orthonormal basis of the subspace of $L^2(SU(2))$ composed of central functions. To prove this, consider a generic central function $f \in$

$L^2(SU(2))$, and assume that f is orthogonal to $\{\chi_l : l \geq 0\}$. This is equivalent to the fact that, for every semi-integer l,

$$0 = \int_{SU(2)} f(g)\overline{\chi_l(g)}dg = \frac{1}{\pi}\int_0^{2\pi} f(t)U_{2l}\left(\cos\frac{t}{2}\right)\sin^2\frac{t}{2}\,dt.$$

After the change of variable $x = \cos\frac{t}{2}$, the last relation yields that $f(t) = 0$ for almost every t, since Chebyshev polynomials constitute an orthonormal basis of the Hilbert space of the functions on $[-1, 1]$ that are square-integrable with respect to $(2/\pi)(1 - x^2)^{1/2}dx$. The conclusion follows. $\qquad\square$

3.3.3 Further properties

In general, as l grows the expressions of Wigner's D matrices and functions become hardly manageable and it is difficult to derive any property from analytic manipulations. On the other hand, several properties of these objects can be derived as special cases of the general theory of representations for compact groups outlined in the previous Chapter 2.

- *Additive Properties.* Since each D^l is an (irreducible) matrix representation of $SU(2)$, for all $g_1, g_2 \in SU(2)$ we have

$$D^l(g_1)D^l(g_2) = D^l(g_1g_2)\,,\, l = 0, \frac{1}{2}, 1...,$$

which is equivalent to the relation

$$D^l_{mn}(g_1g_2) = \sum_{k=-l}^{l} D^l_{mk}(g_1)D^l_{kn}(g_2)\,.$$

This property follows immediately from the definition of a matrix representation, and as we shall see it represents an extremely valuable tool for the analysis of stochastic properties of isotropic spherical random fields. In terms of Euler angles, the expression becomes more complicated, and we have

$$D^l_{mn}(\varphi, \vartheta, \psi) = \sum_{k=-l}^{l} D^l_{mk}(\varphi_1, \vartheta_1, \psi_1)D^l_{kn}(\varphi_2, \vartheta_2, \psi_2) \qquad (3.25)$$

where the angles of the resultant rotation $(\varphi, \vartheta, \psi)$ can be expressed in terms of the angles of the original rotations $(\varphi_1, \vartheta_1, \psi_1), (\varphi_2, \vartheta_2, \psi_2)$ by means of the implicit equations (when they are well-defined – see [195], page 32)

$$\cot(\varphi - \varphi_2) = \cos\vartheta_2\cot(\varphi_1 + \psi_2) + \cot\vartheta_1\frac{\sin\vartheta_2}{\sin(\varphi_1 + \psi_2)}\,, \qquad (3.26)$$

$$\cos\vartheta = \cos\vartheta_1\cos\vartheta_2 - \sin\vartheta_1\sin\vartheta_2\cos(\varphi_1 + \psi_2)\,, \qquad (3.27)$$

$$\cot(\psi - \psi_1) = \cos\beta_1 \cot(\varphi_1 + \psi_2) + \cot\vartheta_2 \frac{\sin\vartheta_1}{\sin(\varphi_1 + \psi_2)} . \qquad (3.28)$$

These relationships can be established using 3.4 to write down an analytic form for the equality $R = R_1 R_2$ and then exploting standard trigonometric identities.

- *Orthonormality.* As an immediate consequence of the Schur's orthogonality relations of Theorem 2.33, we have that

$$\int_{SU(2)} D_{mn}^l(g)\overline{D}_{m'n'}^{l'}(g)dg = \frac{1}{2l+1}\delta_l^{l'}\delta_m^{m'}\delta_n^{n'} , \qquad (3.29)$$

see also Remark 2.35.

- *Unitary Properties.* Since the matrices D^l are unitary, we have that

$$\sum_m D_{mn}^l(g)\overline{D}_{mn'}^l(g) = \delta_n^{n'} . \qquad (3.30)$$

3.3.4 The dual of $SO(3)$

We will now provide an explicit relation between the representations of $SU(2)$ and the representations of $SO(3)$. We start with a simple result.

Lemma 3.15 *(1) For every integer $l = 0, 1, 2, ...$, and for every $g \in SU(2)$,*
$$D^l(g) = D^l(-g).$$
(2) For every $l = \frac{1}{2}, \frac{3}{2}, \frac{5}{2}, ...$, and for every $g \in SU(2)$, $D^l(g) = -D^l(-g)$.

Proof It is convenient to work with Euler angles. For every

$$(\varphi, \vartheta, \psi) \in [0, 2\pi) \times [0, \pi] \times [-2\pi, 2\pi),$$

denote by $\Phi(\varphi, \vartheta, \psi)$ the element of $SU(2)$ defined by (3.1). Then, for every $\psi \in [0, 2\pi)$ we have that $-\Phi(\varphi, \vartheta, \psi) = \Phi(\varphi, \vartheta, \psi - 2\pi)$, and one can use (3.13) to deduce that $D_{mn}^l(\varphi, \vartheta, \psi) = D_{mn}^l(\varphi, \vartheta, \psi - 2\pi)$ whenever l, m, n are integers (thus proving Part (1)), and

$$D_{mn}^l(\varphi, \vartheta, \psi) = -D_{mn}^l(\varphi, \vartheta, \psi - 2\pi),$$

whenever l, m, n are equal to odd multiples of $\frac{1}{2}$ (thus proving Part (2)). To deal with the case $\psi \in [-2\pi, 0)$, replace $\psi - 2\pi$ with $\psi + 2\pi$ and repeat the same argument. □

Now recall that, in view of the results of Section 3.2.1, there exists a surjective homomorphism T of $SU(2)$ onto $SO(3)$ such that

$$SU(2) \xrightarrow{T} SO(3), \quad T(A) = T(B) \text{ if and only if } A = \pm B.$$

Equivalently, for every $R \in SO(3)$, we have that $T^{-1}(R) = \{A, -A\}$ for some $A \in SU(2)$. By virtue of Part (1) of Lemma 3.15, for every integer $l \geq 0$, the mapping on $SO(3)$ given by $g \mapsto D^l \circ T^{-1}(g) := D^l(T^{-1}(g))$ is well-defined and yields a representation of $SO(3)$.

The next statement follows immediately from Theorem 3.14.

Lemma 3.16 *For every integer $l = 0, 1, 2, ...,$ the pair $(D^l \circ T^{-1}, \mathcal{H}_l)$ defines an irreducible representation of $SO(3)$.*

Remark 3.17 As before, denote by $R = R(\varphi, \vartheta, \psi)$ the rotation associated with the Euler angles $(\varphi, \vartheta, \psi) \in [0, 2\pi) \times [0, \pi] \times [0, 2\pi)$, via Proposition 3.1. Then, for every $l = 0, 1, 2, ...,$ and every $m, n = -l, ..., l,$

$$(D^l \circ T^{-1})_{mn}(R) = D^l_{mn}(T^{-1}(R(\varphi, \vartheta, \psi))) = D^l_{mn}(\varphi, \vartheta, \psi).$$

In this sense, for every integer l the restriction of the matrices D^l to the set $[0, 2\pi) \times [0, \pi] \times [0, 2\pi)$ is a representation of $SO(3)$.

Theorem 3.18 *Every irreducible unitary representation of $SO(3)$ is equivalent to one of the representations $D^l \circ T^{-1}$, $l = 0, 1, 2, ...$ In particular,*

$$\widehat{SO(3)} = \left\{ [D^l \circ T^{-1}] : l = 0, 1, 2, ... \right\}.$$

Proof Let σ be any irreducible representation of $SO(3)$ on some vector space W with finite dimension. The composite mapping $\sigma \circ T$, where T is the homomorphism discussed above, is obviously a representation of $SU(2)$ on W. Since σ is irreducible, $\sigma \circ T$ is irreducible and moreover, since $\sigma \circ T(g) = \sigma \circ T(-g)$, Part (1) of Lemma 3.15 implies that $\sigma \circ T$ must be equivalent to some representation D^l, for $l = 0, 1, 2, ...$ Since in this case $D^l(g) = D^l \circ T^{-1}(T(g))$, the conclusion follows immediately. $\square$

Using Remark 3.17 together with Schur's orthogonality relations of Section 2.4.4 and the characterization of the Haar measure of $SO(3)$ (see (3.5)), we deduce the following orthonormality relations for integer Wigner D matrices.

Corollary 3.19 *For every integers $l, l' = 0, 1, 2, ...$*

$$\int_{SO(3)} (D^l \circ T^{-1})_{mn}(g)\overline{(D^{l'} \circ T^{-1})}_{m'n'}(g)dg \tag{3.31}$$

$$= \frac{1}{8\pi^2} \int_0^{2\pi} \int_0^{\pi} \int_0^{2\pi} D^l_{mn}(\varphi, \vartheta, \psi)\overline{D}^{l'}_{m'n'}(\varphi, \vartheta, \psi)d\varphi \sin \vartheta d\vartheta d\psi$$

$$= \frac{1}{2l+1}\delta^{l'}_l \delta^{m'}_m \delta^{n'}_n.$$

Remark 3.20 The second equality in (3.31) can alternatively be deduced from (3.29), by using the characterization of the Haar measure on $SU(2)$ given in (2.7), as well as the fact that, for $\psi \in [0, 2\pi)$ and for l integer, $D^l(\varphi, \vartheta, \psi) = D^l(\varphi, \vartheta, \psi - 2\pi)$.

We have shown that Wigner's D matrices provide a complete set of unitary irreducible representations for $SO(3)$. Building on Theorem (2.47), this fact will allow us to deduce a crucial result for this monograph, namely the *spectral representation theorem* for square integrable functions on the sphere. To this issue we devote the next Section.

3.4 Spherical harmonics and Fourier analysis on S^2

3.4.1 Spherical harmonics and Wigner's D^l matrices

We recall first the standard spherical coordinates: for every $x = (x_1, x_2, x_3) \in \mathbb{R}^3$,

$$x_1 = r \sin \vartheta \cos \varphi , \tag{3.32}$$

$$x_2 = r \sin \vartheta \sin \varphi , \tag{3.33}$$

$$x_3 = r \cos \vartheta , \tag{3.34}$$

where $r \in [0, \infty)$, $\vartheta \in [0, \pi]$, $\varphi \in [0, 2\pi)$.

Remark 3.21 The three parameters r, ϑ, φ are provided by the following relations:

(a) If $x = 0$, then $r = 0$, and one can choose any value for ϑ, φ.
(b) if $x_1 = x_2 = 0$ and $x_3 \neq 0$, then $r = \|x_3\|$, $\vartheta = \pi, 0$, according as $x_3 < 0$ or $x_3 > 0$, and one can choose any value for φ.
(c) In any other case, the three parameters r, ϑ, φ are uniquely determined as:
(i) $r = \|x\|_{\mathbb{R}^3}$, (ii) $\vartheta = \arccos(x_3/\|x\|_{\mathbb{R}^3})$, and

$$\text{(iii)} \; \varphi = \arccos \left(\frac{x_1}{\sqrt{x_1^2 + x_2^2}} \right).$$

Remark 3.22 Using (3.32)-(3.34), every triple $(r, \vartheta, \varphi) \in [0, \infty) \times [0, \pi] \times [0, 2\pi)$ uniquely determines one point $(x_1, x_2, x_3) \in \mathbb{R}^3$.

Throughout this book, we shall be concerned with the unit sphere $S^2 = \{x_1, x_2, x_3 : x_1^2 + x_2^2 + x_3^2\}$; here, the case (a) considered in Remark 3.21 is immaterial, and in the case (c) the change of variables is trivially provided by the expressions

$$r = 1, \quad \vartheta = \arccos x_3 = \arccos \left[\text{sign}(x_3) \sqrt{1 - x_1^2 - x_2^2} \right],$$

where $\text{sign}(x) = 1$ for $x \geq 0$ and $= -1$ otherwise, and

$$\varphi = \arccos \left(\frac{x_1}{\sqrt{x_1^2 + x_2^2}} \right).$$

Let us now introduce the class of *spherical harmonics*

$$\{Y_{lm}(\vartheta, \varphi) : l = 0, 1, 2, \ldots; \ m = -l, \ldots, l\}.$$

Definition 3.23   For every integer $l = 0, 1, 2, \ldots$, and every $m = -l, \ldots, l,$, the **spherical harmonic function** of index (l, m), written $Y_{lm} : S^2 \to \mathbb{C}$, is defined (in spherical coordinates) as the mapping

$$(\vartheta, \varphi) \mapsto Y_{lm}(\vartheta, \varphi) := \sqrt{\frac{2l+1}{4\pi}} \overline{D}_{m0}^l(\varphi, \vartheta, \cdot) \tag{3.35}$$

$$= \sqrt{\frac{2l+1}{4\pi}} D_{0-m}^l(\cdot, \vartheta, \varphi) = \sqrt{\frac{2l+1}{4\pi}} d_{0-m}^l(\vartheta) e^{im\varphi} \tag{3.36}$$

$$= (-1)^m \sqrt{\frac{2l+1}{4\pi}} D_{-m0}^l(\varphi, \vartheta, \cdot) = (-1)^m \sqrt{\frac{2l+1}{4\pi}} \overline{D}_{0m}^l(\cdot, \vartheta, \varphi),$$

where $(\vartheta, \varphi) \in [0, \pi] \times [0, 2\pi)$.

In other words, the spherical harmonics correspond (up to a normalization factor) to the elements of the "central" column (or row) in the Wigner's D matrix. In all the previous expressions, the "dot" stands for an arbitrary angle.

Remark 3.24   The following definition is equivalent to 3.23 (see for instance [195, p. 133] or [127, p. 330]):

$$Y_{lm}(\vartheta, \varphi) = \sqrt{\frac{2l+1}{4\pi} \frac{(l-m)!}{(l+m)!}} P_{lm}(\cos \vartheta) \exp(im\varphi), \ m \geq 0,$$

$$Y_{lm}(\vartheta, \varphi) = (-1)^m \overline{Y}_{l-m}(\vartheta, \varphi), \ m < 0.$$

where $\{P_{lm}\}$ denote the *associated Legendre function* (see Appendix), which is defined in terms of the *Legendre polynomials* $\{P_l : l = 0, 1, \ldots\}$ by the equation

$$P_{lm}(\mu) = (-1)^m (1 - \mu^2)^{m/2} \frac{d^m}{d\mu^m} P_l(\mu), \ m = 0, 1 .., l, \ l = 0, 1, 2, \ldots.$$

(in particular, $P_{l0} = P_l$). The following inverse relations hold:

$$D_{0m}^l(\cdot, \vartheta, \varphi) = \sqrt{\frac{4\pi}{2l+1}} Y_{l-m}(\vartheta, \varphi) = (-1)^m \sqrt{\frac{4\pi}{2l+1}} \overline{Y}_{lm}(\vartheta, \varphi). \qquad (3.37)$$

Remark 3.25 (On notation.) In the sequel, we shall use the same notation Y_{lm} to denote spherical harmonics as a function of both the spherical and the Euclidean coordinates, i.e. for all $x \in S^2$, we shall write (with a slight abuse of notation)

$$Y_{lm}(x) := Y_{lm}(\vartheta, \varphi), \quad x = (\sin \vartheta \cos \varphi, \sin \vartheta \sin \varphi, \cos \vartheta).$$

We shall also use $d\sigma(x)$ to denote the *Lebesgue measure* on the sphere, which, in spherical coordinates is defined as

$$d\sigma(x) := \sin \vartheta d\vartheta d\varphi,$$

meaning that, for every bounded function on S^2,

$$\int_{S^2} f(x) d\sigma(x) = \int_0^{2\pi} \int_0^\pi f(\sin \vartheta \cos \varphi, \sin \vartheta \sin \varphi, \cos \vartheta) \sin \vartheta d\vartheta d\varphi$$

Using the content of Remark 3.24, it is immediate to check that

$$Y_{l0}(\vartheta, \varphi) = \sqrt{\frac{2l+1}{4\pi}} P_l(\cos \vartheta), \text{ for all } \varphi,$$

and because $P_l(1) \equiv 1$, for all $l = 1, 2, \ldots$

$$Y_{l0}(0, 0) = \sqrt{\frac{2l+1}{4\pi}} \text{ for all } l = 0, 1, 2, \ldots.$$

3.4.2 Some properties of spherical harmonics

In the sequel, for any $x, y \in S^2$ we shall label $d(x, y) := \arccos(\langle x, y \rangle)$ the usual spherical distance, i.e. the angle between x, y, where $\langle ., . \rangle$ denotes the Euclidean inner product

$$\langle x, y \rangle := \langle x, y \rangle_{\mathbb{R}^3} = x_1 y_1 + x_2 y_2 + x_3 y_3 \qquad (3.38)$$

$$= \sin \vartheta_x \sin \vartheta_y \{\cos \varphi_x \cos \varphi_y + \sin \varphi_x \sin \varphi_y\} + \cos \vartheta_x \cos \vartheta_y$$

$$= \sin \vartheta_x \sin \vartheta_y \{\cos (\varphi_x - \varphi_y)\} + \cos \vartheta_x \cos \vartheta_y,$$

$$x = (\sin \vartheta_x \cos \varphi_x, \sin \vartheta_x \sin \varphi_x, \cos \vartheta_x), \ y = (\sin \vartheta_y \cos \varphi_y, \sin \vartheta_y \sin \varphi_y, \cos \vartheta_y).$$

In view of (3.35) and (3.37), it is straightforward to derive a set of properties of the spherical harmonics, which we shall exploit heavily in the sequel of this book.

- *Orthonormality*: for all l, l', m, m'

$$\int_0^\pi \int_0^{2\pi} Y_{lm}(\vartheta, \varphi)\overline{Y_{l'm'}}(\vartheta, \varphi) \sin\vartheta d\varphi d\vartheta \tag{3.39}$$

$$= \frac{\sqrt{(2l+1)(2l'+1)}}{8\pi^2} \int_0^{2\pi} \int_0^\pi \int_0^{2\pi} \overline{D}_{m0}^l(\varphi, \vartheta, \psi)D_{m'0}^{l'}(\varphi, \vartheta, \psi) \sin\vartheta d\psi d\vartheta d\varphi$$

$$= \delta_l^{l'}\delta_m^{m'},$$

in view of (3.37) and the orthogonality properties of elements in the irre-ducible matrix representations of $SO(3)$ given in (3.31). In particular,

$$\int_0^\pi \int_0^{2\pi} Y_{lm}(\vartheta, \varphi) \sin\vartheta d\varphi d\vartheta = \sqrt{4\pi} \int_0^\pi \int_0^{2\pi} Y_{lm}(\vartheta, \varphi)\overline{Y}_{00}(\vartheta, \varphi) \sin\vartheta d\varphi d\vartheta$$

$$= 0, \quad \text{for all } l > 0.$$

- *Symmetry*: for all $x \in S^2$

$$\overline{Y}_{lm}(x) \equiv (-1)^m Y_{l-m}(x), \tag{3.40}$$

and in spherical coordinates

$$\overline{Y}_{lm}(\vartheta, \varphi) = Y_{lm}(\vartheta, -\varphi). \tag{3.41}$$

These are immediate consequences of the definition of spherical harmonics and (3.16).

- *Addition Formula*: for all $x, y \in S^2$

$$\sum_{m=-l}^l \overline{Y}_{lm}(x)Y_{lm}(y) = \frac{2l+1}{4\pi}P_l(\langle x, y\rangle), \tag{3.42}$$

where P_l is the lth Legendre polynomial (see Remark 3.24). In particular, for all $x \in S^2$,

$$\sum_{m=-l}^l Y_{lm}(x)\overline{Y}_{lm}(x) = \frac{2l+1}{4\pi}.$$

To prove this relationships, we define an application $\Phi_{S^2} : S^2 \to SO(3)$, such that

$$\Phi_{S^2}(\sin\vartheta\cos\varphi, \sin\vartheta\sin\varphi, \cos\vartheta) = (\varphi, \vartheta, 0).$$

Set $g_x := \Phi_{S^2}(\varphi_x, \vartheta_x, 0)$, $g_y := \Phi_{S^2}(\varphi_y, \vartheta_y, 0)$ to obtain

$$\sum_{m=-l}^l \overline{Y}_{lm}(x)Y_{lm}(y) = \frac{2l+1}{4\pi}\sum_{m=-l}^l D_{m0}^l(g_x)\overline{D}_{m0}^l(g_y)$$

$$= \frac{2l+1}{4\pi} \sum_{m=-l}^{l} D_{m0}^l(g_x) D_{0m}^l(g_y^{-1})$$

$$= \frac{2l+1}{4\pi} D_{00}^l(g_y^{-1} g_x) . \qquad (3.43)$$

Now,

$$D_{m0}^l(g_x) = D_{m0}^l(\varphi_x, \vartheta_x, 0) , \; D_{0m}^l(g_y^{-1}) = D_{0m}^l(0, -\vartheta_y, -\varphi_y) ,$$

so that from (3.27) we obtain

$$D_{00}^l(g_y^{-1} g_x) = P_l(\cos \beta),$$

where

$$\cos \beta = \cos \vartheta_x \cos \vartheta_y + \sin \vartheta_x \sin \vartheta_y \cos(\varphi_x - \varphi_y) = \langle x, y \rangle ,$$

as desired.

- *Behaviour under rotations*: for all $g \in SO(3)$ and all $x \in S^2$, denote by $g \cdot x$ the image of x through the rotation g (that is, $g \cdot x$ indicates the position of x after the rotation g is applied to the sphere). Then, one has

$$Y_{lm}(g \cdot x) = \sum_{m'} D_{m'm}^l(g^{-1}) Y_{lm'}(x) . \qquad (3.44)$$

Again, this is just an immediate consequence of (3.37) and the properties of the matrices D^l. Indeed

$$Y_{lm}(g \cdot x) = \sqrt{\frac{2l+1}{4\pi}} \overline{D}_{m0}^l(g_{g \cdot x})$$

$$= \sqrt{\frac{2l+1}{4\pi}} \sum_{m'} \overline{D}_{mm'}^l(g) \overline{D}_{m'0}^l(g_x)$$

$$= \sqrt{\frac{2l+1}{4\pi}} \sum_{m'} D_{m'm}^l(g^{-1}) \overline{D}_{m'0}^l(g_x) = \sum_{m'} D_{m'm}^l(g^{-1}) Y_{lm'}(x) ,$$

where the second step follows from

$$D_{m0}^l(g_{g \cdot x}) = D_{m0}^l(\Phi_{S^2}(g \cdot x)) = D_{m0}^l(g \Phi_{S^2}(x)) = D_{m0}^l(g g_x) .$$

Remark 3.26 Write $Y_{l.}(x)$ for the $(2l+1) \times 1$ column vector s.t. $Y_{l.}' = \{Y_{l,-l}, ..., Y_{ll}\}$, so that (3.44) becomes

$$Y_{l.}(g \cdot x) = D^l(g^{-1})' Y_{l.}(x) .$$

Now fix $l = 0, 1, ...,$ let $a'_{l.} = \{a_{l,-l}, ..., a_{l,l}\}$ be a vector of complex valued coefficients, and consider the function

$$f_l(x) = Y'_{l.}(x)a_{l.} = \sum_{m=-l}^{l} a_{lm} Y_{lm}(x).$$

We write V_l to indicate the vector space of complex-valued functions on the unit sphere, having the form of f_l. For a generic (square-integrable) function f on S^2, the *left action* of $SO(3)$ on f is given by

$$L(g)f(x) = f(g^{-1} \cdot x), \quad g \in SO(3), \quad x \in S^2.$$

It is easily seen that L defines a representation of $SO(3)$ on $L^2(S^2, d\sigma) = L^2(S^2)$, known as the *left regular representation* of $SO(3)$ on $L^2(S^2)$. A crucial fact is that, if $f_l \in V_l$ as above, then

$$L(g)f_l(x) = f_l(g^{-1} \cdot x) = Y'_{l.}(g^{-1} \cdot x)a_{l.}$$
$$= Y'_{l.}(x)D^l(g)a_{l.} = Y'_{l.}(x)\widetilde{a}_{l.},$$

for

$$\widetilde{a}_{l.} := D^l(g)a_{l.}. \tag{3.45}$$

This yields the following fundamental result, providing a useful alternative description of the dual of $SO(3)$.

Proposition 3.27 *For every integer l, the pair (L, V_l) defines a representation of $SO(3)$. The matrix of the representation (L, V_l) with respect to the basis $Y_{l.}$ is given by the Wigner's matrix D^l, and the representation is therefore irreducible and equivalent to $(D^l, \mathcal{H}_l)$.*

Proposition 3.27 will play a crucial role for the characterization of the spectral properties of isotropic random fields in Chapters 6-8.

Example 3.28 For convenience, we report exact expressions of the first few spherical harmonics in the following table.

$l\backslash m$	0	1	2
$l = 0$	$(\frac{1}{4\pi})^{\frac{1}{2}}$	–	–
$l = 1$	$(\frac{3}{4\pi})^{\frac{1}{2}} \cos \vartheta$	$-(\frac{3}{8\pi})^{\frac{1}{2}} \sin \vartheta e^{i\varphi}$	–
$l = 2$	$(\frac{5}{16\pi})^{\frac{1}{2}} (3\cos^2 \vartheta - 1)$	$-(\frac{15}{8\pi})^{\frac{1}{2}} \sin \vartheta \cos \vartheta e^{i\varphi}$	$(\frac{15}{32\pi})^{\frac{1}{2}} \sin^2 \vartheta e^{2i\varphi}$

For our purposes, though, the most important result is the following

Proposition 3.29 (Peter-Weyl Theorem on the sphere) *For all complex-valued functions $f \in L^2(S^2)$, we have (in spherical coordinates)*

$$f(\vartheta, \varphi) = \sum_{lm} a_{lm} Y_{lm}(\vartheta, \varphi),$$

where

$$a_{lm} := \int_{S^2} f(\vartheta, \varphi) \overline{Y_{lm}}(\vartheta, \varphi) \sin \vartheta d\vartheta d\varphi = (-1)^m \overline{a}_{l,-m},$$

and convergence holds in the $L^2(S^2)$ sense, i.e.

$$\lim_{L \to \infty} \int_0^\pi \int_0^{2\pi} \left\{ f(\vartheta, \varphi) - \sum_{l=0}^L \sum_{m=-l}^l a_{lm} Y_{lm}(\vartheta, \varphi) \right\}^2 \sin \vartheta d\vartheta d\varphi = 0.$$

Proof By exploiting directly the Peter-Weyl Theorem 2.47, we have that for any square integrable function $f \in L^2(SO(3))$

$$f(g) = f(\varphi, \vartheta, \psi) = \sum_l (2l + 1) \sum_{m,n} b_{mn}^l D_{mn}^l(\varphi, \vartheta, \psi),$$

$$b_{mn}^l = b_{mn}^l(f) = \int_{SO(3)} f(g) \overline{D}_{mn}^l(g) dg.$$

As discussed above, we can view the sphere as a quotient space, namely

$$S^2 \simeq SO(3)/SO(2);$$

we can hence identify the functions on S^2 as applications in $L^2(SO(3))$ which are constant on the cosets or, more explicitly, constant with respect to the Euler angle ψ (compare Proposition 2.12). More precisely, take as before $\mathbf{e}_3 = (0, 0, 1)$ (the "North Pole"); clearly $R_{\mathbf{e}_3}(\psi) \cdot \mathbf{e}_3 = \mathbf{e}_3$. For any function $f : S^2 \to \mathbb{R}$ we can define $f_{SO(3)} : SO(3) \to \mathbb{R}$ by

$$\begin{aligned} f_{SO(3)}(g) = f_{SO(3)}(R(\varphi, \vartheta, \psi)) &:= f(R_{\mathbf{e}_3}(\varphi) R_{\mathbf{e}_2}(\vartheta) R_{\mathbf{e}_3}(\psi) \mathbf{e}_3) \\ &= f(R_{\mathbf{e}_3}(\varphi) R_{\mathbf{e}_2}(\vartheta) \mathbf{e} x_3) \\ &= f(\vartheta, \varphi), \end{aligned}$$

for all $\psi \in [0, 2\pi]$, where $(\varphi, \vartheta, \psi) = R^{-1}(g)$, and we write for brevity

$$f(\vartheta, \varphi) = f(\sin \vartheta \cos \varphi, \sin \vartheta \sin \varphi, \cos \vartheta).$$

We have that

$$\begin{aligned} (2l + 1) b_{mn}^l(f_{SO(3)}) &= \frac{2l + 1}{8\pi^2} \int_{SO(3)} f_{SO(3)}(R(\varphi, \vartheta, \psi)) \overline{D}_{mn}^l(\varphi, \vartheta, \psi) \sin \vartheta d\varphi d\vartheta d\psi \\ &= \frac{2l + 1}{8\pi^2} \int_{S^2} f(\vartheta, \varphi) \left\{ \int_0^{2\pi} e^{in\psi} d\psi \right\} d_{mn}^l(\vartheta) e^{im\varphi} \sin \vartheta d\varphi d\vartheta \end{aligned}$$

$$= \begin{cases} \frac{2l+1}{4\pi} \int f(\vartheta,\varphi) d^l_{mn}(\vartheta) e^{im\varphi} \sin \vartheta d\varphi d\vartheta \text{ , for } n=0 \\ 0 \text{ , otherwise.} \end{cases}$$

Thus

$$f_{SO(3)} = \sum_l (2l+1) \sum_{m=-l}^{l} b^l_{m0}(f_{SO(3)}) D^l_{m0} = \sum_l \sum_{m=-l}^{l} a_{lm}(f) Y_{lm} , \qquad (3.46)$$

where

$$a_{lm}(f) = \int_{S^2} f(\vartheta,\varphi) \overline{Y_{lm}}(\vartheta,\varphi) \sin \vartheta d\vartheta d\varphi.$$

Note that, in the second equality of (3.46) we have used the relation $Y_{lm} = (-1)^m \overline{Y}_{l-m}$, along with the fact that the sum $\sum_{m=-l}^{l}$ is symmetric around zero.

□

Remark 3.30 The elements of the "central" column $n = 0$ in Wigner's D matrices are thus seen to play a special role in Fourier expansions on the sphere. A very natural question then arises, concerning the geometrical meaning of the spaces spanned by different columns, other than $n = 0$. This issue will turn out to have deep implications and great practical relevance from the point of view of applications. Indeed, in the last chapter, we shall address this issue in the framework of spin spherical harmonics and spin random fields, which are strongly motivated by the analysis of CMB polarization and gravitational lensing data. This will lead us to the analysis of random sections of fiber bundles over the sphere, which can be viewed as an extension of the notion of standard (scalar) functions. For the time being (and for most of the book), however, we shall stick to the scalar case.

In the following subsection, we shall discuss the spectral expansion from a different point of view, and focus more direct on the analytic properties of the spherical harmonics basis $\{Y_{lm}\}$. In particular, we shall introduce the spherical harmonics $\{Y_{lm}\}$ as a complete orthonormal system for the $L^2(S^2)$ space on the sphere.

3.4.3 An alternative characterization of spherical harmonics

Let us recall that the standard Laplacian operator in $\mathbb{R}^3$ is given by

$$\Delta = \frac{\partial^2}{\partial x_1^2} + \frac{\partial^2}{\partial x_2^2} + \frac{\partial^2}{\partial x_3^2} . \qquad (3.47)$$

Our first aim is to discuss how the Laplacian transforms under spherical coordinates (3.32-3.34). For this purpose, we present the following result.

Proposition 3.31 *In spherical coordinates* (r, ϑ, φ), *the Laplacian is provided by*

$$\Delta = \frac{1}{r^2}\frac{\partial}{\partial r}\left(r^2\frac{\partial}{\partial r}\right) + \frac{1}{r^2\sin\vartheta}\frac{\partial}{\partial\vartheta}\left(\sin\vartheta\frac{\partial}{\partial\vartheta}\right) + \frac{1}{r^2\sin^2\vartheta}\frac{\partial^2}{\partial\varphi^2}. \qquad (3.48)$$

Proof A direct proof can be given by simply using standard rules for multi-variate change of variables. More precisely, from (3.32-3.34) we have

$$\begin{pmatrix}\frac{\partial}{\partial r}\\[4pt]\frac{\partial}{\partial\vartheta}\\[4pt]\frac{\partial}{\partial\varphi}\end{pmatrix} = \begin{pmatrix}\sin\vartheta\cos\varphi & \sin\vartheta\sin\varphi & \cos\vartheta\\ r\cos\vartheta\cos\varphi & r\cos\vartheta\sin\varphi & -r\sin\vartheta\\ -r\sin\vartheta\sin\varphi & r\sin\vartheta\cos\varphi & 0\end{pmatrix}\begin{pmatrix}\frac{\partial}{\partial x_1}\\[4pt]\frac{\partial}{\partial x_2}\\[4pt]\frac{\partial}{\partial x_3}\end{pmatrix}$$

leading to

$$\begin{pmatrix}\frac{\partial}{\partial r}\\[4pt]\frac{1}{r}\frac{\partial}{\partial\vartheta}\\[4pt]\frac{1}{r\sin\vartheta}\frac{\partial}{\partial\varphi}\end{pmatrix} = \begin{pmatrix}\sin\vartheta\cos\varphi & \sin\vartheta\sin\varphi & \cos\vartheta\\ \cos\vartheta\cos\varphi & \cos\vartheta\sin\varphi & -\sin\vartheta\\ -\sin\varphi & \cos\varphi & 0\end{pmatrix}\begin{pmatrix}\frac{\partial}{\partial x_1}\\[4pt]\frac{\partial}{\partial x_2}\\[4pt]\frac{\partial}{\partial x_3}\end{pmatrix}$$

and because the matrix on the right is orthogonal, we get

$$\begin{pmatrix}\frac{\partial}{\partial x_1}\\[4pt]\frac{\partial}{\partial x_2}\\[4pt]\frac{\partial}{\partial x_3}\end{pmatrix} = \begin{pmatrix}\sin\vartheta\cos\varphi & \cos\vartheta\cos\varphi & -\sin\varphi\\ \sin\vartheta\sin\varphi & \cos\vartheta\sin\varphi & \cos\varphi\\ \cos\vartheta & -\sin\vartheta & 0\end{pmatrix}\begin{pmatrix}\frac{\partial}{\partial r}\\[4pt]\frac{1}{r}\frac{\partial}{\partial\vartheta}\\[4pt]\frac{1}{r\sin\vartheta}\frac{\partial}{\partial\varphi}\end{pmatrix}.$$

The proof can then be concluded by substituting into (3.47), expanding the squares and (trivial, but lengthy and tedious) algebraic manipulations. □

Remark 3.32 A more elegant proof of the previous result, which requires a greater background in differential geometry, can instead be sketched as follows. Given any set of coordinates y_j, $j = 1, ..., d$, the (tensor) Laplacian is intrinsically defined as (see for instance Bishop and Goldberg [25])

$$\frac{1}{\sqrt{\det\{E\}}}\sum_{i,j=1}^{3}\frac{\partial}{\partial y^i}\left(\sqrt{\det\{E\}}E^{ij}\frac{\partial f}{\partial y^j}\right), \qquad (3.49)$$

where the matrix with elements $E^{-1} = \left\{E^{ij}\right\}_{i,j=1,2,3}$ is the inverse of $E = \left\{E_{ij}\right\}_{i,j=1,2,3}$, the *metric tensor* corresponding to the choice of coordinates. For standard Cartesian coordinates on $\mathbb{R}^3$, we have the Euclidean metric tensor

$$\left\{E_{ij}\right\}_{i,j=1,2,3} = \begin{pmatrix}1 & 0 & 0\\ 0 & 1 & 0\\ 0 & 0 & 1\end{pmatrix},$$

and (3.49) immediately gives (3.47). Under spherical coordinates, the metric tensor becomes (see [25])

$$\left\{E_{ij}\right\}_{i,j=1,2,3} = \begin{pmatrix} 1 & 0 & 0 \\ 0 & r & 0 \\ 0 & 0 & r\sin^2\vartheta \end{pmatrix},$$

and (3.49) provides (3.48).

It is natural to decompose the Laplacian into a radial part and a spherical part, denoted respectively by Δ_r and Δ_{S^2}, that is:

$$\Delta = \Delta_r + \frac{1}{r^2}\Delta_{S^2},$$

$$\Delta_r = \frac{1}{r^2}\frac{\partial}{\partial r}\left(r^2\frac{\partial}{\partial r}\right),$$

$$\Delta_{S^2} = \frac{1}{\sin\vartheta}\frac{\partial}{\partial\vartheta}\left(\sin\vartheta\frac{\partial}{\partial\vartheta}\right) + \frac{1}{\sin^2\vartheta}\frac{\partial^2}{\partial\varphi^2};$$

the operator Δ_{S^2} is usually called the *spherical Laplacian*.

Let us now consider the harmonic homogeneous polynomials of degree l, of the type

$$\sum_{\alpha,\beta} a_{\alpha\beta}x_1^\alpha x_2^\beta x_3^{l-(\alpha+\beta)},$$

and satisfying

$$\Delta\left(\sum_{\alpha,\beta} c_{\alpha\beta}x_1^\alpha x_2^\beta x_3^{l-(\alpha+\beta)}\right) = 0,$$

i.e. under spherical coordinates,

$$\Delta\left(r^l H_l(\vartheta,\varphi)\right) = 0, \qquad (3.50)$$

with

$$H_l(\vartheta,\varphi) = \sum_{\alpha,\beta\geq 0, \alpha+\beta\leq l} c_{\alpha\beta}(\sin\vartheta\cos\varphi)^\alpha(\sin\vartheta\sin\varphi)^\beta(\cos\vartheta)^{l-(\alpha+\beta)}.$$

Combining (3.48) and (3.50), after some manipulations, we obtain:

$$\Delta\left(r^l H_l(\vartheta,\varphi)\right) = \Delta_r\left(r^l H_l(\vartheta,\varphi)\right) + \frac{1}{r^2}\Delta_{S^2}\left(r^l H_l(\vartheta,\varphi)\right)$$
$$= r^{l-2}l(l+1)H_l(\vartheta,\varphi) + r^{l-2}\Delta_{S^2}H_l(\vartheta,\varphi) = 0, \qquad (3.51)$$

from which it follows immediately that

$$\Delta_{S^2}H_l(\vartheta,\varphi) = -l(l+1)H_l(\vartheta,\varphi). \qquad (3.52)$$

We have then established a very important fact, that is, that the space of harmonic polynomials of degree l on the sphere is included into the space $\mathcal{H}_l\left(S^2\right) = \mathcal{H}_l$ spanned by the eigenfunctions of the spherical Laplacian, with eigenvalue $-l(l+1)$.

We recall that a *weak k-composition* of an integer $l \geq 0$ is an ordered vector $(n_1, ..., n_k)$ of nonnegative integers whose sum is equal to l, and that every l admits exactly $\binom{l+k-1}{k-1}$ distinct weak k-compositions. Now call $\mathbf{P}_l$ the space of homogeneous polynomials of degree l on $\mathbb{R}^3$, and fix $l \geq 2$. There are as many linearly independent homogeneous monomials of degree l as there are weak three-compositions of l, that is $\dim \mathbf{P}_l = \binom{l+2}{2} = \frac{1}{2}(l+1)(l+2)$. Using (3.51), we deduce that the dimension (say $d(l)$) of the space spanned by harmonic polynomials of degree l on S^2 coincides with the dimension of the kernel of the operator $\Delta : \mathbf{P}_l \to \mathbf{P}_{l-2}$. Since this operator is onto (as shown by an easy induction argument), we deduce that

$$d(l) = \dim \mathbf{P}_l - \dim \mathbf{P}_{l-2} = \binom{l+2}{2} - \binom{l}{2} = 2l + 1,$$

showing that $\dim \mathcal{H}_l\left(S^2\right) \geq d(l) = 2l+1$. Of course, one has also that $\dim \mathcal{H}_0 = 1$, and $\dim \mathcal{H}_1 = \dim \mathbf{P}_1 = 3$.

The above estimates are indeed much sharper, as made clear by the following statement.

Proposition 3.33 (1) *The spaces* $\mathcal{H}_l\left(S^2\right)$ *are dense in* $L^2(S^2)$.

(2) $\mathcal{H}_l\left(S^2\right) = \text{span}\,\{Y_{lm}, \ m = -l, ..., l\}$ *for all* $l = 0, 1, 2, 3, ...$, *so that in particular* $\dim \mathcal{H}_l\left(S^2\right) = 2l + 1$.

Proof (1) The fact that the spaces $\mathcal{H}_l\left(S^2\right)$ are dense in $L^2(S^2)$ is standard in functional analysis. More precisely, a standard application of the Weierstrass Theorem implies that a continuous function $f(x_1, x_2, x_3)$ which is defined in the compact domain $-1 \leq x_1, x_2, x_3 \leq 1$, can be uniformly approximated by polynomials of the type

$$\sum_{\alpha, \beta, \gamma} c_{\alpha\beta\gamma} x_1^\alpha x_2^\beta x_3^\gamma .$$

The monomials $x_1^\alpha x_2^\beta x_3^\gamma$ therefore form a basis for the function space of square-integrable functions over this domain; that is, they form a complete set of functions. Because this space is spanned by homogeneous polynomials, the result follows, and Point 1 in the statement is proved.

(2) We know that the set span $\{Y_{lm}, \ m = -l, ..., l\}$ has dimension $(2l + 1)$. In view of the previous discussion, the result will follow if we can prove that these

spherical harmonics are a basis of the class of eigenfunctions of the spherical Laplacian with eigenvalue $-l(l + 1)$. To do so, we need to solve the differential equations (3.52). We use the method of separation of variables, and look for solutions of the form $\Theta_{lm}(\vartheta)\Psi_{lm}(\varphi)$. For $\Psi_{lm}(\varphi)$, we shall take the functions $\exp(im\varphi)$, which are easily seen to form an orthogonal system over the sphere. With this choice for Ψ_{lm}, we find that Θ_{lm} satisfies the equation

$$\left\{ \frac{1}{\sin\vartheta}\frac{\partial}{\partial\vartheta}\left(\sin\vartheta\frac{\partial}{\partial\vartheta}\right) + \frac{1}{\sin^2\vartheta}\frac{\partial^2}{\partial\varphi^2}\right\}\Theta_{lm}(\vartheta)\exp(im\varphi) + l(l+1)\Theta_{lm}(\vartheta)\exp(im\varphi)$$

$$= \frac{1}{\sin\vartheta}\frac{\partial}{\partial\vartheta}\left(\sin\vartheta\frac{\partial}{\partial\vartheta}\right)\Theta_{lm}(\vartheta)\exp(im\varphi)$$

$$+ \frac{\Theta_{lm}(\vartheta)}{\sin^2\vartheta}\frac{\partial^2}{\partial\varphi^2}\exp(im\varphi) + l(l+1)\Theta_{lm}(\vartheta)\exp(im\varphi)$$

$$= \left\{ \frac{1}{\sin\vartheta}\frac{\partial}{\partial\vartheta}\left(\sin\vartheta\frac{\partial}{\partial\vartheta}\right) + \left[l(l+1) - \frac{m^2}{\sin^2\vartheta}\right]\right\}\Theta_{lm}(\vartheta)\exp(im\varphi)$$

which leads to

$$\left\{ \frac{1}{\sin\vartheta}\frac{\partial}{\partial\vartheta}\left(\sin\vartheta\frac{\partial}{\partial\vartheta}\right) + \left[l(l+1) - \frac{m^2}{\sin^2\vartheta}\right]\right\}\Theta_{lm}(\vartheta) = 0 ,$$

i.e., after the change of variable $\mu = \cos\vartheta$,

$$\left\{ \frac{d}{d\mu}\left((1-\mu^2)\frac{d}{d\mu}\right) + \left[l(l+1) - \frac{m^2}{1-\mu^2}\right]\right\}\Theta_{lm}(\arccos\mu) = 0 . \qquad (3.53)$$

Equation (3.53) is known as the *Legendre equation*. For $m = 0$ it becomes

$$\frac{d}{d\mu}\left((1-\mu^2)\frac{d}{d\mu}\right)\Theta_{l0}(\arccos\mu) + l(l+1)\Theta_{l0}(\arccos\mu) = 0 ,$$

and a solution is given by

$$P_l(\mu) = \frac{1}{2^l l!}\frac{d^l}{d\mu^l}(\mu^2 - 1)^l , \qquad (3.54)$$

where we recognize from (13.1), the Legendre polynomial of the lth degree. For general $m = -l, ..., l$, it can be shown (see for instance [127]) that a complete set of solutions is obtained by introducing the Associated Legendre function (see again the Appendix for more discussion and details)

$$P_{lm}(\mu) = (-1)^m(1-\mu^2)^{m/2}\frac{d^m}{d\mu^m}P_l(\mu) , \quad m = -l, ..., l, \quad l = 0, 1, 2, ...$$

and setting $\Theta_{lm}(\vartheta) = P_{lm}(\cos\vartheta)$. $\qquad\qquad\qquad\qquad\qquad\qquad\qquad\qquad\square$

Remark 3.34 Note that, once we have shown that the spherical harmonics are eigenfunctions of the spherical Laplacian corresponding to different eigenvalues, their orthogonality properties across different multipoles l follow easily. Indeed, by an application of the Stokes' Theorem and standard manipulations (see [190]), it is possible to show that the spherical Laplacian is a self-adjoint operator, namely

$$\int_{S^2} \Delta_{S^2} Y_{l_1 m_1} \overline{Y}_{l_2 m_2} \sin \vartheta d\vartheta d\varphi = \int_{S^2} Y_{l_1 m_1} \Delta_{S^2} \overline{Y}_{l_2 m_2} \sin \vartheta d\vartheta d\varphi \,.$$

Now we known that spherical harmonics satisfy $\Delta_{S^2} Y_{lm} = -l(l+1) Y_{lm}$, whence

$$-l_1 (l_1 + 1) \int_{S^2} Y_{l_1 m_1} \overline{Y}_{l_2 m_2} \sin \vartheta d\vartheta d\varphi = \int_{S^2} \Delta_{S^2} Y_{l_1 m_1} \overline{Y}_{l_2 m_2} \sin \vartheta d\vartheta d\varphi$$

$$= \int_{S^2} Y_{l_1 m_1} \Delta_{S^2} \overline{Y}_{l_2 m_2} \sin \vartheta d\vartheta d\varphi$$

$$= -l_2 (l_2 + 1) \int_{S^2} Y_{lm} \overline{Y}_{l_2 m_2} \sin \vartheta d\vartheta d\varphi \,.$$

The orthogonality property for $l_1 \neq l_2$ follows immediately.

Remark 3.35 An important property of the spherical Laplacian is that it commutes with the action of the group $SO(3)$, that is, for all $g \in SO(3)$

$$(\Delta_{S^2} f)(g^{-1} \cdot x) = (\Delta_{S^2}(f \circ g^{-1}))(x) \,.$$

An immediate consequence is that the spaces $\mathcal{H}_l$ are rotationally invariant, that is $(f \circ g^{-1})(.) \in \mathcal{H}_l$ if and only if $f(.) \in \mathcal{H}_l$. In fact,

$$(\Delta_{S^2}(f \circ g^{-1}))(x) = (\Delta_{S^2} f)(g^{-1} \cdot x) = -l(l+1) f(g^{-1} \cdot x) = -l(l+1)(f \circ g^{-1})(x) \,,$$

whence if f is an eigenfunctions of Δ_{S^2} with eigenvalue $-l(l+1)$, so is $f \circ g^{-1})(.)$, and the result follows easily.

Example 3.36 The set of homogeneous polynomials of degree one (which coincides with $\mathcal{H}_1(S^2)$) is obviously obtained by taking linear combination of the (trivially) harmonic monomials x_1, x_2 and x_3. In spherical coordinates, $\mathcal{H}_1(S^2)$ is the three dimensional space of polynomials of the form

$$H_2(\vartheta, \varphi) = \sum_{\alpha, \beta \geq 0, \alpha + \beta \leq 1} c_{\alpha\beta} (\sin \vartheta \cos \varphi)^\alpha (\sin \vartheta \sin \varphi)^\beta (\cos \vartheta)^{1-(\alpha+\beta)}$$

$$= c_{00} \cos \vartheta + c_{01} \sin \vartheta \sin \varphi + c_{10} \sin \vartheta \cos \varphi \,.$$

It is immediate to see that $\mathcal{H}_1(S^2)$ is spanned by

$$\{Y_{1,-1}, Y_{1,0}, Y_{1,1}\} = \left\{ \left(\frac{3}{8\pi}\right)^{\frac{1}{2}} \sin \vartheta e^{-i\varphi}, \left(\frac{3}{4\pi}\right)^{\frac{1}{2}} \cos \vartheta, -\left(\frac{3}{8\pi}\right)^{\frac{1}{2}} \sin \vartheta e^{i\varphi} \right\} \,.$$

For $l = 2$, homogeneous polynomials on the unit sphere are provided by

$$H_1(\vartheta, \varphi) = \sum_{\alpha=0}^{2} c_{\alpha 1} x_1^{\alpha} x_2^{2-\alpha} + \sum_{\alpha=0}^{1} c_{\alpha 2} x_1^{\alpha} x_2^{1-\alpha} x_3$$

$$= c_{01} (\sin \vartheta \sin \varphi)^2 + c_{11} \sin^2 \vartheta \cos \varphi \sin \varphi + c_{21} (\sin \vartheta \cos \varphi)^2$$

$$+ c_{02} \sin \vartheta \sin \varphi \cos \vartheta + c_{12} \sin \vartheta \cos \varphi \cos \vartheta .$$

As before, it is simple to verify that $\mathcal{H}_2(S^2)$ is a five-dimensional vector space which is spanned by

$$\{Y_{2,-2}, Y_{2,-1}, Y_{2,0}, Y_{2,1}, Y_{2,2}\}$$

$$= \sqrt{\frac{15}{8\pi}} \left\{ \frac{1}{2} \sin^2 \vartheta e^{-2i\varphi}, - \sin \vartheta \cos \vartheta e^{-i\varphi}, \right.$$

$$\left. \frac{1}{6^{\frac{1}{2}}} (3 \cos^2 \vartheta - 1), - \sin \vartheta \cos \vartheta e^{i\varphi}, \frac{1}{2} \sin^2 \vartheta e^{2i\varphi} \right\} .$$

Remark 3.37   We recall that, for real valued functions f,

$$a_{lm}(f) = (-1)^m \overline{a_{l,-m}}(f) ,$$

whence

$$\sum_{m=-l}^{l} a_{lm} Y_{lm} = a_{l0} Y_{l0} + 2 \sum_{m=1}^{l} [\{\operatorname{Re} a_{lm}\} \{\operatorname{Re} Y_{lm}\} - \{\operatorname{Im} a_{lm}\} \{\operatorname{Im} Y_{lm}\}] ,$$

so that we can rewrite

$$\mathcal{H}_l(S^2) = span\left\{ \sqrt{2} \operatorname{Re} Y_{ll}, ..., \sqrt{2} \operatorname{Re} Y_{l1}, Y_{l0}, \sqrt{2} \operatorname{Im} Y_{l1}, ..., \sqrt{2} \operatorname{Im} Y_{ll} \right\} .$$

The real basis of spherical harmonics is widely used in the literature (see for instance Kim and Koo [113], Wigman [206, 207]); the elements of this basis are obviously real-valued harmonic polynomials of degree l in $\{x_1, x_2, x_3\}$, $x_1^2 + x_2^2 + x_3^2 = 1$. Throughout this book, however, we stick to the complex-valued system, which has simpler symmetry properties and a more direct connection with the Wigner's matrices D^l.

In general the analytic manipulation of spherical harmonics is troublesome. As we mentioned before, most of our result below will be established by a different route, i.e., by exploiting the connection of spherical harmonics with group representation theory. A remarkable example is given in the next section below, where we discuss how to evaluate multiple integrals of spherical harmonics by exploiting representation theory for the group of rotations $SO(3)$.

3.5 The Clebsch-Gordan coefficients

3.5.1 Clebsch-Gordan matrices

Clebsch-Gordan matrices and coefficients (to be defined below) are the fundamental tools for the evaluation of multiple integrals of spherical harmonics. Because of this connection, their role is ubiquitous throughout this monograph, ranging from the characterizations of isotropy (Chapter 6), to the analysis of nonlinear transforms of Gaussian fields (Chapter 7), and from the asymptotic theory for quadratic statistics (Chapter 8), to the definition of angular polyspectra (Chapter 9), and the asymptotic theory for nonlinear functions of wavelets coefficients (Chapter 11).

The role of Clebsch-Gordan coefficients can be explained as follows. According to the discussion contained in Section 2.4.3, we can exploit the family of Wigner D matrices $\left\{ D^l : l = 0, 1, 2, \ldots \right\}$ in order to build alternative (reducible) representations, either by forming the tensor products

$$\left\{ D^{l_1} \otimes D^{l_2} : l_1, l_2 \geq 0 \right\},$$

or by considering direct sums

$$\left\{ \oplus_{l=|l_2-l_1|}^{l_2+l_1} D^l : l_1, l_2 \geq 0 \right\}.$$

For a fixed pair (l_1, l_2), the tensor product representation $D^{l_1} \otimes D^{l_2}$ and the direct sum representation $\oplus_{l=|l_2-l_1|}^{l_2+l_1} D^l$ have common dimension

$$(2l_1 + 1)(2l_2 + 1) \times (2l_1 + 1)(2l_2 + 1)$$

and are unitarily equivalent (this fact will become clear from the subsequent discussion). From formula (2.8), we infer that there exists a unitary matrix $C_{l_1 l_2}$, known as a *Clebsch-Gordan matrix*, such that

$$\left\{ D^{l_1} \otimes D^{l_2} \right\} = C_{l_1 l_2} \left\{ \oplus_{l=|l_2-l_1|}^{l_2+l_1} D^l \right\} C^*_{l_1 l_2}. \tag{3.55}$$

The matrix $C_{l_1 l_2}$ is a $\{(2l_1 + 1)(2l_2 + 1) \times (2l_1 + 1)(2l_2 + 1)\}$ block matrix, whose blocks, of dimensions $(2l_2 + 1) \times (2l + 1)$, are customarily denoted by $C^l_{l_1(m_1)l_2}$, $l = |l_2 - l_1|, \ldots, l_1 + l_2$, $m_1 = -l_1, \ldots, l_1$. The elements of the lth block are indexed by m_2 (over rows) and m (over columns; note that $m = -(2l + 1), \ldots, 2l + 1$). More precisely,

$$C_{l_1 l_2} = \left[C^{l.}_{l_1(m_1)l_2.} \right]_{m_1 = -l_1, \ldots, l_1 ; l = |l_2 - l_1|, \ldots, l_2 + l_1} \tag{3.56}$$

$$C^{l.}_{l_1(m_1)l_2.} = \left\{ C^{lm}_{l_1 m_1 l_2 m_2} \right\}_{m_2 = -l_2, \ldots, l_2 ; m = -l, \ldots, l}. \tag{3.57}$$

Remark 3.40 below provides further details on the structure of the matrix $C_{l_1 l_2}$.

Remark 3.38 (Triangle conditions.)   Given nonnegative integers l_1, l_2, l_3, the relations $|l_2 - l_1| \le l_3 \le l_1 + l_2$ are equivalent to $l_{\rho(1)} \le l_{\rho(2)} + l_{\rho(3)}$, for every permutation ρ of $\{1, 2, 3\}$. As customary, we refer to these constraints on l_1, l_2, l_3 as the *triangle conditions*. Whenever, l_1, l_2, l_3 do not verify the triangle conditions, we set $C_{l_1 m_1 l_2 m_2}^{l_3 m_3} = 0$ by convention.

Remark 3.39 (On dimensions.)   The fact that the two matrices $D^{l_1} \otimes D^{l_2}$ and $\oplus_{l=|l_2-l_1|}^{l_2+l_1} D^l$ have the same dimension follows from the elementary relation (valid for any integers $l_1, l_2 \ge 0$):

$$\sum_{l=|l_2-l_1|}^{l_1+l_2} (2l + 1) = (2l_1 + 1)(2l_2 + 1). \tag{3.58}$$

By induction, we also obtain that, for every $n \ge 3$,

$$\sum_{\lambda_1=|l_2-l_1|}^{l_1+l_2} \sum_{\lambda_2=|l_3-\lambda_1|}^{\lambda_1+l_3} \cdots \sum_{\lambda_{n-1}=|l_n-\lambda_{n-2}|}^{\lambda_{n-2}+l_n} (2\lambda_{n-1} + 1) = \prod_{j=1}^{n} (2l_j + 1), \tag{3.59}$$

for any integers $l_1, ..., l_n \ge 0$ (relation (3.59) is needed in Section 6.5.2).

The *Clebsch-Gordan coefficients* for $SO(3)$ are then defined as the collection $\left\{ C_{l_1 m_1 l_2 m_2}^{lm} \right\}$ of the elements of the unitary matrices $C_{l_1 l_2}$, extended according to the convention described in Remark 3.38. These coefficients were introduced in Mathematics in the XIX century, as motivated by the analysis of invariants in Algebraic Geometry; in the XX century, they have gained an enormous importance in the quantum theory of angular momentum, where $C_{l_1 m_1 l_2 m_2}^{lm}$ represents the *probability amplitude* that two particles with total angular momentum l_1, l_2 and momentum projection on the z-axis m_1 and m_2 are coupled to form a system with total angular momentum l and projection m (see e.g. [127]). Their use in the analysis of isotropic random fields is much more recent, see for instance [102] and the references therein.

Remark 3.40 (Detailed description of a Clebsch-Gordan matrix.)   We now provide an alternate presentation of the Clebsch-Gordan matrices $C_{l_1 l_2}$, based on a combinatorial description of the labels of its entries. Fix integers $l_1, l_2 \ge 0$ such that $l_1 \le l_2$ (this is just for notational convenience), and consider the Clebsch-Gordan coefficients $\left\{ C_{l_1 m_1 l_2 m_2}^{lm} \right\}$. According to the above discussion, we have that: (i) $-l_i \le m_i \le l_i$ for $i = 1, 2$, (ii) $l_2 - l_1 \le l \le l_1 + l_2$, (iii) $-l \le m \le l$, and (iv) the symbols (l_1, m_1, l_2, m_2) label rows, whereas the pairs (l, m) are attached to columns. Now introduce the total order $\prec_c$ on the "column pairs" (l, m), by setting that $(l, m) \prec_c (l', m')$, whenever either $l < l'$ or $l = l'$ and $m < m'$. Analogously, introduce a total order $\prec_r$ over the "row symbols" (l_1, m_1, l_2, m_2), by setting that $(l_1, m_1, l_2, m_2) \prec_r (l'_1, m'_1, l'_2, m'_2)$, if either $m_1 < m'_1$, or $m_1 = m'_1$ and $m_2 < m'_2$ (recall that l_1 and l_2 are fixed). It can be checked that the set of column pairs (resp. row symbols) can now be written as a *saturated chain* with respect to $\prec_c$ (resp. $\prec_r$) with a least element given by $(l_2 - l_1, -(l_2 - l_1))$ (resp. $(l_1, -l_1, l_2, -l_2)$) and a maximal element given by $(l_2 + l_1, l_2 + l_1)$ (resp. (l_1, l_1, l_2, l_2)); recall that, given a finite set $A = \{a_j : j = 1, ..., N\}$ and an order $\prec$ on A, one says that A is a saturated chain with respect to $\prec$ if there exists a permutation π of $\{1, ..., N\}$ such that

$$a_{\pi(1)} \prec a_{\pi(2)} \prec \cdots \prec a_{\pi(N-1)} \prec a_{\pi(N)}.$$

In this case, $a_{\pi(1)}$ and $a_{\pi(N)}$ are called, respectively, the least and the maximal elements of the chain. Then, (A) dispose the columns from west to east, increasingly according to $\prec_c$, (B) dispose the rows from north to south, increasingly according to $\prec_r$. For instance, by setting $l_1 = 0$ and $l_2 \geq 1$, one obtains that $C_{l_1 l_2}$ is the $(2l_2 + 1) \times (2l_2 + 1)$ square matrix $\{C_{00 l_2 m_2}^{l_2 m}\}$ with column indices $m = -(2l_2 + 1), ..., (2l_2 + 1)$ and row indices $m_2 = -(2l_2 + 1), ..., (2l_2 + 1)$ (from the subsequent discussion, one also deduces that, in general, $C_{00 l_2 m_2}^{lm} = \delta_l^{l_2} \delta_m^{m_2}$). By selecting $l_1 = l_2 = 1$, one sees that C_{11} is the 9×9 matrix with elements $C_{1 m_1 1 m_2}^{lm}$ (for $m_1, m_2 = -1, 0, 1; l = 0, 1, 2, m = -l, ..., l$) arranged as follows:

$$
\begin{pmatrix}
C_{1,-1;1,-1}^{0,0} & C_{1,-1;1,-1}^{1,-1} & C_{1,-1;1,-1}^{10} & C_{1,-1;1,-1}^{11} & C_{1,-1;1,-1}^{2,-2} & C_{1,-1;1,-1}^{2,-1} & C_{1,-1;1,-1}^{2,0} & C_{1,-1;1,-1}^{2,1} & C_{1,-1;1,-1}^{2,2} \\
C_{1,-1;1,0}^{0,0} & \cdots & \cdots & \cdots & \cdots & \cdots & \cdots & \cdots & \cdots \\
C_{1,-1;1,1}^{0,0} & \cdots & \cdots & C_{1,-1;1,1}^{11} & \cdots & \cdots & \cdots & \cdots & \cdots \\
C_{1,0;1,-1}^{0,0} & \cdots & \cdots & \cdots & \cdots & \cdots & C_{1,0;1,-1}^{2,0} & \cdots & \cdots \\
C_{1,0;1,0}^{0,0} & \cdots & \cdots & \cdots & C_{1,0;1,0}^{2,-2} & \cdots & \cdots & \cdots & \cdots \\
C_{1,0;1,1}^{0,0} & \cdots & \cdots & \cdots & \cdots & \cdots & \cdots & C_{1,0;1,1}^{2,1} & \cdots \\
C_{1,1;1,-1}^{0,0} & \cdots & \cdots & \cdots & \cdots & \cdots & \cdots & \cdots & \cdots \\
C_{1,1;1,0}^{0,0} & \cdots & \cdots & \cdots & \cdots & \cdots & \cdots & \cdots & \cdots \\
C_{1,1;1,1}^{0,0} & C_{1,1;1,1}^{1,-1} & C_{1,1;1,1}^{1,0} & C_{1,1;1,1}^{1,1} & C_{1,1;1,1}^{2,-2} & C_{1,1;1,1}^{2,-1} & C_{1,1;1,1}^{2,0} & C_{1,1;1,1}^{2,1} & C_{1,1;1,1}^{2,2}
\end{pmatrix}.
$$

Remark 3.41 Using the previous description of Clebsch-Gordan matrices, one can reformulate "pointwise" the algebraic relation (3.55) as follows: for every $g \in SO(3)$,

$$D_{m_1 n_1}^{l_1}(g) D_{m_2 n_2}^{l_2}(g) = \sum_{l=|l_1-l_2|}^{l_1+l_2} \sum_{\mu_1 \mu_2} C_{l_1 m_1 l_2 m_2}^{l \mu_1} D_{\mu_1 \mu_2}^{l}(g) C_{l_1 n_1 l_2 n_2}^{l \mu_2}. \tag{3.60}$$

See [195, formula 4.6.1.1].

It can be proved (see e.g. [195, Ch. 8], or Remark 3.45 below) that *Clebsch-Gordan coefficients are real-valued*. Their analytic expression is known, but is in general hardly manageable. We have e.g. (see [195, formula 8.2.1.5])

$$C^{l_3 - m_3}_{l_1 m_1 l_2 m_2} = (-1)^{l_1 + l_3 + m_2} \sqrt{2l_3 + 1} \left[\frac{(l_1 + l_2 - l_3)!(l_1 - l_2 + l_3)!(l_1 - l_2 + l_3)!}{(l_1 + l_2 + l_3 + 1)!} \right]^{1/2}$$

$$\times \left[\frac{(l_3 + m_3)!(l_3 - m_3)!}{(l_1 + m_1)!(l_1 - m_1)!(l_2 + m_2)!(l_2 - m_2)!} \right]^{1/2}$$

$$\sum_z \frac{(-1)^z (l_2 + l_3 + m_1 - z)!(l_1 - m_1 + z)!}{z!(l_2 + l_3 - l_1 - z)!(l_3 + m_3 - z)!(l_1 - l_2 - m_3 + z)!},$$

where the summation runs over all z's such that the factorials are non-negative. This expression becomes neater for $m_1 = m_2 = m_3 = 0$, where we have

$$C^{l_3 0}_{l_1 0 l_2 0} = \begin{cases} 0, \text{ for } l_1 + l_2 + l_3 \text{ odd} \\ \frac{(-1)^{\frac{l_1 + l_2 - l_3}{2}} \sqrt{2l_3 + 1} [(l_1 + l_2 + l_3)/2]!}{[(l_1 + l_2 - l_3)/2]! [(l_1 - l_2 + l_3)/2]! [(-l_1 + l_2 + l_3)/2]!} \left\{ \frac{(l_1 + l_2 - l_3)!(l_1 - l_2 + l_3)!(-l_1 + l_2 + l_3)!}{(l_1 + l_2 + l_3 + 1)!} \right\}^{1/2}, \\ \text{for } l_1 + l_2 + l_3 \text{ even,} \end{cases}$$

(3.61)

(see Proposition 3.44 for a partial explanation of these formulae). Clebsch-Gordan coefficients enjoy a nice set of symmetry and orthogonality properties, playing a crucial role in our results to follow. For instance, from unitary equivalence we have the two relations:

$$\sum_{m_1, m_2} C^{lm}_{l_1 m_1 l_2 m_2} C^{l' m'}_{l_1 m_1 l_2 m_2} = \delta^{l'}_l \delta^{m'}_m, \tag{3.62}$$

$$\sum_{l, m} C^{lm}_{l_1 m_1 l_2 m_2} C^{lm}_{l_1 m'_1 l_2 m'_2} = \delta^{m'_1}_{m_1} \delta^{m'_2}_{m_2}; \tag{3.63}$$

in particular, (3.62) is a consequence of the orthonormality of row vectors, whereas (3.63) comes from the orthonormality of columns.

Further properties are provided in Section 4.5 on the so-called "graphical method".

Remark 3.42 Clebsch-Gordan coefficients have an elementary form in the case of a commutative compact group, whose irreducible representations have all dimension one (see Proposition 2.30). Take for instance the torus $\mathbb{T} = [0, 2\pi)$. Then, every irreducible representation of $\mathbb{T}$ is equivalent to a one-dimensional matrix representation of the type $\vartheta \mapsto \exp(i2\pi k\vartheta)$, $k \in \mathbb{Z}$, $\vartheta \in \mathbb{T}$. Hence, the associated one-dimensional Clebsch-Gordan matrices satisfy

$$\exp(i2\pi k_1 \vartheta) \exp(i2\pi k_2 \vartheta) = C^{k_3}_{k_1 k_2} \exp(i2\pi k_3 \vartheta)$$

which gives immediately $C^{k_3}_{k_1 k_2} = \delta^{k_3}_{k_1 + k_2}$. In this case the Clebsch-Gordan coefficients are therefore given by the Kronecker delta function. It follows that the role of Clebsch-Gordan coefficients is just as important in standard Fourier analysis and its statistical applications as it is in this book. However, due to their very simple form, the role of Clebsch-Gordan coefficients in the commutative case is somewhat hidden, and manipulation of Fourier transforms and coefficients can be performed without mentioning them at all.

3.5.2 Integrals of multiple spherical harmonics

Evaluating integrals involving products of spherical harmonics is a difficult task, and it is often useful to perform this task by using Clebsch-Gordan matrices. We will explore this venue in Chapter 6 below. In this section, we shall only provide a brief illustration of this approach.

The basic idea is to exploit the fact that spherical harmonics are proportional to the elements of representation matrices, and then use the connection (3.55) between tensor product and direct sum representations. A first example is given by the following, simple result.

Proposition 3.43 *For all l_1, l_2, l_3,*

$$\int_{S^2} Y_{l_1 m_1}(x) Y_{l_2 m_2}(x) \overline{Y}_{l_3 m_3}(x) d\sigma(x) \tag{3.64}$$

$$= \sqrt{\frac{(2l_1 + 1)(2l_2 + 1)}{4\pi(2l_3 + 1)}} C^{l_3 m_3}_{l_1 m_1 l_2 m_2} C^{l_3 0}_{l_1 0 l_2 0},$$

where we used the convention of Remark 3.38 for those integers l_1, l_2, l_3 not verifying the triangle conditions. In particular, the previous relation implies that $C^{l_3 m_3}_{l_1 m_1 l_2 m_2} = 0$ whenever $m_1 + m_2 \neq m_3$.

Proof We start by assumping that l_1, l_2, l_3 verify the triangle conditions. By virtue of Definition 3.23, we have

$$\int_{S^2} Y_{l_1 m_1}(x) Y_{l_2 m_2}(x) \overline{Y}_{l_3 m_3}(x) d\sigma(x)$$

$$= \sqrt{\frac{(2l_1 + 1)(2l_2 + 1)(2l_3 + 1)}{(4\pi)^3}} \times \tag{3.65}$$

$$\times \int_0^\pi d^{l_1}_{m_1 0}(\vartheta) d^{l_2}_{m_2 0}(\vartheta) d^{l_3}_{m_3 0}(\vartheta) \left[\int_0^{2\pi} \exp\{i(m_1 + m_2 - m_3)\varphi\} \, d\varphi \right] \sin\vartheta d\vartheta$$

$$= \sqrt{\frac{(2l_1 + 1)(2l_2 + 1)(2l_3 + 1)}{(4\pi)^3}} \times \tag{3.66}$$

$$\times \frac{1}{2\pi} \int_0^{2\pi} \int_0^\pi \int_0^{2\pi} \overline{D}^{l_1}_{m_1 0}(\varphi, \vartheta, \psi) \overline{D}^{l_2}_{m_2 0}(\varphi, \vartheta, \psi) D^{l_3}_{m_3 0}(\varphi, \vartheta, \psi) \sin\vartheta d\varphi d\vartheta d\psi$$

$$= \sqrt{\frac{(2l_1 + 1)(2l_2 + 1)(2l_3 + 1)}{(4\pi)^3}} 4\pi \int_{SO(3)} \overline{D}^{l_1}_{m_1 0}(g) \overline{D}^{l_2}_{m_2 0}(g) D^{l_3}_{m_3 0}(g) dg \ .$$

Using (3.55) (or, equivalently, (3.60)) we have that, for all $g \in SO(3)$,

$$D^{l_1}_{m_1 0}(g) D^{l_2}_{m_2 0}(g) = \sum_{l=|l_2-l_1|}^{l_2+l_1} \sum_{\mu_1 \mu_2} C^{l\mu_1}_{l_1 m_1 l_2 m_2} C^{l\mu_2}_{l_1 0 l_2 0} D^{l}_{\mu_1 \mu_2}(g),$$

yielding

$$\int_{SO(3)} D^{l_1}_{m_1 0}(g) D^{l_2}_{m_2 0}(g) \overline{D}^{l_3}_{m_3 0}(g) dg$$

$$= \sum_{l=|l_2-l_1|}^{l_2+l_1} \sum_{\mu_1 \mu_2} C^{l\mu_1}_{l_1 m_1 l_2 m_2} C^{l\mu_2}_{l_1 0 l_2 0} \int_{SO(3)} D^{l}_{\mu_1 \mu_2}(g) \overline{D}^{l_3}_{m_3 0}(g) dg$$

$$= \sum_{l=|l_2-l_1|}^{l_2+l_1} \sum_{\mu_1 \mu_2} C^{l\mu_1}_{l_1 m_1 l_2 m_2} C^{l\mu_2}_{l_1 0 l_2 0} \frac{\delta^{l_3}_l \delta^{m_3}_{\mu_1} \delta^0_{\mu_2}}{2l_3 + 1} = \frac{1}{2l_3 + 1} C^{l_3 m_3}_{l_1 m_1 l_2 m_2} C^{l_3 0}_{l_1 0 l_2 0}.$$

If l_1, l_2, l_3 do not verify the triangle conditions, then a slight modification of the above arguments shows the mapping $g \mapsto Y_{l_1 m_1}(g) Y_{l_2 m_2}(g)$ is orthogonal to $Y_{l_3 m_3}$ in $L^2(S^2, d\sigma)$ thus proving the first part of the statement. To prove the last part, just observe that specializing the above computations to the case $m_1 = m_2 = m_3 = 0$ show that $C^{l_3 0}_{l_1 0 l_2 0} \neq 0$. The proof is concluded by using the equality (3.65), showing that the LHS of (3.64) is zero whenever $m_1 + m_2 \neq m_3$.   □

Expression (3.64) is known as a *Gaunt integral*. This simple argument, and its extension to higher-order integrals (which is only notationally more complicated) will lie at the heart of several computations throughout the book. In particular, iterating the previous argument we shall obtain an expression for the so-called *generalized Gaunt integrals*, i.e.

$$\mathcal{G}\{l_1, m_1; ...; l_r, m_r\} := \int_{S^2} Y_{l_1, m_1}(x) \cdots Y_{l_r, m_r}(x) \, d\sigma(x).$$

Explicit formulae and properties of generalized Gaunt integrals are discussed in Chapter 6. As a special case, we mention

$$\mathcal{G}\{l_1, m_1; l_2, m_2; l, -m\} = (-1)^m \sqrt{\frac{(2l_1 + 1)(2l_2 + 1)}{4\pi(2l + 1)}} C^{lm}_{l_1 m_1 l_2 m_2} C^{l0}_{l_1 0 l_2 0},$$

which is just a consequence of Proposition 3.43 and (3.40).

3.5.3 Wigner $3j$ coefficients

It is often convenient to use the so-called *Wigner $3j$ coefficients*, which are related to the Clebsch-Gordan coefficients by the identities (see [195, Section 8.1.2])

$$\begin{pmatrix} l_1 & l_2 & l_3 \\ m_1 & m_2 & m_3 \end{pmatrix} := (-1)^{l_3+m_3} \frac{1}{\sqrt{2l_3+1}} C^{l_3 m_3}_{l_1-m_1 l_2-m_2} \tag{3.67}$$

$$C^{l_3 m_3}_{l_1 m_1 l_2 m_2} = (-1)^{l_1-l_2+m_3} \sqrt{2l_3+1} \begin{pmatrix} l_1 & l_2 & l_3 \\ m_1 & m_2 & -m_3 \end{pmatrix}. \tag{3.68}$$

Some elementary properties of Wigner coefficients are listed in the next proposition.

Proposition 3.44 *Fix integers l_1, l_2, l_3, and consider the associated Wigner coefficient*

$$\begin{pmatrix} l_1 & l_2 & l_3 \\ m_1 & m_2 & m_3 \end{pmatrix}, \ -(2l_i+1) \le m_i \le 2l_i+1, \ i = 1, 2, 3.$$

(1) The Wigner coefficients are zero unless the triangle conditions $|l_i - l_j| \le l_k \le l_i + l_j$, $i, j, k = 1, 2, 3$ are satisfied and $m_1 + m_2 + m_3 = 0$.

(2) When the sum $l_1 + l_2 + l_3$ is even, the Wigner coefficients are invariant under permutations of any two columns.

(3) For any l_1, l_2, l_3, the following upper bound holds:

$$\left| \begin{pmatrix} l_1 & l_2 & l_3 \\ m_1 & m_2 & m_3 \end{pmatrix} \right| \le [\max \{2l_1+1, 2l_2+1, 2l_3+1\}]^{-1/2}$$

(4) When the triangle conditions are verified and $m_1 = m_2 = m_3 = 0$, the Wigner coefficients are different from zero only when the sum $l_1 + l_2 + l_3$ is even.

(5) The following sign inversion rule holds for every choice of l_1, l_2, l_3

$$\begin{pmatrix} l_1 & l_2 & l_3 \\ m_1 & m_2 & m_3 \end{pmatrix} = (-1)^{l_1+l_2+l_3} \begin{pmatrix} l_1 & l_2 & l_3 \\ -m_1 & -m_2 & -m_3 \end{pmatrix}.$$

Proof Points (1), (2) and (3) follow directly from (3.67) and the properties of Clebsch-Gordan coefficients. To prove Point (4), we first observe that

$$\int_{S^2} Y_{l_1 0}(x) Y_{l_2 0}(x) Y_{l_3 0}(x) d\sigma(x) \tag{3.69}$$

$$= \sqrt{\frac{(2l_1+1)(2l_2+1)(2l_3+1)}{(4\pi)^3}} 2\pi \int_0^\pi \sin \vartheta d^{l_1}_{00}(\vartheta) d^{l_2}_{00}(\vartheta) d^{l_3}_{00}(\vartheta) d\vartheta$$

$$= \sqrt{\frac{(2l_1 + 1)(2l_2 + 1)(2l_3 + 1)}{4\pi}} \begin{pmatrix} l_1 & l_2 & l_3 \\ 0 & 0 & 0 \end{pmatrix}^2.$$

Performing the change of variable $\vartheta \mapsto \pi - \vartheta$ and using (3.18), we infer that

$$\int_0^\pi \sin \vartheta d_{00}^{l_1}(\vartheta) d_{00}^{l_2}(\vartheta) d_{00}^{l_3}(\vartheta) d\vartheta = (-1)^{l_1+l_2+l_3} \int_0^\pi \sin \vartheta d_{00}^{l_1}(\vartheta) d_{00}^{l_2}(\vartheta) d_{00}^{l_3}(\vartheta) d\vartheta,$$

from which the result follows immediately. Point (5) in the statement follows from similar manipulations. $\qquad\square$

Remark 3.45 Using formula (3.69) in the previous proof, one can deduce a quick argument showing that Clebsch-Gordan coefficients are indeed real-valued. Using the definition of spherical harmonics, we can in fact prove (after some standard computations) that

$$\int_{S^2} Y_{l_1 0}(x) Y_{l_2 0}(x) Y_{l_3 0}(x) d\sigma(x) \geq 0,$$

from which it follows immediately that

$$\begin{pmatrix} l_1 & l_2 & l_3 \\ 0 & 0 & 0 \end{pmatrix}^2 \geq 0,$$

from which we infer (using (3.68) in the case $m_1 = m_2 = m_3 = 0$) that $C_{l_1 0 l_2 0}^{l_3 0}$ is real-valued. Since the LHS of (3.64) is real, we conclude that $C_{l_1 m_1 l_2 m_2}^{l_3 m_3}$ is real for every choice of $l_i, m_i, i = 1, 2, 3$.

Remark 3.46 In terms of the Wigner $3j$ coefficients, the integral of three spherical harmonics gives

$$\int_{S^2} Y_{l_1 m_1}(x) Y_{l_2 m_2}(x) Y_{l_3 m_3}(x) d\sigma(x)$$

$$= \sqrt{\frac{(2l_1 + 1)(2l_2 + 1)(2l_3 + 1)}{(4\pi)^3}} \begin{pmatrix} l_1 & l_2 & l_3 \\ m_1 & m_2 & m_3 \end{pmatrix} \begin{pmatrix} l_1 & l_2 & l_3 \\ 0 & 0 & 0 \end{pmatrix}.$$

4

Background Results in Probability and Graphical Methods

4.1 Introduction

This chapter collects several probabilistic results that are used throughout the text. All random objects appearing in this chapter and in the rest of the book are defined on a suitable common probability space $(\Omega, \mathcal{F}, P)$, and we shall denote by E the mathematical expectation with respect to P. We use the standard notation

$$Z \sim N(\mu, \sigma^2)$$

to indicate that Z is a one-dimensional Gaussian random variable with mean μ and variance σ^2. For a general integer k, the notation

$$(Z_1, ..., Z_k) \sim N_k(m, V)$$

indicates that $(Z_1, ..., Z_k)$ is a k-dimensional Gaussian vector with mean $m \in \mathbb{R}^k$ and $k \times k$ covariance matrix V. Of course, the symbols N_1 and N indicate the same objects. To avoid ambiguity, when specifying the distribution of a given random object, we shall sometimes use the symbol " $\overset{law}{\sim}$ " instead of " $\sim$ ". Standard references for probability theory and stochastic calculus are, respectively, Dudley [56] and Revuz and Yor [169].

   Section 4.2 contains some basic results about Brownian motion and stochastic integration. Section 4.3 concerns moments, cumulants and diagram formulae. Section 4.4 deals with a recent simplified version of the so-called "method of moments and cumulants". Finally, in Section 4.5 we shall establish a connection with the material in the previous chapter by discussing the so-called *graphical method*, which is usually adopted in order to deal with large convolutions of Clebsch-Gordan coefficients (again, arising from multiple integrals of spherical harmonics).

4.2 Brownian motion and stochastic calculus

In this monograph, we will often study random objects (for instance, random fields on the sphere) that are square-integrable functionals of some infinite-dimensional collection of Gaussian random variables. A basic (but useful) remark is that every collection of Gaussian random variables can be uniquely embedded in some $L^2(P)$-closed Gaussian space, and also that all these Gaussian spaces are isomorphic (they are indeed separable Hilbert spaces). It follows that, as long as one is merely interested in the Hilbertian structure of the Gaussian space, there is no loss in generality in assuming that the underlying field is indeed a real-valued standard Brownian motion on a finite time interval. This is exactly the point of view adopted in this monograph, where we will systematically represent every Gaussian random object in terms of some underlying (possibly complex-valued) Brownian motion on $[0, 1]$. Some useful results and definitions on Brownian motions are presented in the subsequent sections.

Example 4.1   (1) We anticipate some notions introduced in the next subsection. Suppose that $\xi = \{\xi_i : i \geq 1\}$ is a sequence of independent Gaussian random variables, such that $E\left(\xi_i^2\right) = \sigma_i^2 > 0$. Then, it is possible to represent the law of ξ in terms of a Brownian motion W on $[0, 1]$, as follows. First, select an orthonormal basis $\{e_i : i \geq 1\}$ of $L^2([0, 1])$, and then set

$$\zeta_i = \sigma_i \times \int_0^1 e_i(s)\,dW_s, \quad i \geq 1.$$

By definition, one has that $\zeta = \{\zeta_i : i \geq 1\}$ is a centered Gaussian family such that

$$E\left[\zeta_i\zeta_j\right] = \sigma_i\sigma_j \int_0^1 e_i(s)\,e_j(s)\,ds = \begin{cases} 0 & \text{if } i \neq j \\ \sigma_i^2 & \text{otherwise.} \end{cases}$$

Since the distribution of a Gaussian family is completely determined by its means and covariances, it follows that ζ and ξ have the same law.

(2) Let ξ_t be a Gaussian zero-mean and stationary process, with covariance function $\gamma_{t-s} := E\xi_t\xi_s$ admitting a spectral density (see for instance [30]), i.e. a nonnegative function $f : [-\pi, \pi] \to \mathbb{R}^+$ such that

$$\gamma_{t-s} = \int_0^{2\pi} \exp(i(t-s)\lambda)f(\lambda)d\lambda\,.$$

Then, the sequence $\{\xi_t\}$ can be represented as

$$\xi_t = \int_0^{2\pi} \exp(it\lambda)\sqrt{f(\lambda)}dW_\lambda\,, t = 1, 2, \ldots\,.$$

This is just a special case of the spectral representation theorem by Cramer and Leadbetter (see again [30]). Here, $W_\lambda = \text{Re}(W_\lambda) + i\text{Im}(W_\lambda)$ is complex-valued Brownian motion, i.e. $\text{Re}(W_\lambda), \text{Im}(W_\lambda)$ are independent realizations of standard Brownian motion in $[0, 2\pi]$, see the discussion below for definitions and further results. Of course, Example (1) is just a special case of Example (2).

(3) Generalizing the previous example, we will see in Section 5.3 that every real-valued stationary Gaussian field on $\mathbb{R}^m$, say

$$Z = \{Z(x) : x \in \mathbb{R}^m\}$$

can be represented in the form

$$Z(x) = \int_{\mathbb{R}^m} e^{i\langle x, \lambda \rangle} M(d\lambda), \tag{4.1}$$

where M is a complex-valued Gaussian measure and $\langle \cdot, \cdot \rangle$ is the usual inner product in $\mathbb{R}^m$. We will also show that one can always reduce the representation (4.1) to the form

$$Z(x) = \int_0^1 \Psi\left(e^{i\langle x, \cdot \rangle}\right) dW,$$

where W is again a standard Brownian motion, and $\Psi(\cdot)$ is an appropriate isomorphism between spaces of deterministic functions.

4.2.1 Brownian motion and stochastic integrals

A *standard Brownian motion* (initialized at zero) on $[0, 1]$ is a centered Gaussian process of the type

$$W = \{W_t : t \in [0, 1]\},$$

such that the following properties are verified: **(1)** $E(W_t W_s) = \min(s, t)$, **(2)** $W_0 = 0$, **(3)** the paths of W are continuous. It is well-known that such an object exists (see e.g. [169, p. 15]). We shall assume that $\mathcal{F} = \sigma(W)$ (that is, our reference σ-field is generated by W), and we will write

$$\mathcal{F}_s = \sigma\{W_u : u \le s\} \vee \mathcal{N}, \quad s \in [0, 1],$$

in order to indicate the canonical augmented filtration of W, where $\mathcal{N}$ stands for the class of P-null sets of $\mathcal{F}$. In other words, for every s, the σ-field $\mathcal{F}_s$ is the σ-field generated by $\sigma\{W_u : u \le s\}$ and by the subsets of the elements of $\mathcal{F}$ having P-measure zero. We recall that property **(1)** above ensures that, for every $s < t$, the increment $W_t - W_s$ is independent of $\mathcal{F}_s$, and also that $W_t - W_s$

has a centered Gaussian law with variance $t - s$. Throughout the following, we assume that the reader is familiar with the construction and the properties of stochastic integrals of the type $\int_0^1 \phi_s dW_s$, where the stochastic process $s \mapsto \phi_s$ is $\mathcal{F}_s$-adapted and such that $E \int_0^1 \phi_s^2 ds < \infty$ – see e.g. [169, Section IV.2]. For instance, we shall often use the fact that the random variable $\int_0^1 \phi_s dW_s$ has mean zero and that the following isometric property is in order: for every pair of $\mathcal{F}_s$-adapted stochastic processes (ϕ_s, ψ_s), verifying $E \int_0^1 (\phi_s^2 + \psi_s^2) ds < \infty$, one has that

$$E\left[\int_0^1 \phi_s dW_s \int_0^1 \psi_s dW_s \right] = E\left[\int_0^1 \phi_s \psi_s ds \right]. \tag{4.2}$$

From now on, we will denote by $L^2([0, 1], ds) = L^2([0, 1])$ the space of real-valued functions on $[0, 1]$ that are square-integrable with respect to the Lebesgue measure. Also, for $d \geq 2$, we shall write $L_s^2([0, 1]^d)$ for the space of the *symmetric* functions on $[0, 1]^d$ that are square integrable with respect to the product Lebesgue measure; we will sometimes use the notation $L_s^2([0, 1]^1) = L^2([0, 1])$. The following definition introduces a class of random variables that are the basic building blocks of our analysis, that is, single and multiple stochastic integrals of deterministic kernels with respect to W. We shall see later on that these objects are indeed the infinite-dimensional analogous of Hermite polynomials defined on the real line.

Definition 4.2 (Stochastic integrals.)   (1) (Single Wiener-Itô integrals) For every $f \in L^2([0, 1])$, we will write $I_1(f)$ in order to indicate the single stochastic integral $\int_0^1 f(s) dW_s$, of f with respect to W.

(2) (Multiple Wiener-Itô integrals.) For every $d \geq 2$ and for every $f \in L_s^2([0, 1]^d)$, we write $I_d(f)$ in order to indicate the iterated multiple stochastic integral

$$I_d(f) = d! \int_0^1 \int_0^{s_1} \cdots \int_0^{s_{d-1}} f(s_1, ..., s_d) dW_{s_d} \cdots dW_{s_1}, \tag{4.3}$$

of f with respect to W.

Note that the stochastic integrals $I_d(f)$ appearing in (4.3) is a usual adapted stochastic integral, with respect to W, of the $\mathcal{F}_{s_1}$-adapted process

$$s_1 \mapsto \int_0^{s_1} \cdots \int_0^{s_{d-1}} f(s_1, ..., s_d) dW_{s_d} \cdots dW_{s_2}, \tag{4.4}$$

that is,

$$I_d(f) = \int_0^1 \left[d! \int_0^{s_1} \cdots \int_0^{s_{d-1}} f(s_1, ..., s_d) \, dW_{s_d} \cdots dW_{s_2} \right] dW_{s_1}. \qquad (4.5)$$

Relation (4.5) also yields that $E[I_d(f)] = 0$.

The discussion here needs a little more precision. Indeed, the process (4.4) is in principle well-defined only for kernels f that are *elementary*, that is, that are linear combinations of products of indicators of intervals. In the general case, (4.4) must be defined by an approximation argument involving sequences of elementary kernels. See [159, Chapter 5] or [169, Chapter V] for more details.

Remark 4.3   (1) The collection $C_1 = \{I_1(f) : f \in L^2([0, 1])\}$ is called the **first Wiener chaos** associated with W. It is a $L^2(P)$-closed Gaussian space whose covariance structure is given by

$$E[I_1(f) I_1(g)] = \int_0^1 f(s) g(s) \, ds. \qquad (4.6)$$

(2) For $d \geq 2$, the class $C_d = \{I_d(f) : f \in L_s^2([0, 1]^2)\}$ is called the d**th Wiener chaos** associated with W. Each space C_d is a Hilbert subspace of $L^2(P)$. Multiple integrals also verify the isometric property

$$E[I_d(f) I_{d'}(g)] \qquad (4.7)$$
$$= \begin{cases} 0 & \text{if } d \neq d' \\ d! \int_{[0,1]^d} f(s_1, ... s_d) g(s_1, ..., s_d) \, ds_1 \cdots ds_d & \text{if } d = d' \end{cases}.$$

(3) Note that (4.6) is just a special case of (4.2). On the other hand, relation (4.7) can be derived from (4.2) by iteration, as well as by exploiting the fact that f and g are symmetric by assumption, and that $E[I_d(f)] = E[I_d(g)] = 0$.

4.2.2 From multiplication to Hermite polynomials

Our aim is now to show that (multiple) Wiener-Itô integrals are the infinite-dimensional analogues of Hermite polynomials, and that they therefore provide a natural orthogonal decomposition of the space $L^2(\sigma(W))$ of square-integrable functionals of W. This point is crucial for understanding our approach to Central Limit Theorems. Indeed, we will see that, since every random variable in $L^2(\sigma(W))$ can be represented as an infinite series of multiple integrals, the characterization of the joint convergence to Gaussian of vectors of multiple integrals is the key element in order to derive general CLTs on $L^2(\sigma(W))$.

The concept of *contraction* plays a fundamental role in the theory developed in this section.

Definition 4.4   For every $q, p \geq 1$, $f \in L_s^2([0,1]^p)$, $g \in L_s^2([0,1]^q)$ and every $r = 0, ..., q \wedge p$, the **contraction of order** r *of* f and g is the function $f \otimes_r g$ of $p + q - 2r$ variables defined as follows: for $r = 1, ..., p \wedge q$ and $(t_1, ..., t_{p-r}, s_1, ..., s_{q-r}) \in [0,1]^{p+q-2r}$,

$$
\begin{aligned}
&f \otimes_r g(t_1, ..., t_{p-r}, s_1, ..., s_{q-r}) \\
&= \int_{[0,1]^r} f(z_1, ..., z_r, t_1, ..., t_{p-r}) g(z_1, ..., z_r, s_1, ..., s_{q-r}) dz_1 ... dz_r, \quad (4.8)
\end{aligned}
$$

and, for $r = 0$,

$$
\begin{aligned}
f \otimes_r g(t_1, ..., t_p, s_1, ..., s_q) &= f \otimes g(t_1, ..., t_p, s_1, ..., s_q) \qquad (4.9) \\
&= f(t_1, ..., t_{p-r}) g(s_1, ..., s_{q-r}).
\end{aligned}
$$

Note that, if $p = q$, then $f \otimes_p g = \langle f, g \rangle_{L^2([0,1]^p)}$. For instance, if $p = q = 2$, one has

$$
f \otimes_1 g(t, s) = \int_{[0,1]} f(z, t) g(z, s) \, dz, \qquad (4.10)
$$

$$
f \otimes_2 g = \int_{[0,1]^2} f(z_1, z_2) g(z_1, z_2) \, dz_1 dz_2. \qquad (4.11)
$$

By an application of the Cauchy-Schwartz inequality, it is straightforward to prove that, for every $r = 0, ..., q \wedge p$, the function $f \otimes_r g$ is an element of $L^2\left([0,1]^{p+q-2r}\right)$. Note that $f \otimes_r g$ is in general not symmetric (although f and g are): we shall denote by $f \widetilde{\otimes}_r g$ the canonical symmetrization of $f \otimes_r g$, that is,

$$
f \widetilde{\otimes}_r g(t_1, ..., t_{p+q-2r}) = \frac{1}{(p+q-2r)!} \sum_{\sigma \in S_{p+q-2r}} f \otimes_r g(t_{\sigma(1)}, ..., t_{\sigma(p+q-2r)}),
$$

where σ runs over all permutations of $\{1, ..., p + q - 2r\}$.

The next result is a multiplication formula for (multiple) Wiener-Itô integrals. It will be crucial for the rest of this chapter.

Theorem 4.5   *For every $p, q \geq 1$ and every $f \in L_s^2([0,1]^p)$, $g \in L_s^2([0,1]^q)$,*

$$
I_p(f) I_q(g) = \sum_{r=0}^{p \wedge q} r! \binom{p}{r} \binom{q}{r} I_{p+q-2r}\left(f \widetilde{\otimes}_r g\right). \qquad (4.12)
$$

Theorem 4.5, whose proof is omitted, can be established by at least two routes, namely by induction (see e.g. [151, Proposition 1.1.3]), or by using the concept of "diagonal measure" (see [159, Section 6.4]).

Example 4.6 Consider two centered independent Gaussian random variables with unit variance of the type $(I_1(e_1), I_1(e_2))$, where e_1, e_2 are two orthonormal elements of $L^2([0, 1], dt)$ and

$$I_1(e_i) = \int_0^1 e_i(t)\, dW_t, \quad i = 1, 2.$$

The multiplication formula (4.12) implies that

$$
\begin{aligned}
I_1(e_1)I_1(e_2) &= \sum_{r=0}^1 r! \binom{1}{r}\binom{1}{r} I_{p+q-2r}\left(e_1 \widetilde{\otimes}_r e_2\right) \\
&= I_2\left(e_1 \widetilde{\otimes}_0 e_2\right) \\
&= \int_0^1 \int_0^1 e_1(s)\, e_2(t)\, dW_s dW_t.
\end{aligned}
$$

Remark 4.7 Formula (4.12) implies that, for every $m \geq 1$, a random variable belonging to the space $\oplus_{j=0}^m C_j(G)$ (where here "$\oplus$" stands for an orthogonal sum in $L^2(P)$) has finite moments of any order. This result can be made more precise. Indeed, for every $p > 2$ and every $n \geq 1$, one can prove that there exists a universal constant $c_{p,n} > 0$, such that the following hypercontractivity relation takes place

$$E\left[|I_n(f)|^p\right]^{1/p} \leq c_{n,p} E\left[I_n(f)^2\right]^{1/2}, \tag{4.13}$$

$\forall f \in L^2(\mu^n)$ (see e.g. [105, Ch. V]). We also observe that, on every finite sum of Wiener chaoses $\oplus_{j=0}^m C_j(G)$ and for every $p \geq 1$, the topology induced by the $L^p(\mathbb{P})$ convergence is equivalent to the L^0-topology induced by convergence in probability, that is, convergence in probability is equivalent to convergence in L^p, for every $p \geq 1$. This fact has been first proved by Schreiber in [177] – see also [105, Chapter VI]. One can also prove that *the law of a non-zero random variable living in a finite sum of Wiener chaoses always admits a density* (see e.g. [185]).

The properties of Hermite polynomials will be constantly employed in this monograph.

Definition 4.8   The sequence of **Hermite polynomials** $\{H_q : q \geq 0\}$ on $\mathbb{R}$, is defined via the following relations: $H_0 \equiv 1$ and, for $q \geq 1$,

$$H_q(x) = (-1)^q \, e^{\frac{x^2}{2}} \, \frac{d^q}{dx^q} e^{-\frac{x^2}{2}}, \quad x \in \mathbb{R}. \tag{4.14}$$

For instance, $H_1(x) = 1$, $H_2(x) = x^2 - 1$, $H_3(x) = x^3 - 3x$ and $H_4(x) = x^4 - 6x^2 + 3$.

Recall that the sequence $\{(q!)^{-1/2} H_q : q \geq 0\}$ is an orthonormal basis of the space $L^2(\mathbb{R}, (2\pi)^{-1/2} e^{-x^2/2}dx)$. Several relevant properties of Hermite polynomials can be deduced from the following formula, valid for every $t, x \in \mathbb{R}$,

$$\exp\left(tx - \frac{t^2}{2}\right) = \sum_{n=0}^{\infty} \frac{t^n}{n!} H_n(x). \tag{4.15}$$

For instance, one deduces immediately from the previous expression that

$$\frac{d}{dx} H_n(x) = nH_{n-1}(x), \quad n \geq 1, \tag{4.16}$$

$$H_{n+1}(x) = xH_n(x) - nH_{n-1}(x), \quad n \geq 1. \tag{4.17}$$

The next result uses (4.12) and (4.17) in order to establish the announced explicit relation between multiple stochastic integrals and Hermite polynomials.

Proposition 4.9   *Let $h \in L^2([0,1])$ be such that $\|h\|_{L^2([0,1])} = 1$, and, for $n \geq 2$, define*

$$h^{\otimes n}(z_1, .., z_n) = h(z_1) \times \cdots \times h(z_n), \quad (z_1, ..., z_n) \in [0,1]^n.$$

Then,

$$I_n\left(h^{\otimes n}\right) = H_n(I_1(h)). \tag{4.18}$$

Proof   Of course, $H_1(I_1(h)) = I_1(h)$. By the multiplication formula (4.12), one has therefore that, for $n \geq 2$,

$$I_n\left(h^{\otimes n}\right) I_1(h) = I_{n+1}\left(h^{\otimes n+1}\right) + nI_{n-1}\left(h^{\otimes n-1}\right),$$

and the conclusion is obtained from (4.17), and by recursion on n.   □

Remark 4.10   By using the relation $\mathbb{E}\left[I_n(h^{\otimes n}) I_n(g^{\otimes n})\right] = n! \langle h^{\otimes n}, g^{\otimes n} \rangle_{L^2(\mu^n)} = n! \langle h, g \rangle^n_{L^2(\mu)}$, we infer from (4.18) that, for every jointly Gaussian random variables (U, V) with zero mean and unitary variance,

$$E[H_n(U) H_m(V)] = \begin{cases} 0 & \text{if } m \neq n \\ n! E[UV]^n & \text{if } m = n. \end{cases}$$

4.2.3 Chaotic decompositions

By combining (4.15) and (4.18), one obtains the following fundamental decomposition of the square-integrable functionals of W.

Theorem 4.11 (Chaotic decomposition)   *For every $F \in L^2(\sigma(W))$ (that is, F is a square-integrable functional of W), there exists a unique sequence $\{f_n : n \geq 1\}$, with $f_n \in L_s^2([0,1]^n)$, such that*

$$F = E[F] + \sum_{n=1}^{\infty} I_n(f_n), \qquad (4.19)$$

where the series converges in $L^2(P)$.

Proof   Fix $h \in L^2([0,1])$ such that $\|h\|_{L^2([0,1])} = 1$, as well as $t \in \mathbb{R}$. By using (4.15) and (4.18), one obtains that

$$\exp\left(tI_1(h) - \frac{t^2}{2}\right) = \sum_{n=0}^{\infty} \frac{t^n}{n!} H_n(I_1(h)) = 1 + \sum_{n=1}^{\infty} \frac{t^n}{n!} I_n\left(h^{\otimes n}\right). \qquad (4.20)$$

Since $\mathbb{E}\left[\exp\left(tI_1(h) - \frac{t^2}{2}\right)\right] = 1$, one deduces that (4.19) holds for every random variable of the form $F = \exp\left(tI_1(h) - \frac{t^2}{2}\right)$, with $f_n = \frac{t^n}{n!} h^{\otimes n}$. The conclusion is obtained by observing that the linear combinations of random variables of this type are dense in $L^2(\sigma(W))$. $\qquad \square$

Remark 4.12   (1) Relations (4.6) and (4.7), together with (4.19), imply that

$$E\left[F^2\right] = E[F]^2 + \sum_{n=1}^{\infty} n! \, \|f_n\|_{L^2([0,1]^n)}^2. \qquad (4.21)$$

(2) The statement of Theorem 4.11 can be reformulated as follows:

$$L^2(\sigma(W)) = \bigoplus_{n=0}^{\infty} C_n,$$

where "$\oplus$" indicates, as before, an orthogonal sum.

(3) By inspection of the proof of Theorem 4.11, we deduce that the linear combinations of random variables of the type $I_n(h^{\otimes n})$, with $n \geq 1$ and $\|h\|_{L^2([0,1])} = 1$, are dense in $L^2(\sigma(W))$. This implies in particular that the random variables $I_n(h^{\otimes n}) = H_n(I_1(h))$ generate the nth Wiener chaos C_n.

(4) The first proof of (4.19) dates back to Wiener [205]. See [151] and [159] for further references and results on chaotic decompositions.

4.3 Moments, cumulants and diagram formulae

We recall here the definition of *cumulant*, and we present several of its properties. A complete and self-contained combinatorial analysis of moments and cumulants is provided in the monograph [159], from which we borrow some notation. Another classic reference for cumulants is contained in the book by Shiryaev [186]. In what follows, we shall sometimes use the notation $[n] = \{1, ..., n\}$, where n is an integer.

For $n \geq 1$, we consider a vector of real-valued random variables $\mathbf{X}_{[n]} = (X_1, ..., X_n)$ such that $E|X_j|^n < +\infty$, $\forall j = 1, ..., n$. For every subset

$$b = \{j_1, ..., j_k\} \subseteq [n] = \{1, ..., n\},$$

we write

$$\mathbf{X}_b = \left(X_{j_1}, ..., X_{j_k}\right) \quad \text{and} \quad \mathbf{X}^b = X_{j_1} \times \cdots \times X_{j_k}. \qquad (4.22)$$

For instance, $\forall m \leq n$,

$$\mathbf{X}_{[m]} = (X_1, .., X_m) \quad \text{and} \quad \mathbf{X}^{[m]} = X_1 \times \cdots \times X_m.$$

For every $b = \{j_1, ..., j_k\} \subseteq [n]$ and $(z_1, ..., z_k) \in \mathbb{R}^k$, we let $g_{\mathbf{X}_b}(z_1, .., z_k) = E\left[\exp\left(i \sum_{\ell=1}^{k} z_\ell X_{j_\ell}\right)\right]$ be the joint characteristic function of the components of $\mathbf{X}_b$. The *joint cumulant* of the components of the vector $\mathbf{X}_b$ is defined as

$$Cum(\mathbf{X}_b) = (-i)^k \frac{\partial^k}{\partial z_1 \cdots \partial z_k} \log g_{\mathbf{X}_b}(z_1, ..., z_k)|_{z_1 = ... = z_k = 0}. \qquad (4.23)$$

We recall the following basic facts.

(i) The application $\mathbf{X}_b \mapsto Cum(\mathbf{X}_b)$ is *homogeneous*, that is, for every $\mathbf{h} = (h_1, ..., h_k) \in \mathbb{R}^k$,

$$Cum\left(h_1 X_{j_1}, ..., h_k X_{j_k}\right) = (\Pi_{\ell=1}^{k} h_\ell) \times Cum(\mathbf{X}_b);$$

(ii) The application $\mathbf{X}_b \mapsto Cum(\mathbf{X}_b)$ is invariant with respect to the permutations of b;

(iii) $Cum(\mathbf{X}_b) = 0$, if the vector $\mathbf{X}_b$ has the form $\mathbf{X}_b = \mathbf{X}_{b'} \cup \mathbf{X}_{b''}$, with $b', b'' \neq \varnothing$, $b' \cap b'' = \varnothing$ and $\mathbf{X}_{b'}$ and $\mathbf{X}_{b''}$ independent;

(iv) if $\mathbf{Y} = \{Y_j : j \in J\}$ is a Gaussian family and if $\mathbf{X}_{[n]}$ is a vector obtained by juxtaposing $n \geq 3$ elements of $\mathbf{Y}$ (with possible repetitions), then $Cum(\mathbf{X}_{[n]}) = 0$.

Properties (i) and (ii) follow immediately from (4.23). To see how to deduce (iii) from (4.23), just observe that, if $\mathbf{X}_b$ has the structure described in (iii), then

$$\log g_{\mathbf{X}_b}(z_1, .., z_k) = \log g_{\mathbf{X}_{b'}}(z_\ell : j_\ell \in b') + \log g_{\mathbf{X}_{b''}}(z_\ell : j_\ell \in b'')$$

(by independence), so that

$$\frac{\partial^k}{\partial z_1 \cdots \partial z_k} \log g_{\mathbf{X}_b}(z_1, .., z_k)$$

$$= \frac{\partial^k}{\partial z_1 \cdots \partial z_k} \log g_{\mathbf{X}_{b'}}(z_\ell : j_\ell \in b') + \frac{\partial^k}{\partial z_1 \cdots \partial z_k} \log g_{\mathbf{X}_{b''}}(z_\ell : j_\ell \in b'') = 0.$$

Finally, property (iv) is proved by using the fact that, if $\mathbf{X}_{[n]}$ is obtained by juxtaposing $n \geq 3$ elements of a Gaussian family (even with repetitions), then $\log g_{\mathbf{X}_b}(z_1, .., z_k)$ has necessarily the form $\sum_l a(l) z_l + \sum_{i,j} b(i, j) z_i z_j$, where $a(k)$ and $b(i, j)$ are coefficients not depending on the z_l's.

When $|b| = n$, one says that the cumulant $Cum(\mathbf{X}_b)$, given by (4.23), *has order n*. When $\mathbf{X}_{[n]}$ is such that $X_j = X, \forall j = 1, ..., n$, where X is a random variable in $L^n(P)$, we write

$$Cum(\mathbf{X}_{[n]}) = Cum_n(X),$$

and we say that $Cum_n(X)$ is the *nth cumulant* (or the *cumulant of order n*) of X. Note that, if $X, Y \in L^n(\mathbb{P})$ $(n \geq 1)$ are independent random variables, then (4.23) implies that

$$Cum_n(X + Y) = Cum_n(X) + Cum_n(Y),$$

since $Cum_n(X + Y)$ involve the derivative of $E\left[\exp\left(i(X + Y)\sum_{j=1}^n z_j\right)\right]$ with respect to $z_1, ..., z_n$.

We recall that a *partition* of a finite set b is a collection of r subsets of b, say $\{b_1, ..., b_r\}$, such that each b_j is not empty and $\cup_{j=1, ..., r} b_j = b$. The next result contains two crucial relations (known as *Leonov-Shiryaev identities*), linking the cumulants and the moments associated with a random vector $\mathbf{X}_{[n]}$. See Peccati and Taqqu [159] for a proof.

Proposition 4.13 *For every non empty $b \subseteq [n]$, denote by $\mathcal{P}(b)$ the class of all partitions of b. Then,*

(1)

$$E\left[\mathbf{X}^b\right] = \sum_{\pi=\{b_1, ..., b_k\} \in \mathcal{P}(b)} Cum(\mathbf{X}_{b_1}) \cdots Cum(\mathbf{X}_{b_k}); \qquad (4.24)$$

(2)

$$Cum\,(\mathbf{X}_b) = \sum_{\sigma=\{a_1,\,...,\,a_r\}\in\mathcal{P}(b)} (-1)^{r-1}\,(r-1)!\,E\,(\mathbf{X}^{a_1})\cdots E\,(\mathbf{X}^{a_r}).\quad (4.25)$$

4.3.1 Diagram formulae

To complete our background, we need a quick review of some *diagram formulae* for the computation of moments and cumulants (see for instance [159] or [186]). The *diagrams* we are interested in are essentially mnemonic devices used for the computation of the moments and cumulants associated with polynomial forms in Gaussian random variables. See [159, Chapters 2–4] for a self-contained presentation using integer partitions and Möbius inversion formulae. We start with formal definitions. Fig. 4.1 will provide an illustration.

Let p and $l_i \geq 1$, $i = 1, ..., p$, be given integers. A *diagram* γ *of order* $(l_1, ..., l_p)$ is a set of points $\{(j, l) : 1 \leq j \leq p, 1 \leq l \leq l_j\}$ (called *vertices* and represented as a table $W = \overrightarrow{l_1} \otimes \cdots \otimes \overrightarrow{l_p}$) and a partition of these points into pairs

$$\{((j, l), (k, s)) : 1 \leq j \leq k \leq p; 1 \leq l \leq l_j, 1 \leq s \leq l_k\},$$

called *edges*, such that $(j, l) \neq (k, s)$ (that is, *no vertex can be linked with itself*), and *each pair (a, b) appears in one and only one edge*. Plainly, if the integer $l_1 + \cdots + l_p$ is odd, then W does not admit any diagram. We denote by $\Gamma(l_1, ..., l_p)$ the set of all diagrams of order $(l_1, ..., l_p)$. If the order is such that $l_1 = \cdots = l_p = q$, for simplicity, we also write $\Gamma(p, q)$ instead of $\Gamma(l_1, ..., l_p)$. With this notation, one has that $\Gamma(p, q) = \emptyset$ whenever pq is odd.

Remark 4.14 The table W described above is composed of p rows, the jth $(j = 1, ..., p)$ row being composed of the pairs $(j, 1), (j, 2), ..., (j, l_j)$. One can graphically represent W by arranging $l_1 + \cdots + l_p$ dots into p rows, the jth row containing l_j dots. The lth dot (from left to right) of the jth row corresponds to the pair (j, l). Once the table W has been drawn, the edges of the diagram are represented as (possibly curved) lines connecting the two corresponding dots. In the language of graph theory, the resulting graph is called a *perfect matching*. See Fig. 4.1 for some examples.

We say that:

- A diagram has a *flat edge* if there is at least one pair $((i, j), (i', j'))$ such that $i = i'$. We write Γ_F for the set of diagrams having at least one flat edge, and $\Gamma_{\overline{F}}$ for the collection of all diagrams with no flat edges. For example, the diagrams in Fig. 4.1-(b, d) have no flat edges. The diagram appearing in Fig. 4.1(c) has two flat edges, namely $((1, 1), (1, 2))$ and $((4, 1), (4, 2))$.

- A diagram γ is *connected* if it is not possible to partition the rows $\overrightarrow{l_1}, \cdots, \overrightarrow{l_p}$ of the table W into two non-connected subdiagrams. This means that one cannot find a partition $K_1 \cup K_2 = \{1, ..., p\}$ such that, for each member V_k of the set of edges $(V_1, ..., V_r)$ in a diagram γ, either V_k links vertices in $\cup_{j \in K_1} \overrightarrow{l_j}$, or V_k links vertices in $\cup_{j \in K_2} \overrightarrow{l_j}$. We write Γ_C for the collection of all connected diagrams, and $\Gamma_{\overline{C}}$ for the class of all diagrams that are not connected. For instance, the diagrams in Fig. 4.1(b,c) are not connected (for both, one can choose the partition $K_1 = \{1, 4\}$, $K_2 = \{2, 3\}$). The diagram in Fig. 4.1-(d) is connected.

- A diagram is *paired* if, considering any two edges $((i_1, j_1), (i_2, j_2))$, and $((i_3, j_3), (i_4, j_4))$, then $i_1 = i_3$ implies $i_2 = i_4$; in words, the rows are completely coupled two by two. We write Γ_P for the set of diagrams for paired diagrams, and $\Gamma_{\overline{P}}$ for the set of diagrams that are not paired. For example, the diagrams in Fig. 4.1(b, c) are paired (in both diagrams, the first row is coupled with the fourth, and the second row is coupled with the third).

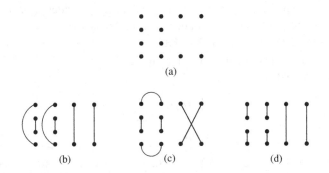

(a)

(b)   (c)   (d)

Figure 4.1 A table (a) and three different diagrams (b, c, d). (a) A representation of a table W of order $(4, 2, 2, 4)$, that is, $p = 4$, $l_1 = l_4 = 4$ and $l_2 = l_3 = 2$. (b) A representation of $\{((1, j), (4, j)), j = 1, ..., 4; ((2, k), (3, k)), k = 1, 2\}$, where the diagram is paired, non-flat and not connected. (c) A representation of $\{((i, 1), (i, 2)), i = 1, 4; ((1, 4), (4, 3)); (1, 3), (4, 4); ((2, k), (3, k)), k = 1, 2\}$, where the diagram is paired, with two flat edges and not connected. (c) A representation of $\{((1, j), (2, j)), j = 1, 2; ((1, k), (4, k)), k = 3, 4; ((3, l), (4, l)), l = 1, 2\}$, where the diagram is not paired, non-flat and connected.

The next statement provides a well-known combinatorial description of the moments and cumulants associated with Hermite transformations of (possibily correlated) Gaussian random variables. In view of Proposition 4.9, one can see such a result of the diagram formulae associated with general multiple stochastic integrals. See [159, Section 7.3] for statements and proofs.

Proposition 4.15 (Diagram formulae for Hermite polynomials) *Let $(Z_1, ..., Z_p)$ be a centered Gaussian vector, and let $\gamma_{ij} = E[Z_i Z_j], i, j = 1, ..., p$ be its covariance. Let $H_{l_1}, ..., H_{l_p}$ be Hermite polynomials of degrees $l_1, ..., l_p \ (\geq 1)$ respectively. As above, let $\Gamma_{\overline{F}}(l_1, ..., l_p)$ (resp. $\Gamma_C(l_1, ..., l_p)$) be the collection of all diagrams with no flat edges (resp. connected diagrams) of order $l_1, ..., l_p$. Then,*

$$E[\Pi_{j=1}^{p} H_{l_j}(Z_j)] = \sum_{G \in \Gamma_{\overline{F}}(l_1, ..., l_p)} \Pi_{1 \leq i \leq j \leq p} \gamma_{ij}^{\eta_{ij}(G)} \qquad (4.26)$$

$$Cum(H_{l_1}(Z_1), ..., H_{l_p}(Z_p)) = \sum_{G \in \Gamma_{\overline{F}}(l_1, ..., l_p) \cap \Gamma_C(l_1, ..., l_p)} \Pi_{1 \leq i \leq j \leq p} \gamma_{ij}^{\eta_{ij}(G)} \quad (4.27)$$

where, for each diagram G, $\eta_{ij}(G)$ is the exact number of edges between the ith row and the jth row of the diagram G.

Example 4.16 (1) Consider the quantity EZ^{2p}, where $Z \sim N(0, \sigma^2)$. Proposition 4.15 implies that this expected value can be written as a sum over all diagrams with $2p$ rows and one column. Standard combinatorial arguments show that there exist $(2p-1)!! = (2p-1) \times (2p-3) \times \cdots \times 1$ different diagrams of this type, so that, selecting for $Z_i = Z_j = Z$, $(1 \leq i \leq j \leq 2p)$ in (4.26) we deduce:

$$EZ^{2p} = E[\Pi_{j=1}^{p} H_1(Z_j)] = \sum_{G \in \Gamma(1, ..., 1)} \Pi_{1 \leq i \leq j \leq p} EZ_i Z_j$$

$$= (2p - 1)!! \sigma^{2p}.$$

(2) Analogously, we see that $EZ^{2p+1} = 0$, where again $Z \sim N(0, \sigma^2)$ and p is any nonnegative integer. Indeed, as already observed, there cannot be any diagrams with an odd number of vertices.

(3) Now consider the quantity $EH_m(Z_1)H_n(Z_2)$, $Z_i \sim N(0, \sigma_i^2)$, $i = 1, 2$. This corresponds to the case $p = 2$, $l_1 = m$, $l_2 = n$. If $p \neq q$, then $\Gamma_{\overline{F}}(l_1, ..., l_p) = \emptyset$. If or $p = q$ then $\Gamma_{\overline{F}}(l_1, ..., l_p)$ contains $p!$ different diagrams (one for every permutation of $\{1, ..., p\}$). Formula (4.26), therefore yields

$$EH_p(Z_1)H_q(Z_2) = \delta_p^q p! \{EZ_1 Z_2\}^p .$$

This is consistent with the content of Remark 4.10.

(4) Consider $EZ_1^2 Z_2^2 Z_3^2$, where $Z_i \sim N(0, \sigma_i^2)$, $i = 1, 2, 3$. Using the fact that $H_2(x) = x^2 - 1$ and (4.26), we obtain

$$
\begin{aligned}
EZ_1^2 Z_2^2 Z_3^2 &= E\{H_2(Z_1)H_2(Z_2)H_2(Z_3)\} \\
&\quad + E\{H_2(Z_2)H_2(Z_3)\} \\
&\quad + E\{H_2(Z_1)H_2(Z_2)\} + E\{H_2(Z_1)H_2(Z_3)\} \\
&\quad + E\{H_2(Z_1)\} + E\{H_2(Z_2)\} \\
&\quad + E\{H_2(Z_3)\} + 1 \\
&= 8\gamma_{12}\gamma_{23}\gamma_{31} + 2\left\{\gamma_{12}^2 + \gamma_{23}^2 + \gamma_{13}^2\right\} + 1,
\end{aligned}
$$

$\gamma_{12} = EZ_1 Z_2$, $\gamma_{23} = EZ_2 Z_3$, $\gamma_{13} = EZ_1 Z_3$.

4.4 The simplified method of moments on Wiener chaos

4.4.1 Real kernels

The following statement provides a classic application of the method of moments (and cumulants) to the derivation of multivariate CLTs on Wiener chaos.

Proposition 4.17 (The method of moments and cumulants) *Fix integers $k \geq 1$ and $1 \leq d_1, ..., d_k < \infty$, and let $\mathbf{Z} = (Z_1, ..., Z_k) \sim N_k(0, V)$ be a k-dimensional centered Gaussian vector with nonnegative covariance matrix*

$$
V = \{V(i, j) : i, j = 1, ..., k\}.
$$

Suppose that

$$
\mathbf{F}_n = \left(F_1^{(n)}, ..., F_k^{(n)}\right) = \left(I_{d_1}\left(f_1^{(n)}\right), ..., I_{d_k}\left(f_k^{(n)}\right)\right), \quad n \geq 1, \tag{4.28}
$$

is a sequence of k-dimensional vectors of chaotic random variables such that $f_j^{(n)} \in L_s^2\left([0, 1]^{d_j}\right)$ and, for every $1 \leq i, j \leq k$,

$$
\lim_{n \to \infty} E\left[F_i^{(n)} F_j^{(n)}\right] = V(i, j). \tag{4.29}
$$

Then, the following three conditions are equivalent, as $n \to \infty$.

(1) The vectors $\mathbf{F}_n$ converge in law to Z.

(2) For every choice of nonnegative integers $p_1, ..., p_k$ such that $\sum p_j \geq 3$

$$
E\left[\prod_{j=1}^{k}\left[F_i^{(n)}\right]^{p_j}\right] \longrightarrow E\left[\prod_{j=1}^{k} Z_i^{p_j}\right]. \tag{4.30}
$$

(3) For every $r \geq 3$ and for every vector $(j_1, ..., j_r) \in \{1, ..., k\}^r$ (note that such a vector may contain repeated indices)

$$Cum\left(F_{j_1}^{(n)}, ..., F_{j_r}^{(n)}\right) \to Cum\left(Z_{j_1}, ..., Z_{j_r}\right) = 0. \qquad (4.31)$$

Proof To prove that Point (1) and Point (2) are equivalent, combine (4.29) with (4.13) and use uniform integrability, as well as the fact that the Gaussian distribution is determined by its moments. To prove the equivalence of Point (3), use the Leonov-Shyriaev relation (4.25). □

The use of results analogous to Proposition 4.17 in the proof of CLTs for non-linear functionals of Gaussian fields, is classic. See e.g. Breuer and Major [27] or Chambers and Slud [38]. It can be however technically quite demanding to verify conditions (4.30)–(4.31), since they involve an infinity of asymptotic relations of increasing complexity. In recent years, great efforts have been made in order to obtain drastic simplifications of the convergence criteria implied by Proposition 4.17. The most powerful achievements in this direction (at least, in the case of one-dimensional CLTs) are given in the subsequent statement.

We recall that the **total variation distance**, between the law of two real-valued random variables X and Y, is given by the quantity

$$d_{TV}(X, Y) = \sup_A |P(X \in A) - P(Y \in A)|,$$

where the supremum is taken over the class of all Borel sets. Note that the topology induced by d_{TV} (on the class of all probability measures on $\mathbb{R}$) is strictly stronger than the topology of convergence in distribution (see e.g. [56, Chapter 11]).

Recall the notion of "contraction" of order p, and the associated operator $\otimes_p$, as introduced in Definition 4.4.

Theorem 4.18 (Total variation bounds – see [147]) *Let $\sigma^2 > 0$ and let $Z \sim N\left(0, \sigma^2\right)$ be a centered Gaussian random variable with variance σ^2. For $d \geq 2$, let $F = I_d(f) \in C_d$ be an element of the dth chaos of W. Then, $Cum_4(F) = E\left(F^4\right) - 3E\left(F^2\right)^2$, and*

$$d_{TV}(Z, F) \leq \frac{2}{\sigma^2}\left|\sigma^2 - E\left[F^2\right]\right| \qquad (4.32)$$

$$+ \frac{2}{\sigma^2}\sqrt{d^2 \sum_{p=1}^{d-1}(p-1)!^2\binom{d-1}{p-1}^4(2d-2p)! \left\|f\widetilde{\otimes}_p f\right\|_{L^2([0,1]^{2(d-p)})}^2}$$

$$\leq \frac{2}{\sigma^2}\left[\left|\sigma^2 - E\left[F^2\right]\right| + \sqrt{\frac{q-1}{3q}}\,Cum_4\left(F\right)\right].$$

In particular, if $\{F_n : n \geq 1\} \subset C_d$ is a sequence of chaotic random variables such that $F_n = I_d\left(f_n\right)$ and $E\left(F_n^2\right) \to \sigma^2$, then the following three conditions are equivalent as n diverges to infinity

(1) F_n *converges in law to Z.*

(2) F_n *converges to Z in the total variation distance.*

(3) $Cum_4\left(F_n\right) \to 0$ *(or, equivalently, $E\left(F_n^4\right) \to 3\sigma^4$).*

(4) For every $p = 1, ..., d - 1$, $\left\|f_n \widetilde{\otimes}_p f_n\right\|_{L^2\left([0,1]^{2(d-p)}\right)} \to 0$.

Remark 4.19 (1) The first proof of the "fourth moment" bound in (4.32) appears in Nourdin, Peccati and Reinert [147]. It is based on Malliavin calculus, and the so-called Lindeberg principle and Stein's method for probabilistic approximations. This result also builds on previous estimates by Nourdin and Peccati [148].

(2) Note that the bound (4.32) provides an immediate proof of the equivalence of Points (1)–(4) in the statement of Theorem 4.18. Indeed, if Point (3) is in order, then (4.32) yields that Points (4) and (2) necessarily hold, which also gives Point (1), since convergence in total variation implies convergence in law. On the other hand, if $E\left(F_n^2\right) \to 1$ and F_n converges to $Z \sim N\left(0, \sigma^2\right)$ in law, then one can use (4.13) together with a uniform integrability argument, in order to deduce that $Cum_4\left(F_n\right) = E\left(F_n^4\right) - 3E\left(F_n^2\right)^2 \to 0$.

(3) The equivalence between Points 1, 3 and 4 in the statement of Theorem 4.18 was first proved by Nualart and Peccati in [152], by means of stochastic calculus techniques.

(4) Observe that $\left\|f_n \widetilde{\otimes}_p f_n\right\|_{L^2\left([0,1]^{2(d-p)}\right)} \leq \left\|f_n \otimes_p f_n\right\|_{L^2\left([0,1]^{2(d-p)}\right)}$. In [152] it is also proved that $\left\|f_n \widetilde{\otimes}_p f_n\right\|_{L^2\left([0,1]^{2(d-p)}\right)} \to 0$ for every $p = 1, ..., d - 1$ if and only if $\left\|f_n \otimes_p f_n\right\|_{L^2\left([0,1]^{2(d-p)}\right)} \to 0$ for every $p = 1, ..., d - 1$

The next result, first proved in [160] allows to deduce joint CLTs from one dimensional convergence results.

Theorem 4.20 (Joint CLTs on Wiener chaos – see [160]) *Keep the assumptions and notation of Proposition 4.17 (in particular, the sequence $\{\mathbf{F}_n : n \geq 1\}$ verifies (4.29)). Then, the vectors $\mathbf{F}_n$ converge in law to Z if and only if $F_j^{(n)}$ converges in law to Z_j for every $j = 1, ..., k$.*

The original proof of Theorem 4.20, as given in [160], used stochastic time-changes and other tools from continuous-time stochastic calculus. A more direct proof is now available, using the following estimate, taken again from [147].

Theorem 4.21 (Bounds on vectors)   *Let*

$$\mathbf{F} = (I_{d_1}(f_1), ..., I_{d_k}(f_k))$$

be a k-dimensional vector of multiple integrals, where $d_1, ..., d_k \geq 1$ and $f_j \in L_s^2\left([0,1]^{d_j}\right)$. Let V denote the (nonnegative definite) covariance matrix of $\mathbf{F}$. Let Z be a centered k-dimensional Gaussian vector with the same covariance V. Then, for every twice differentiable function $\phi : \mathbb{R}^k \to \mathbb{R}$,

$$\left|E\left[\phi(\mathbf{F})\right] - E\left[\phi(Z)\right]\right| \leq C \left\|\phi''\right\|_\infty \sum_{j=1}^{k} \sum_{p=1}^{d_j-1} \left\|f_j \widetilde{\otimes}_p f_j\right\|, \qquad (4.33)$$

where $C = C(k; d_1, ..., d_k)$ is a strictly positive constant depending uniquely on k and $d_1, ..., d_k$, and $\|\phi''\|_\infty$ is defined as

$$\|\phi''\|_\infty := \sup_{\alpha; x_1, ..., x_k} \frac{\partial^{|\alpha|}}{\partial x_1^{\alpha_1} \cdots \partial x_d^{\alpha_k}} \phi(x_1, ..., x_k), \qquad (4.34)$$

and the symbol α runs over all integer-valued multiindex $\alpha = (\alpha_1, ..., \alpha_k)$ such that $|\alpha| := \alpha_1 + \cdots + \alpha_k = 2$.

It is also clear the combination of Theorem 4.18 and Theorem 4.20 yields an important simplification of the method of moments and cumulants, as stated in Proposition 4.17. This fact is so useful that we prefer to write it as a separate statement.

Proposition 4.22 (The simplified method of moments)   *Keep the assumptions and notation of Proposition 4.17 (in particular, the sequence $\{\mathbf{F}_n : n \geq 1\}$ verifies (4.29)). Then, the following two conditions are equivalent:*

(1) The vectors $\mathbf{F}_n$ converge in law to Z.
(2) For every $j = 1, ..., k$,

$$E\left[\left(F_j^{(n)}\right)^4\right] \to 3\sigma^4. \qquad (4.35)$$

We conclude this section with a useful estimate on moments, taken from the survey paper [149].

Proposition 4.23 *Let $d \geq 2$ be an integer, and let $F = I_d(f)$ be a multiple integral of order d of some kernel $f \in L_s^2([0,1]^d)$. Assume that $\mathrm{Var}(F) = E(Z^2) = 1$, and let $Z \sim N(0,1)$. Then, for all integer $k \geq 3$,*

$$\left| E(F^k) - E(Z^k) \right| \leq c_{k,d} \sqrt{E(F^4) - E(Z^4)}, \tag{4.36}$$

where the constant $c_{k,d}$ is given by

$$c_{k,d} = (k-1)2^{k-\frac{5}{2}} \sqrt{\frac{d-1}{3d}} \left(\sqrt{\frac{(2k-4)!}{2^{k-2}(k-2)!}} + (2k-5)^{\frac{kd}{2}-d} \right).$$

4.4.2 Further results on complex kernels

Since we will work with harmonic decompositions based on complex-valued Hilbert spaces, we will sometimes need criteria for CLTs involving random quantities taking values in $\mathbb{C}$. Note that, in general, CLTs for complex-valued random variables are simply joint CLTs for the real and imaginary parts, so that the subsequent results can be regarded as corollaries of Theorem 4.20 of the previous section. However, the direct formulation in terms of complex quantities is very useful, and we believe that it deserves a separate discussion.

For every integer $d \geq 1$, $L_{\mathbb{C}}^2([0,1]^d)$ and $L_{s,\mathbb{C}}^2([0,1]^d)$ are the Hilbert spaces, respectively, of square integrable and square integrable and symmetric complex-valued functions with respect to the product Lebesgue measure. For every $g \in L_{s,\mathbb{C}}^2([0,1]^d)$ with the form $g = a + ib$, where $a, b \in L_s^2([0,1]^d)$, we set $I_d(g) = I_d(a) + iI_d(b)$. Note that, by isometry (4.7),

$$E\left[I_d(f) \overline{I_{d'}(g)} \right] \tag{4.37}$$

$$= \begin{cases} 0 & \text{if } d \neq d' \\ d! \int_{[0,1]^d} f(s_1, \dots s_d) \overline{g(s_1, \dots, s_d)} ds_1 \cdots ds_d & \text{if } d = d'. \end{cases}$$

Also, a random variable such as $I_d(g)$ is real-valued if and only if g is real valued. For every pair and $g_k = a_k + ib_k \in L_{s,\mathbb{C}}^2([0,1]^d)$, $k = 1, 2$, and every $q = 1, \dots, d-1$, we set

$$g_1 \otimes_q g_2 = a_1 \otimes_q a_2 - b_1 \otimes_q b_2 + i\left(a_1 \otimes_q b_2 + b_1 \otimes_q a_2 \right). \tag{4.38}$$

The following result has been proved in [134, Proposition 5].

Proposition 4.24 *Suppose that the sequence $g_l = a_l + ib_l \in L_{s,\mathbb{C}}^2(\mu^d)$, $l \geq 1$, is such that*

$$\lim_{l \to +\infty} d! \|a_l\|_{L^2([0,1]^d)}^2 = \lim_{l \to +\infty} d! \|b_l\|_{L^2([0,1]^d)}^2 \to \frac{1}{2} \quad \text{and} \quad \langle a_l, b_l \rangle_{L^2([0,1]^d)} = 0. \tag{4.39}$$

Then, the following conditions are equivalent: as $l \to \infty$,

(1) $I_d(g_l) \overset{law}{\to} N + iN'$, *where* $N, N' \sim N(0, 1/2)$ *are independent;*

(2) $g_l \otimes_q \overline{g_l} \to 0$ *and* $g_l \otimes_q g_l \to 0$ *in* $L^2_{\mathbb{C}}\left([0, 1]^{2(d-q)}\right)$ *for every* $q = 1, ..., d - 1;$

(3) $g_l \otimes_q \overline{g_l} \to 0$ *in* $L^2_{\mathbb{C}}\left([0, 1]^{2(d-q)}\right)$ *for every* $q = 1, ..., d - 1;$

(4) $a_l \otimes_q a_l \to 0$, $b_l \otimes_q b_l \to 0$ *and* $a_l \otimes_q b_l \to 0$ *in* $L^2\left([0, 1]^{2(d-q)}\right)$ *for every* $q = 1, ..., d - 1;$

(5) $a_l \otimes_q a_l \to 0$, $b_l \otimes_q b_l \to 0$ *in* $L^2\left([0, 1]^{2(d-q)}\right)$ *for every* $q = 1, ..., d - 1;$

(6) $E\left[I_d(a_l)^4\right] \to 3/4$, $E\left[I_d(b_l)^4\right] \to 3/4$ *and* $E\left[I_d(a_l)^2 I_d(b_l)^2\right] \to 1/4;$

(7) $E\left[I_d(a_l)^4\right] \to 3/4$, $E\left[I_d(b_l)^4\right] \to 3/4$.

4.5 The graphical method for Wigner coefficients

We now focus on some combinatorial results, known as "the graphical method", involving the class of Clebsch-Gordan and Wigner coefficients introduced in Section 3.5. These formulae enter quite naturally into the computation of moments and cumulants associated with non-linear transformations of Gaussian fields defined on the sphere S^2.

Graphical methods are well-known to the physicists' community, and go far beyond the results described in this section. An exhaustive discussion of these techniques can be found in textbooks such as [195, Chapter 11] and [17]; see also [23]. Here, we shall simply present the main staples of the graphical method, in a form which is suitable for the applications developed in the subsequent chapters.

4.5.1 From diagrams to graphs

Recall that a *graph* is a pair (I, E), where I is a set of vertices and E is a collection of edges, that is, of unordered pairs $\{x, y\}$, where $x, y \in I$. Considering unordered pairs $\{x, y\}$ makes the graph *not directed*, that is, $\{x, y\}$ and $\{y, x\}$ identify the same edge. Also, we allow for repetitions, meaning that the edge $\{x, y\}$ may appear more than once into E (or, equivalently, every edge is counted with a multiplicity possibly greater than one). Due to this circumstance, the term *multigraph* might be more appropriate, but we shall avoid this terminology.

In what follows, we will exclusively deal with graphs that are obtained by "compression" of diagrams of the class $\Gamma(p, 3)$, as defined in Section 4.3.1. Recall that the elements of $\Gamma(p, 3)$ are associated with a table W composed of three columns and p rows. In standard matrix notation, we shall denote by (i, j), $(i = 1, ..., p, j = 1, 2, 3)$ the element of W corresponding to the ith row and jth column. Since we are focussing on diagrams with a specific structure, in this section we switch from G to γ, in order to indicate a generic element of $\Gamma(p, 3)$.

Now fix an even integer $p \geq 2$. Given a diagram $\gamma \in \Gamma(p, 3)$ on a $p \times 3$ table W, we build a graph $\hat{\gamma}$ with p vertices and $3p/2$ edges by implementing the following procedure: (i) identify the ith row of W with the ith vertex of $\hat{\gamma}$, $i = 1, ..., p$ and (ii) draw one edge linking the vertex i_1 and the vertex i_2 for every pair of the type $((i_1, j_1), (i_2, j_2))$ appearing in γ. If one counts loops twice (recall that a *loop* is an edge connecting one vertex to itself), then $\hat{\gamma}$ is such that there are exactly three edges that are incident to each vertex.

Example 4.25   Fig. 4.2 provides two illustrations of the construction described above. For instance, on the left of (a) one has $\gamma = \{((1, 1), (1, 3)),$ $((1, 2), (2, 1)), ((2, 2), (3, 1)), ((2, 3), (3, 2))((3, 3), (4, 2))((4, 1), (4, 3))\}$, which is in $\Gamma(4, 3)$. The graph $\hat{\gamma}$ (on the right) is given by

$$\hat{\gamma} = \{\{1, 1\}, \{1, 2\}, \{2, 3\}, \{2, 3\}, \{3, 4\}, \{4, 4\}\}.$$

Analogously, the diagram on the left of (b) is $\gamma = \{((1, j), (2, j)), ((3, j), (4, j)) :$ $j = 1, 2, 3\}$, and

$$\hat{\gamma} = \{\{1, 2\}, \{1, 2\}, \{1, 2\}, \{3, 4\}, \{3, 4\}, \{3, 4\}\}.$$

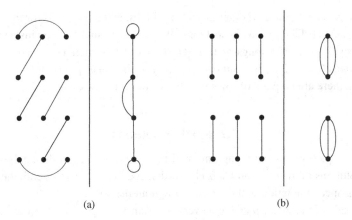

(a)   (b)

Figure 4.2 (a) The diagram on the left is flat and connected, and generates a graph (right) with two loops and one two-loop. (b) The diagram on the left is paired and not connected, and so is the generated graph on the right.

Remark 4.26 The following facts are elementary, and can be checked by the reader.

(1) To every flat edge of the diagram γ, there corresponds a loop of $\hat{\gamma}$.
(2) The diagram γ is connected (in the specific sense of Section 4.3.1) if and only if $\hat{\gamma}$ is a connected graph (in the usual sense of graph theory, that is, $\hat{\gamma}$ cannot be partitioned into two disjoint subgraphs).
(3) The diagram γ is paired if and only if $\hat{\gamma}$ is paired. The fact that $\hat{\gamma}$ is paired means that two vertices are either not linked, or linked by exactly three edges. Observe that the graph on the right of Fig. 4.2(b) is paired.

Definition 4.27 We shall say that a diagram γ has a *k-loop* if there exists a sequence of k edges

$$((i_1, j_1), (a_1, b_1)), ((i_2, j_2), (a_2, b_2))..., ((i_k, j_k), (a_k, b_k)),$$

such that the integers i_j, $j = 1, ..., k$, are all different and $a_x = i_{x+1}$ for $x = 1, ..., k$, with $i_{k+1} = i_1$ by definition. We write $\gamma \in \Gamma_{L(k)}(p, 3)$ for diagrams with a k-loop and no loop of order smaller than k. Plainly, to every k-loop of γ as above, there corresponds a k-loop of $\hat{\gamma}$, which is given by the sequence of edges $\{i_1, i_2\}, \{i_2, i_3\}, ..., \{i_k, i_1\}$. Note that a one-loop of γ is just a flat edge (so that $\Gamma_{L(1)}(p, 3) = \Gamma_F(p, 3)$), whereas a one-loop of $\hat{\gamma}$ is a loop in plain graph theoretical language. Also, we write

$$\Gamma_{CL(k)}(p, 3) = \Gamma_C(p, 3) \cap \Gamma_{L(k)}(p, 3),$$

for the class of connected diagrams with k-loops, and $\Gamma_{\overline{CL(k)}}(p, 3)$ for the class of connected diagrams with no loops of order k or smaller. For instance, a connected diagram belongs to $\Gamma_{\overline{CL(2)}}(p, 3)$ if there are neither flat edges nor two edges $((i_1, j_1), (i_2, j_2))$ and $((i_3, j_3), (i_4, j_4))$ such that $i_1 = i_3$ and $i_2 = i_4$; in words, there are no pairs of rows which are connected twice.

4.5.2 Further notation

In the sequel, graphs and diagrams will be crucial devices in the analysis of convolutions of Wigner and Clebsch-Gordan coefficients. To this end, the following special notation will be used throughout the book.

For any even integer $p \geq 2$ and every diagram $\gamma \in \Gamma(p, 3)$ we shall write

$$D[\gamma; l_1, l_2, l_3] = \left\{ \prod_{i=1}^{p} \prod_{j=1}^{3} \sum_{m_{ij}=-l_j}^{l_j} \right\} \prod_{i=1}^{p} \begin{pmatrix} l_1 & l_2 & l_3 \\ m_{i1} & m_{i2} & m_{i3} \end{pmatrix} \delta(\gamma; l_1, l_2, l_3),$$

(4.40)

where

$$\delta(\gamma; l_1, l_2, l_3) = \prod_{((i_u, j_u),(i'_u, j'_u)) \in \gamma} (-1)^{m_{i_u j_u}} \delta_{m_{i_u j_u}}^{-m'_{i'_u j'_u}} \delta_{l_{j_u}}^{l'_{j'_u}},$$

and

$$\left\{ \prod_{i=1}^{p} \prod_{j=1}^{3} \sum_{m_{ij}=-l_j}^{l_j} \right\} = \sum_{m_{11}} \sum_{m_{12}} \sum_{m_{13}} \cdots \sum_{m_{p1}} \sum_{m_{p2}} \sum_{m_{p3}}.$$

This symbol implies therefore that there are exactly $3 \times p$ summations to compute.

Remark on notation. In what follows, when the range of a sum is not specified, the sum has to be regarded as over all possible values of the parameters.

For instance, when $p = 2$ we get

$$\left\{ \prod_{i \in \{1,2\}} \prod_{j=1}^{3} \sum_{m_{ij}=-l_j}^{l_j} \right\} = \sum_{m_{11}=-l_1}^{l_1} \sum_{m_{22}=-l_2}^{l_2} \sum_{m_{13}=-l_3}^{l_3} \sum_{m_{21}=-l_1}^{l_1} \sum_{m_{22}=-l_2}^{l_2} \sum_{m_{23}=-l_3}^{l_3},$$

so that, if $\gamma = \{((1, j), (2, j)) : j = 1, 2, 3\}$, then

$$D[\gamma; l_1, l_2, l_3] =$$

$$= \sum_{m_{11}=-l_1}^{l_1} \sum_{m_{22}=-l_2}^{l_2} \sum_{m_{13}=-l_3}^{l_3} \sum_{m_{21}=-l_1}^{l_1} \sum_{m_{22}=-l_2}^{l_2} \sum_{m_{23}=-l_3}^{l_3} \delta(\gamma; l_1, l_2, l_3)$$

$$\times \begin{pmatrix} l_1 & l_2 & l_3 \\ m_{11} & m_{12} & m_{13} \end{pmatrix} \begin{pmatrix} l_1 & l_2 & l_3 \\ m_{21} & m_{22} & m_{23} \end{pmatrix}$$

$$= \sum_{m_{11}=-l_1}^{l_1} \sum_{m_{22}=-l_2}^{l_2} \sum_{m_{13}=-l_3}^{l_3} \begin{pmatrix} l_1 & l_2 & l_3 \\ m_{11} & m_{12} & m_{13} \end{pmatrix} \begin{pmatrix} l_1 & l_2 & l_3 \\ -m_{11} & -m_{12} & -m_{13} \end{pmatrix},$$

where we have implicitly used the fact that Wigner coefficients vanish whenever $m_1 + m_2 + m_3 \neq 0$.

4.5.3 First example: sums of squares

The so-called "graphical method" is a collection of mnemonic devices, allowing to represent products of Wigner coefficients in terms of one of the graphs $\hat{\gamma}$ introduced in Section 4.5.1. It is better to focus on Wigner coefficients rather than Clebsch-Gordan, because the former have simpler symmetry properties.

Remark 4.28 To simplify the notation, from now on the integers in the upper row of $3j$ symbols are implicitly assumed to satisfy the "triangle conditions" described in Remark 3.38.

The idea is to view each Wigner coefficient as a vertex with three incident edges, corresponding to $(l_1 m_1), (l_2 m_2), (l_3 m_3)$. Multiplying Wigner coefficients just corresponds to putting the corresponding dots in the same plot, while an edge connects two vertices if and only if the corresponding index m_i is common and summed in the two coefficients. For instance, to represent the product

$$T = \sum_{m_{11} m_{12} m_{13}} \begin{pmatrix} l_{11} & l_{12} & l_{13} \\ m_{11} & m_{12} & m_{13} \end{pmatrix}^2 \sum_{m_{21} m_{22} m_{23}} \begin{pmatrix} l_{21} & l_{22} & l_{23} \\ m_{21} & m_{22} & m_{23} \end{pmatrix}^2 = 1,$$

(4.41)

(the last equality is just a consequence of the unitarity of the Clebsch-Gordan matrices) one uses the graph in Fig. 4.3. Here, the first and second dots (from above) correspond to the two Wigner coefficients juxtaposed in the first sum (so to form a square), whereas the third and fourth correspond to the coefficients in the second sum.

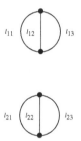

Figure 4.3 A graphical representation of a double sum of squares

It is an important remark that a graph such as the one in Fig. 4.3 *does not provide* all the information that is needed to reconstruct a given sum. Indeed, in what follows we shall use graphs in order to keep track of the pairings of the indices m_{ij}, without indicating possible factors of the type $(-1)^\alpha$ (inside and outside the $3j$ symbols). For instance, the graph in Fig. 4.3 also represents the quantity $T' = D[\gamma; l_{11}, l_{12}, l_{13}] \times D[\gamma; l_{21}, l_{22}, l_{23}]$, where $\gamma = \{((1, j), (2, j)) : j = 1, 2, 3\}$. Note that, by Proposition 3.44(4), one has that $T' = (-1)^{\sum_{ij} l_{ij}} T = (-1)^{\sum_{ij} l_{ij}}$. As explained by [195], it would be possible to take into account these phase terms by considering oriented graphs; in this monograph, however, we shall be concerned either with the absolute values of Wigner's convolutions or with cases where α is even, and hence we stick to a simpler undirected representation.

4.5.4 Cliques and Wigner $6j$ coefficients

A sum of the type

$$\sum_{\alpha,\beta,\gamma,\varepsilon,\delta,\phi}(-1)^{\eta}\begin{pmatrix} a & b & c \\ \alpha & \beta & \gamma \end{pmatrix}\begin{pmatrix} a & e & f \\ \alpha & \epsilon & -\phi \end{pmatrix}\begin{pmatrix} d & b & f \\ -\delta & \beta & \phi \end{pmatrix}\begin{pmatrix} d & e & c \\ \delta & -\epsilon & \gamma \end{pmatrix},$$

where $\eta := d + e + f + \delta + \varepsilon + \phi$, is called a *Wigner $6j$ coefficient* (see [195, Chapter 9]), and it is customarily denoted by the symbol

$$\begin{Bmatrix} a & b & c \\ d & e & f \end{Bmatrix}.$$

In view of the graphical rules discussed above, such a symbol is represented by a *clique*, that is, by a graph in which every pair of vertices is connected. Fig. 4.4 provides an illustration.

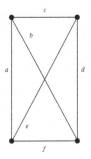

Figure 4.4 A Wigner $6j$ coefficient represented as a clique

Note that, in Fig. 4.4 and in those to follow, the vertices of the graph $\hat{\gamma}$ are not necessarily vertically organized.

Remark 4.29   In [195], Wigner $6j$ coefficients are sometimes represented as triangular cliques. By definition, these cliques are isomorphic to the one in Fig. 4.4. We also recall the following *upper bound on Wigner's $6j$:* for any positive integers a, b, c, d, e and f

$$\left| \left\{ \begin{array}{ccc} a & b & c \\ d & e & f \end{array} \right\} \right|$$

$$\leq \min \left(\frac{1}{\sqrt{(2c+1)(2f+1)}}, \frac{1}{\sqrt{(2a+1)(2d+1)}}, \frac{1}{\sqrt{(2b+1)(2e+1)}} \right).$$

To conclude the chapter, we shall present four useful rules for the computation of sums involving Wigner and Clebsch-Gordan coefficients. All these rules admit some neat representations in terms of the graphical method.

4.5.5 Rule n. 1: loops are zero

The following lemma shows that, when $l_1, l_2, l_2 > 0$, all graphs with a loop generate sums equal to zero.

Lemma 4.30 *Let $p \geq 2$ and consider integers $l_1, l_2, l_3 > 0$. Assume that $\gamma \in \Gamma_F(p, 3)$ (that is, γ is a flat diagram with p rows). Then,*

$$D[\gamma; l_1, l_2, l_3] = 0.$$

Proof The existence of a flat edge implies that there must exist an $i^* = 1, ..., p$ such that $m_{i^* j_1} = -m_{i^* j_2}$, for some $j_1 \neq j_2$. Denote by $\gamma_{R(i^*)}$ the diagram which we obtain from $\gamma \in \Gamma_F(p, 3)$ by deleting the one-loop $[(i^*, j_1), (i^*, j_2)]$. Now because $m_{i1} + m_{i2} + m_{i3} \equiv 0$, for all i's, we obtain (up to a permutation)

$$D[\gamma; l_1, l_2, l_3] = \left\{ \prod_{i=1}^{p} \prod_{j=1}^{3} \sum_{m_{ij}=-l_j}^{l_j} \right\} \prod_{i=1}^{p} \left(\begin{array}{ccc} l_1 & l_2 & l_3 \\ m_{i1} & m_{i2} & m_{i3} \end{array} \right) \delta(\gamma; l_1, l_2, l_3)$$

$$= \sum_{m_{i^*3}=-l_3}^{l_3} \sum_{m_{i^*1}=-l_1}^{l_1} (-1)^{m_{i^*1}} \left(\begin{array}{ccc} l_1 & l_1 & l_3 \\ m_{i^*1} & -m_{i^*2} & m_{i^*3} \end{array} \right)$$

$$\times \left\{ \prod_{i \neq i^*} \prod_{j \in J} \sum_{m_{ij}=-l_j}^{l_j} \right\} \prod_{i \neq i^*} \left(\begin{array}{ccc} l_1 & l_2 & l_3 \\ m_{i1} & m_{i2} & m_{i3} \end{array} \right) \delta(\gamma_{R(i^*)}; l_1, l_2, l_3),$$

where we used a slightly incomplete notation, since in the previous formula we did not indicate which index in the remaining symbols is coupled with m_{i^*} and therefore set to be equal to zero (this is immaterial for the conclusion of the proof). Now, from [195, Section 8.7.1] we learn that

$$\sum_{m_{i^*1}=-l_1}^{l_1} (-1)^{m_{i^*1}} \left(\begin{array}{ccc} l_1 & l_2 & l_3 \\ m_{i^*1} & -m_{i^*1} & m_{i^*3} \end{array} \right) = (-1)^{l_1} \delta_{m_{i^*3}}^{0} \delta_{l_3}^{0} \delta_{l_1}^{l_2} \sqrt{2l_1 + 1}.$$

Since we assumed that $l_1, l_2, l_3 > 0$, it follows immediately that $D[\gamma; l_1, l_2, l_3] = 0$, as claimed. □

Consider for instance the case $p = 2$, as well as the flat diagram $\gamma = \{(1, 3), (2, 3)); ((i, 1), (i, 2) : i = 1, 2\}$. In this case, the associated graph $\hat{\gamma}$ (drawn in Fig. 4.5) has two loops, from which we infer that

$$D[\gamma; l_1, l_2, l_3] = \sum_{m_1 m_2 m_3} (-1)^{m_1 + m_2} \begin{pmatrix} l_1 & l_1 & l_3 \\ m_1 & -m_1 & m_3 \end{pmatrix} \begin{pmatrix} l_2 & l_2 & l_3 \\ m_2 & -m_2 & -m_3 \end{pmatrix} = 0,$$

whenever $l_1, l_2, l_3 > 0$.

Figure 4.5 A graph with two loops generating a zero sum

Note that in Fig. 4.5 each loop has two (identical) labels.

4.5.6 Rule n. 2: paired sums are one

This rule just refers to the already observed fact that

$$\sum_{m_1 m_2 m_3} \begin{pmatrix} l_1 & l_2 & l_3 \\ m_1 & m_2 & m_3 \end{pmatrix}^2 = 1,$$

which is a direct consequence of the unitarity of Clebsch-Gordan matrices. As above, a paired sum can be identified with a two-vertex clique, such as the ones appearing in Fig. 4.3.

4.5.7 Rule n. 3: 2-loops can be cut, and leave a factor

The following statement allows to simplify graphs with a two-loop, by replacing it with an edge multiplied by an appropriate factor. It corresponds e.g. to

[195, formula 12.1.2.3]. Recall that, in this section, we systematically assume that the triangle conditions of Remark 3.38 is verified by the integers in the upper row of $3j$ symbols.

Lemma 4.31 *The following simplification is in order:*

$$\sum_{\psi,\kappa}(-1)^{p+\psi+q+\kappa}\begin{pmatrix} p & q & a \\ \psi & \kappa & \alpha \end{pmatrix}\begin{pmatrix} p & q & a' \\ \psi & \kappa & \alpha' \end{pmatrix} = \frac{(-1)^{a+\alpha}}{2a+1}\delta_a^{a'}\delta_\alpha^{\alpha'}.$$

Proof Using the definition of the $3j$ coefficients, the above relation is equivalent to

$$\sum_{\psi,\kappa} C_{p\psi q\kappa}^{a-\alpha} C_{p\psi q\kappa}^{a'-\alpha'} = \delta_a^{a'}\delta_\alpha^{\alpha'},$$

which is once again a consequence of the unitarity of Clebsch-Gordan matrices. □

An illustration is provided in Fig. 4.6.

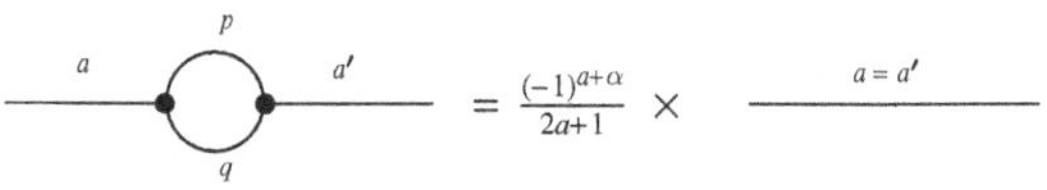

Figure 4.6 A two-loop equals an edge multiplied by a factor

4.5.8 Rule n. 4: three-loops can be cut, and leave a clique

The following statement deals with the simplification of three-loops, that can indeed be replaced by a vertex (with three incident edges) multiplied by a $6j$ coefficient. It corresponds to [195, formula 12.1.3.6]. The proof follows arguments similar to the ones already used in this section (in particular, the unitary properties of Clebsch-Gordan matrices). We stress once again that the triangle conditions of Remark 3.38 are assumed to be in order for every $3j$ coefficient. Also, we shall use the $6j$ notation introduced in Section 4.5.4.

Lemma 4.32 *The following equality holds:*

$$\sum_{\psi,\kappa,\rho}(-1)^{\eta}\begin{pmatrix} p & a & q \\ \psi & \alpha & -\kappa \end{pmatrix}\begin{pmatrix} q & b & r \\ \kappa & \beta & -\rho \end{pmatrix}\begin{pmatrix} r & c & p \\ \rho & \gamma & -\psi \end{pmatrix}$$

$$=\begin{pmatrix} a & b & c \\ -\alpha & -\beta & -\gamma \end{pmatrix}\begin{Bmatrix} a & b & c \\ r & p & q \end{Bmatrix},$$

where $\eta := p + q + r + \psi + \kappa + \rho.$

Fig. 4.7 provides a demonstration of this principle.

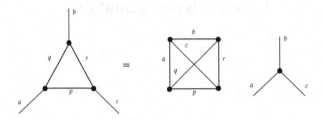

Figure 4.7 A three-loop equals a vertex multiplied by a clique

5

Spectral Representations

5.1 Introduction

In this chapter, we present a number of representation results for random fields defined on $\mathbb{R}^3$, or on the sphere S^2. We shall focus on random fields whose law verifies some invariance properties, namely *isotropy* (i.e., invariance with respect to rotations) and/or *stationarity* (i.e., invariance with respect to translations).

Section 5.2 contains facts that are implicitly used throughout the rest of this book, connecting strong isotropy to group representations and the Peter-Weyl Theorem – with special emphasis on spherical random fields, that is, random fields indexed by the unit sphere. In Sections 5.3 and 5.4 we focus on weak versions of stationarity and isotropy and the associated spectral representations. In particular, the content of Section 5.4 provides alternate proofs (with no group theory involved) of some of the main findings of Section 5.2. Note that Section 5.3 and Section 5.4 are not used in the rest of the book, so that some technical details have been skipped in order to keep the presentation as concise and to the point as possible. Adequate references are given below: in general, for a textbook treatment of stationary and isotropic fields on $\mathbb{R}^m$, one can consult Yadrenko [201], Adler [2], Ivanov and Leonenko [104], Leonenko [125], Adler and Taylor [3] and the references therein.

5.2 The Stochastic Peter-Weyl Theorem

5.2.1 General statements

We will now use notions and definitions that have been introduced in Chapter 2. Let G be a topological compact group, and let dg be the associated Haar measure with unit mass. As in Section 2.5, we denote by $\hat{G}$ the dual of G (that is, $\hat{G}$ is the collection of the equivalence classes of the irreducible representations of G). We recall that $\hat{G}$ is at most countable – see Remark 2.35(3). As before, for every $[\pi] \in \hat{G}$, we use the following notation: (i) d_π (which is a finite integer) is the dimension of any representative element of $[\pi]$, (ii) M_π is the space of the matrix coefficients associated with $[\pi]$, and (iii) χ_π is the character of $[\pi]$. We recall that M_π is a vector space of dimension d_π^2, and that an orthonormal basis of M_π is given by the mappings

$$g \mapsto \sqrt{d_\pi} \pi_{ij}(g), \quad i, j = 1, ..., d_\pi, \tag{5.1}$$

as defined in formula (2.25).

We now consider a *random field* $T = \{T(g) : g \in G\}$ defined on G. As usual, the expression "random field" – as opposed to "random process" – denotes a collection of random variables indexed by a set that is not a subset of the real line. This means that T is a real-valued $\mathcal{F} \otimes \mathcal{G}$ - measurable mapping of the type

$$T : \Omega \times G \to \mathbb{C} : (\omega, g) \mapsto T(\omega, g),$$

where $\mathcal{G}$ is the Borel σ-field generated by the topology on G, and $(\Omega, \mathcal{F}, P)$ is an appropriate probability space. Note that, given a random field T and for every $h \in G$, we can construct a "translated" field ${}^h T$, defined as ${}^h T(\omega, g) = T(\omega, hg)$.

Definition 5.1 Let $\mathbf{T} = \{T_i : i \in I\}$ be a (possibly complex-valued) collection of random fields on G (defined on the same probability space), where I is some set. The class $\mathbf{T}$ is said to be **strongly isotropic** if, for every $h \in G$, every $k \geq 1$ and every $i_1, ..., i_k \in I$, the vectors $\left({}^h T_{i_1}, ..., {}^h T_{i_k}\right)$ and $(T_{i_1}, ..., T_{i_k})$ have the same finite-dimensional distributions. In other words, $\mathbf{T}$ is strongly isotropic if, for every $n, k \geq 1$, every $i_1, ..., i_k \in I$, and every $h, g_1, ..., g_n \in G$,

$$\left\{T_{i_j}(hg_1), ..., T_{i_j}(hg_n) : j = 1, ..., k\right\} \stackrel{law}{=} \left\{T_{i_j}(g_1), ..., T_{i_j}(g_n) : j = 1, ..., k\right\}.$$

The definition of a strongly isotropic field is analogous (it corresponds to the case where I is a singleton).

Definition 5.2 Let $\mathbf{T} = \{T_i : i \in I\}$ be a (possibly complex-valued) collection of random fields on G. Fix $n \geq 2$. The class $\mathbf{T}$ is said to be n-**weakly isotropic** if $E|T_i(g)|^n < \infty$ for every $i \in I$ and every $g \in G$, and if the fields T_i and ${}^h T_i$ have the same joint moments up to the order n. This means that for every $1 \leq k \leq n$, every $i_1, ..., i_k \in I$ (with possible repetitions), every $h, g_1, ..., g_k \in G$, and every $(e_1, ..., e_k) \in \{0, 1\}^k$,

$$E\left[T_{i_1}^{(e_1)}(hg_1) \times \cdots \times T_{i_k}^{(e_k)}(hg_k)\right] = E\left[T_{i_1}^{(e_1)}(g_1) \times \cdots \times T_{i_k}^{(e_k)}(g_k)\right],$$

where $T^{(0)} = T$ and $T^{(1)} = \overline{T}$.

For instance, a real-valued square-integrable field T is two-weakly isotropic if and only if the mapping $g \mapsto E[T(g)]$ is constant over G, and $E[T(g)T(h)] = E[T(kg)T(kh)]$ for every $g, h, k \in G$. Of course, strong isotropy and existence of moments up to the order n imply n-weak isotropy. Also, if $\mathbf{T}$ has square-integrable components and it is Gaussian, then $\mathbf{T}$ is strongly isotropic if and only if it is two-weakly isotropic.

Now suppose that T is a random field on G, and that T verifies the following property: there exists a set $\Omega_0 \subset \Omega$ such that $P(\Omega_0) = 1$ and

$$\forall \omega \in \Omega_0, \text{ the mapping } g \mapsto T(\omega, g) \text{ is in } L^2(G, dg) = L^2(G). \qquad (5.2)$$

Then, for every $[\pi] \in \hat{G}$ and every $i, j = 1, ..., d_\pi$ the mapping

$$(\omega, g) \mapsto \begin{cases} 0, & \text{if } \omega \notin \Omega_0 \\ d_\pi \int_G T(\omega, h) \overline{\pi_{ij}}(h) \, dh \times \pi_{ij}(g), & \text{if } \omega \in \Omega_0 \end{cases}$$

is well-defined. We recall that, according to the second part of the Peter-Weyl Theorem 2.47 , for every $\omega \in \Omega_0$ and every $[\pi] \in \hat{G}$, the function

$$g \mapsto T^\pi(\omega, g) = \sum_{i,j=1}^{d_\pi} d_\pi \int_G T(\omega, h) \overline{\pi_{ij}}(h) \, dh \times \pi_{ij}(g) \qquad (5.3)$$

$$= d_\pi \int_G T(\omega, h) \chi_\pi \left(h^{-1} g \right) dh,$$

provides the projection of $T(\omega, \cdot)$ on M_π. We also define $T^\pi(\omega, \cdot) \equiv 0$ for $\omega \notin \Omega_0$. One simple but useful fact is that the collection $\left\{ T, T^\pi : [\pi] \in \hat{G} \right\}$ inherits the same isotropic properties of the random field T. This is explained in the next two propositions, dealing respectively with strongly and two-weakly isotropic fields (the case of n-weak isotropy for general n can be dealt with similarly, but we do not need such a generality).

Proposition 5.3 *Let T be a strongly isotropic field on G, verifying more-over property (5.2). Then, the class $\left\{ T, T^\pi : [\pi] \in \hat{G} \right\}$ defined in (5.3) is strongly isotropic.*

Proof Fix $a \in G$, and use the left-invariance of the Haar measure to deduce that, for every $[\pi] \in \hat{G}$,

$$T^\pi(ag) = d_\pi \int_G T(h) \chi_\pi \left(h^{-1} ag \right) dh$$

$$= d_\pi \int_G T\left(a(a^{-1}h)\right)\chi_\pi\left(\left(a^{-1}h\right)^{-1}g\right)dh$$

$$= d_\pi \int_G T(ah)\chi_\pi\left(h^{-1}g\right)dh = d_\pi \int_G {}^aT(h)\chi_\pi\left(h^{-1}g\right)dh.$$

To conclude, use the fact that T is strongly isotropic, so that T and ${}^aT$ have the same law. □

Proposition 5.4 *Let T be a 2-weakly isotropic field on G. Then, the following hold.*

(1) The field T verifies property (5.2) and the class $\left\{T, T^\pi : [\pi] \in \hat{G}\right\}$ defined in (5.3) is 2-weakly isotropic.
(2) Denoting by $[\pi_0]$ the equivalence class of the trivial representation, one has that $ET^\pi(g) = 0$ for $[\pi] \neq [\pi_0]$, and $ET^{\pi_0}(g) = ET(g)$.
(3) The fields T^π and $T^{\pi'}$ are not correlated for $[\pi] \neq [\pi']$, in the sense that, for every $a, b \in G$,

$$E[T^\pi(a)\overline{T^{\pi'}(b)}] = 0.$$

Proof Isotropy implies that

$$E\left(\int_G |T(g)|^2 dg\right) = \int_G E|T(g)|^2 dg = E|T(e)|^2 \times \int_G dg = E|T(e)|^2 < \infty,$$

where e is the identity of G, from which we deduce that T verifies (5.2). Arguing as in the previous proof, we see that, for every $a \in G$,

$$T^\pi(ag) = d_\pi \int_G {}^aT(h)\chi_\pi\left(h^{-1}g\right)dh.$$

By a Fubini argument, this shows immediately that $E[T^\pi(g)] = 0$ for every $g \in G$ and every $[\pi] \neq [\pi_0]$, where π_0 is the trivial representation, whereas $E[T^{\pi_0}(g)] = E[T(e)]$. Also, for every $k, g \in G$ and again by isotropy and by a Fubini argument,

$$E[T(ak)T^\pi(ag)] = d_\pi \int_G E[{}^aT(k)\,{}^aT(h)]\chi_\pi\left(h^{-1}g\right)dh$$

$$= d_\pi \int_G E[T(k)T(h)]\chi_\pi\left(h^{-1}g\right)dh = E[T(k)T^\pi(g)], \text{ and}$$

$$E[T^{\pi'}(ak)T^\pi(ag)] = d_\pi d_{\pi'} \int_G \int_G E[{}^aT(b)\,{}^aT(h)]\chi_\pi\left(h^{-1}g\right)\chi_{\pi'}\left(b^{-1}k\right)dhdb$$

$$= d_\pi d_{\pi'} \int_G \int_G E[T(b)T(h)]\chi_\pi\left(h^{-1}g\right)\chi_{\pi'}\left(b^{-1}k\right)dhdb = E[T^{\pi'}(k)T^\pi(g)].$$

The case of mixed moments of order two involving the conjugates of either T

or T^π is dealt with analogously, and we conclude that $\{T, T^\pi : [\pi] \in \hat{G}\}$ is two-weakly isotropic, thus showing Point (1). Point (2) is a consequence of the fact that the space of matrix coefficients M_{π_0} coincides with the class of constant functions on G, whereas M_π is orthogonal to M_{π_0} whenever $[\pi] \neq [\pi_0]$. To prove Point (3), just use the fact that the mappings $g \mapsto T^\pi(ga)$ and $g \mapsto T^{\pi'}(gb)$ are, respectively, in the spaces M_π and $M_{\pi'}$ so that, by weak isotropy and a Fubini argument,

$$E[T^\pi(a)\overline{T^{\pi'}(b)}] = E\left[\int_G T^\pi(ga)\overline{T^{\pi'}(gb)}dg\right] = 0,$$

since M_π and $M_{\pi'}$ are orthogonal in $L^2(G)$, whenever $[\pi] \neq [\pi']$.   □

Proposition 5.3 contains everything we need in order to prove a stochastic analogous of the Peter-Weyl Theorem. Note that, in the following statement, we assume that the field T is 2-weakly isotropic, so that the result applies, in particular, to square-integrable strongly isotropic fields. See also [157] for several statements of a similar nature.

Theorem 5.5 (Stochastic Peter-Weyl Theorem)   *Let T be a square-integrable 2-weakly isotropic field on G. Then, T verifies property (5.2) and*

$$T(g) = \sum_{[\pi] \in \hat{G}} T^\pi(g), \tag{5.4}$$

where we used the notation (5.3). The convergence of the infinite series in (5.4) is both in the sense of $L^2(\Omega \times G, P \otimes dg)$ and $L^2(P)$ for every fixed g, that is, for any enumeration $\{[\pi_k] : k \geq 1\}$ of $\hat{G}$, one has both

$$\lim_{N \to \infty} E\left[\int_G \left|T(g) - \sum_{k=1}^N T^{\pi_k}(g)\right|^2 dg\right] = 0 \tag{5.5}$$

and, for every fixed $g \in G$,

$$\lim_{N \to \infty} E\left|T(g) - \sum_{k=1}^N T^{\pi_k}(g)\right|^2 = 0. \tag{5.6}$$

Proof   Property (5.2) follows from the first part of Proposition 5.4. According to Theorem 2.47, and since (5.2) is in order, for every $\omega \in \Omega_0$ it holds that, as $N \to \infty$,

$$\int_G \left|T(\omega, g) - \sum_{k=1}^N T^{\pi_k}(\omega, g)\right|^2 dg \to 0.$$

By orthogonality, we also deduce that

$$\int_G \left| T(\omega, g) - \sum_{k=1}^{N} T^{\pi_k}(\omega, g) \right|^2 dg = \sum_{k=N+1}^{\infty} \int_G |T^{\pi_k}(\omega, g)|^2 dg$$

$$\leq \sum_{k=1}^{\infty} \int_G |T^{\pi_k}(\omega, g)|^2 dg$$

$$= \int_G |T(\omega, g)|^2 dg.$$

Since $E \int_G |T(g)|^2 dg < \infty$, the proof of (5.5) is concluded by using dominated convergence.

We now turn to the proof of (5.6). Fix $g \in G$. From Proposition 5.4, we deduce that, for every $h \in G$

$$E\left[\left| T(hg) - \sum_{k=1}^{N} T^{\pi_k}(hg) \right|^2\right] = E\left[\left| T(g) - \sum_{k=1}^{N} T^{\pi_k}(g) \right|^2\right],$$

so that, by selecting in particular $h = g^{-1}$, we see that it is sufficient to prove the statement for $g = e$, where e is the identity element of G. Again by weak isotropy (and by the fact that dg has unit mass) we now deduce that

$$E\left[\left| T(e) - \sum_{k=1}^{N} T^{\pi_k}(e) \right|^2\right] = \int_G E\left[\left| T(h) - \sum_{k=1}^{N} T^{\pi_k}(h) \right|^2\right] dh$$

$$= E\left[\int_G \left| T(h) - \sum_{k=1}^{N} T^{\pi_k}(h) \right|^2 dh\right],$$

so that the conclusion is immediately deduced from (5.5). □

Remark 5.6 (1) In order to prove a relation of the type (5.5), isotropy is not needed. Indeed, by inspection of the preceding proof, one sees that relation (5.5) continues to hold for every field T verifying property (5.2) and such that the mapping $g \mapsto E|T(g)|^2$ is bounded (by compactness, it is e.g. sufficient that such a mapping is continuous).

(2) Plainly, the almost sure representation $T = \sum_{[\pi] \in \hat{G}} T^\pi$ may hold even if the random variables $T(x)$ are not square-integrable (and even if the process is not isotropic): indeed, due to the "deterministic" Peter-Weyl Theorem, it is sufficient that, for every ω, the function $T(\omega, \cdot)$ is an element of $L^2(G)$. Of course, under such assumptions, the mean-square convergences at (5.5)–(5.6) may fail. One trivial example is obtained by setting $T(g) \equiv \eta$ for every $g \in G$, where η is some random variable such that $E|\eta| = \infty$.

(3) If G is finite, then $\hat{G}$ is also finite, and the representation $T = \sum_{[\pi] \in \hat{G}} T^\pi$ holds for every random field defined on G.

Remark 5.7 A celebrated result by Cramer and Leadbetter states that for a zero-mean "second-order stationary stochastic processes" $\{X_t : t \geq 1\}$ the following mean square representation holds:

$$X_t = \int_{-\pi}^{\pi} \exp(it\lambda)dW(\lambda) \, , t = 1, 2, 3, \dots .$$

where $dW(\lambda)$ is a complex-valued random measure such that $EdW(\lambda) = 0$,

$$EdW(\lambda)\overline{dW(\mu)} = f(\lambda)d\lambda\delta_\lambda^\mu,$$

$f(\lambda)$ denoting the spectral density of the process, see for instance [30] for details. At this stage, it is certainly tempting to view this as a special case of the stochastic Peter-Weyl Theorem; note indeed that we are dealing with a process defined on the group of integers $\mathbb{N}$, for which a complete set of representations are provided by the functions $\exp(i\lambda.)$, $\lambda \in [-\pi, \pi]$. Some care is needed here, however, as the group of integers $\mathbb{N}$ is not compact, so the Peter-Weyl Theorem does not apply.

We provide now some examples of compact groups where the previous results apply directly.

Example 5.8 (Some Abelian groups)   (1) Let Z_n be the cyclic group. We recall that Z_n is commutative, and that the characters of Z_n are given by the mappings $t \mapsto \exp\left(i\frac{2\pi j}{n}t\right)$, $t \in Z_n$. It follows that every random function $\{T(t) : t \in Z_n\}$ admits the representation

$$T(t) = \sum_{j=1}^{n} a_j \exp\left(i\frac{2\pi j}{n}t\right),$$

where $a_j = \frac{1}{n}\sum_{u=1}^{n} T(u)\exp\left(-i\frac{2\pi j}{n}u\right)$. The mapping

$$j \mapsto \sqrt{n}a_j = \frac{1}{\sqrt{n}}\sum_{u=1}^{n} T(u)\exp\left(-i\frac{2\pi j}{n}u\right), \quad j = 1, \dots, n,$$

is usually called the *discrete Fourier transform* of T.

(2) Let $U > 0$, and let $\mathbb{T} = [0, U)$. We regard $\mathbb{T}$ as an Abelian group (namely, a torus), with group operation given by $xy = (x + y) \bmod U$. Then, the characters of $\mathbb{T}$ are the applications $x \mapsto \exp\left(i\frac{2\pi n}{U}x\right)$, $n \in \mathbb{Z}$, and Theorem 5.5

implies that, for every square-integrable and isotropic field $\{T(x) : x \in \mathbb{T}\}$, one has that

$$\lim_{N \to \infty} E\left[\int_0^U \left|T(x) - \sum_{k=-N}^{N} a_k e^{i\frac{2\pi k}{U}x}\right|^2 dx\right] = 0$$

and, for every fixed x,

$$\lim_{N \to \infty} E\left[\left|T(x) - \sum_{k=-N}^{N} a_k e^{i\frac{2\pi k}{U}x}\right|^2\right] = 0,$$

where

$$a_k = \frac{1}{U}\int_0^U T(x) e^{-i\frac{2\pi k}{U}x}dx.$$

5.2.2 Decompositions on the sphere

We now study on the case where $G = SO(3)$, and use the stochastic Peter-Weyl Theorem 5.5 in order to deduce a spectral representation result for isotropic random fields on the sphere $S^2 \simeq SO(3)/SO(2)$. In what follows, given $g \in SO(3)$ and $x \in S^2$ we shall use the notation $g \cdot x = gx$ to indicate the action of g on x, that is, gx stands for the position of the point x after the rotation g. We start with an analogous of Definitions 5.1 and 5.2. Observe that in this section (mainly in order to simplify the notation) we only focus on real-valued random fields.

Definition 5.9 Let $T = \{T(x) : x \in S^2\}$ be a real-valued spherical random field.

(1) The field T is said to be **strongly isotropic** if, for every $k \in \mathbb{N}$, every $x_1, ..., x_k \in S^2$ and every $g \in SO(3)$ (the group of rotations in $\mathbb{R}^3$) we have

$$\{T(x_1), ..., T(x_k)\} \overset{law}{=} \{T(gx_1), ..., T(gx_k)\}. \tag{5.7}$$

(2) The field T is said to be n-**weakly isotropic** $(n \geq 2)$ if $E|T(x)|^n < \infty$ for every $x \in S^2$, and if, for every $1 \leq k \leq n$, every $x_1, ..., x_k \in S^2$ and every $g \in SO(3)$,

$$E\left[T(x_1) \times \cdots \times T(x_k)\right] = E\left[T(gx_1) \times \cdots \times T(gx_k)\right].$$

As already observed, a field T is two-weakly isotropic if and only if T is square-integrable, $x \mapsto ET(x)$ is constant, and $E(T(x)T(y)) = E(T(gx)T(gy))$ for every $x, y \in S^2$ and every $g \in SO(3)$. The following statement, whose proof is elementary, indicates some further relations between the two notions of isotropy described above.

Proposition 5.10　*(1) A strongly isotropic field with finite moments of some order $n \geq 2$ is also n-weakly isotropic.*

(2) Suppose that the field T is n-weakly isotropic for every $n \geq 2$ (in particular, $E|T(x)|^n < \infty$ for every $n \geq 2$ and every $x \in S^2$) and that, for every $k \geq 1$ and every $(x_1, ..., x_k)$, the law of the vector $\{T(x_1), ..., T(x_k)\}$ is determined by its moments. Then, T is also strongly isotropic.

(3) If T is Gaussian, then T is strongly isotropic if and only if it is two-weakly isotropic.

Remark 5.11　We use the parametrization of $SO(3)$ given by Euler angles, as defined in Chapter 3. This means that to every $g \in SO(3)$ we associate the triple $(\varphi, \vartheta, \psi)$, where $0 \leq \varphi < 2\pi$, $0 \leq \vartheta \leq \pi$ and $0 \leq \psi < 2\pi$. According to the discussion of Chapter 3, we know that each function on $S^2 \simeq SO(3)/SO(2)$ can be identified with a function of $SO(3)$ not depending on the third Euler angle ψ. **Caveat:** as already recalled, in order to stick to the common notational conventions, we write the spherical coordinate representation of every $x \in S^2$ in the form $x = (\vartheta, \varphi)$ where $0 \leq \vartheta \leq \pi$ and $0 \leq \varphi < 2\pi$, that is, in spherical coordinates the angles ϑ and φ appear *in reverse order* with respect to the Euler parametrization of $SO(3)$.

We recall that the unique Haar measure of $SO(3)$ having mass one, noted dg, is such that, in terms of Euler angles,

$$dg = \frac{\sin \vartheta d\varphi d\vartheta d\psi}{8\pi^2},$$

see formula (3.5). As before, we denote by $\{Y_{lm} : l \geq 0, m = -l, ..., l\}$ the collection of the spherical harmonics, introduced in Section 3.4. Our aim is now to use the stochastic Peter-Weyl Theorem 5.5 in order to characterize the spectral decomposition of isotropic spherical fields. Our main tool is the following statement. As before, we shall denote by $\mathbf{e}_3 = (0, 0, 1)$ the North Pole of the unit sphere S^2.

Lemma 5.12　*Let $T = \{T(x) : x \in S^2\}$ be a real valued random field, and assume that T is strongly isotropic (resp. two-weakly isotropic). Then, there exists a strongly isotropic (resp. two-weakly isotropic) random field $U = \{U(g) : g \in SO(3)\}$ such that the representation of U in spherical coordinates, written $U(\varphi, \vartheta, \psi)$, does not depend on the angle ψ, and*

$$T(\vartheta, \varphi)(\omega) = U(\varphi, \vartheta, \psi)(\omega), \tag{5.8}$$

for every ω, and every $\psi \in [0, 2\pi)$.

Proof Assume first that T is strongly isotropic, and observe that, by definition, $T(\vartheta, \varphi) = T(R_{\mathbf{e}_3}(\varphi)R_{\mathbf{e}_2}(\vartheta) \cdot \mathbf{e}_3)$, where we have used the notation of Proposition 3.1. Since T is strongly isotropic, the field U on $SO(3)$ defined by $g \mapsto U(g) := T(g \cdot \mathbf{e}_3)$ is also strongly isotropic. Indeed, for any choice of $h, g_1, ..., g_d \in SO(3)$,

$$\{U(hg_j) : j = 1, ..., d\} = \{T(h \cdot (g_j \cdot \mathbf{e}_3)) : j = 1, ..., d\}$$

has the same law as

$$\{T(g_j \cdot \mathbf{e}_3) : j = 1, ..., d\} = \{U(g_j) : j = 1, ..., d\}.$$

Moreover, the representation of U in terms of the Euler angles is such that

$$U(\varphi, \vartheta, \psi) = T(R_{\mathbf{e}_3}(\varphi)R_{\mathbf{e}_2}(\vartheta)R_{\mathbf{e}_3}(\psi) \cdot \mathbf{e}_3) = T(R_{\mathbf{e}_3}(\varphi)R_{\mathbf{e}_2}(\vartheta) \cdot \mathbf{e}_3) = T(\vartheta, \varphi),$$

thus proving the claim. When T is two-weakly isotropic, we define U as in the first part of the proof, and exploit the relations

$$E[U(hg)] = E[T(h \cdot (g \cdot \mathbf{e}_3))] = E[T(g \cdot \mathbf{e}_3)] = E[U(g)],$$
$$E[U(hg_1)U(hg_2)] = E[T(h \cdot (g_1 \cdot \mathbf{e}_3))T(h \cdot (g_2 \cdot \mathbf{e}_3))]]$$
$$= E[T(g_1 \cdot \mathbf{e}_3)T(g_2 \cdot \mathbf{e}_3)] = E[U(g_1)U(g_2)],$$

holding for every $h, g, g_1, g_2 \in SO(3)$, to conclude the proof. $\square$

The following theorem is one of the staples of the book.

Theorem 5.13 (Spectral representation of isotropic fields on S^2.) *Let* $T = \{T(x) : x \in S^2\}$ *be a two-weakly isotropic field. Then,*

$$\int_{S^2} T(x)^2 \, d\sigma(x) < \infty, \quad a.s.-P, \tag{5.9}$$

and

$$T(x) = \sum_{l=0}^{\infty} \sum_{m=-l}^{l} a_{lm} Y_{lm}(x), \tag{5.10}$$

where

$$a_{lm} = \int_{S^2} T(y) \overline{Y_{lm}(y)} d\sigma(y), \tag{5.11}$$

and $d\sigma = \sin \vartheta d\vartheta d\varphi$ *is the Lebesgue measure on the sphere. The decomposition (5.10) is both in the sense of* $L^2 \left(\Omega \times S^2, P \otimes d\sigma \right)$ *and* $L^2(P)$ *for every* *fixed $x \in S^2$, that is,*

$$\lim_{L \to \infty} E\left[\int_{S^2} \left(T(y) - \sum_{l=0}^{L} \sum_{m=-l}^{l} a_{lm} Y_{lm}(y) \right)^2 d\sigma(y) \right] = 0, \qquad (5.12)$$

and, for every fixed $x \in S^2$,

$$\lim_{L \to \infty} E\left[\left(T(x) - \sum_{l=0}^{L} \sum_{m=-l}^{l} a_{lm} Y_{lm}(x) \right)^2 \right] = 0. \qquad (5.13)$$

Proof Relation (5.9) follows from isotropy and a Fubini argument. To prove the rest of the statement, call U the isotropic field on $SO(3)$ associated with T via Lemma 5.12. Reasoning as in the proof of Proposition 3.29 one infers that, since U does not depend on the Euler angle ψ, then the Peter-Weyl decomposition on U takes necessarily the form

$$U(\varphi, \vartheta, \cdot) = \sum_{l=0}^{\infty} \sum_{m=-l}^{l} a_{lm} Y_{lm}(\vartheta, \varphi).$$

It follows that formulae (5.10)-(5.13) are just a reformulation of the content of Theorem 5.5, as applied to the field U.   □

Remark 5.14 (On adding constants) Suppose that T is a spherical field verifying the assumptions of Theorem 5.13 (hence admitting the decomposition (5.10)), and fix a constant $c \in \mathbb{R}$. Then, the field $T^c := T + c$ also verifies these assumptions, and therefore admits a spectral decomposition of the type (5.10), for some collection of random harmonic coefficients $\{a_{lm}^c\}$. Since spherical harmonics with index $l \geq 1$ are orthogonal to constants, one deduces immediately the relations

$$a_{lm} = a_{lm}^c, \qquad \forall l \geq 1, \ \forall m = -l, ..., l,$$

and $a_{00}^c Y_{00} = c + a_{00} Y_{00}$. In particular, these relations imply that, for $l \geq 1$, the coefficients a_{lm} are always centered random variables (even if the original field T is not centered), whereas $E a_{00} Y_{00} = ET(x)$. In what follows, we shall often consider *centered* isotropic spherical fields: in view of the previous discussion, this is done to simplify the notation, by ensuring that also the trivial coefficient a_{00} in their decomposition (5.10) is a centered random variable. Plainly, the content of the present remark can be seen as a consequence of Proposition 5.4(2).

Remark 5.15 (1) As already observed in connection with the stochastic Peter-Weyl Theorem, isotropy and square integrability are not necessarily needed in order to have a spectral decomposition for a spherical field T. Indeed, if $\int_{S^2} T(x)^2\, dx < \infty$, a.s.-$P$, then the deterministic Peter-Weyl Theorem ensures that, a.s.-P, the decomposition $T = \sum_l \sum_m a_{lm} Y_{lm}$ holds. Plainly, under these minimal assumptions the mean-square convergences at (5.12) and (5.13) may not take place.

(2) A natural question is now the following: under isotropy, can one characterize the joint law (or, at least, the joint moments) of the random harmonic coefficients $\{a_{lm} : l \ge 0, \ m = -l, ..., l\}$? Answering this question is a technically demanding task, to which we shall devote the whole of the forthcoming Chapter 6. However, anticipating some material from the following chapter, we can point out that the spherical harmonic coefficients are necessarily uncorrelated under weak isotropy. More precisely, for a centered and two-weakly isotropic real field T, one has that

$$E a_{l_1 m_1} \bar{a}_{l_2 m_2} = C_{l_1} \delta_{l_1}^{l_2} \delta_{m_1}^{m_2} , \qquad (5.14)$$

for some (nonnegative) sequence $\{C_l\}$ denoting the angular power spectrum of the field. The argument to establish (5.14) can be presented as follows. By isotropy, the covariance function depends only on the Euclidean inner product $\langle x, y \rangle$, and hence we can introduce a function $\Gamma : [-1, 1] \to \mathbb{R}$ by $\Gamma(\langle x, y \rangle) = ET(x)T(y)$. Clearly $\Gamma(.) \in L^2([-1, 1])$ and we can hence write the expansion into Legendre polynomials as

$$\Gamma(\langle x, y \rangle) = \sum_{l=0}^{\infty} \frac{2l+1}{4\pi} C_l P_l(\langle x, y \rangle) = \sum_{l=0}^{\infty} C_l \sum_{m=-l}^{l} Y_{lm}(x) \bar{Y}_{lm}(y) , \qquad (5.15)$$

for some sequence $\{C_l\}$. By an iterated application of Fubini arguments we have easily

$$E a_{l_1 m_1} \bar{a}_{l_2 m_2} = E \left\{ \int_{S^2} T(x) \bar{Y}_{l_1 m_1}(x) d\sigma(x) \int_{S^2} T(y) Y_{l_2 m_2}(y) d\sigma(y) \right\}$$

$$= \int_{S^2} \int_{S^2} \{ET(x)T(y)\} \bar{Y}_{l_1 m_1}(x) Y_{l_2 m_2}(y) d\sigma(x) d\sigma(y)$$

$$= \int_{S^2} \int_{S^2} \Gamma(\langle x, y \rangle) \bar{Y}_{l_1 m_1}(x) Y_{l_2 m_2}(y) d\sigma(x) d\sigma(y)$$

$$= \int_{S^2} \int_{S^2} \sum_{l=0}^{\infty} C_l \sum_{m=-l}^{l} Y_{lm}(x) \bar{Y}_{lm}(y) \bar{Y}_{l_1 m_1}(x) Y_{l_2 m_2}(y) d\sigma(x) d\sigma(y)$$

$$= \sum_{l=0}^{\infty} C_l \sum_{m=-l}^{l} \left\{ \int_{S^2} Y_{lm}(x) \bar{Y}_{l_1 m_1}(x) d\sigma(x) \int_{S^2} \bar{Y}_{lm}(y) Y_{l_2 m_2}(y) d\sigma(y) \right\}$$

$$= \sum_{l=0}^{\infty} C_l \sum_{m=-l}^{l} \delta_{l_1}^{l} \delta_{m_1}^{m} \delta_{l_2}^{l} \delta_{m_2}^{m} = C_{l_1} \delta_{l_1}^{l_2} \delta_{m_1}^{m_2} .$$

The following two statements provide the "spherical counterpart", respectively, of Proposition 5.3 and Proposition 5.4. The proofs are analogous and left to the reader; see also Remark 5.14.

Proposition 5.16 Let $T = \{T(x) : x \in S^2\}$ be a strongly isotropic field such that, with probability one, the mapping $g \mapsto T(g)$ is in $L^2(S^2, d\sigma)$. Set, for $l \geq 0$,

$$T_l(x) = \sum_{m=-l}^{l} a_{lm} Y_{lm}(x), \quad x \in S^2, \tag{5.16}$$

where the notation is the same as in Theorem 5.13. Then, the family

$$\{T, T_l : l \geq 0\}$$

is strongly isotropic, in the sense that, for every $n, k \geq 1$, every $l_1, ..., l_k \geq 0$, every $x_1, ..., x_n \in S^2$ and every $g \in SO(3)$,

$$\{T_{l_j}(gx_1), ..., T_{l_j}(gx_n) : j = 1, ..., k\} \overset{law}{=} \{T_{l_j}(x_1), ..., T_{l_j}(x_n) : j = 1, ..., k\}.$$

Proposition 5.17 Let T be a two-weakly isotropic field on S^2. Then, the class $\{T, T^l : l \geq 1\}$ defined in Proposition 5.16 is two-weakly isotropic. Also,

$$Ea_{00} = ET(x), \quad Ea_{lm} = 0, \quad \forall l \geq 1,$$

and the fields T^l and $T^{l'}$ are not correlated for $l \neq l'$, that is: for every $x, y \in S^2$,

$$E[T^l(x)\overline{T^{l'}(y)}] = 0.$$

Remark 5.18 See the forthcoming Proposition 6.6 for a refinement of Proposition 5.17, providing a characterization of the covariance structure of the coefficients a_{lm}.

5.3 Weakly stationary random fields in $\mathbb{R}^m$

We shall now focus on some classic spectral representations of weakly stationary fields. The facts discussed below are used in the subsequent Section 5.4 in order to deduce a spectral representation for stationary and isotropic fields on $\mathbb{R}^3$. Our presentation is strongly influenced by Section 5.4 in [3], to which the reader is referred for further details. Throughout this section, we shall fix an integer $m \geq 1$. We consider centered, square-integrable and weakly stationary stochastic processes on $\mathbb{R}^m$.

Definition 5.19 (Weakly stationary fields)   Let $Z = \{Z(x) : x \in \mathbb{R}^m\}$ be a real-valued centered random field on $\mathbb{R}^m$, and assume that the following two properties are verified.

(1) For every $x \in \mathbb{R}^m$, $E(Z(x)^2) < \infty$ and $E(Z(x)) = 0$.
(2) For every $x, y \in \mathbb{R}^m$ the covariance $C(x, y) = E[Z(x)Z(y)]$ depends uniquely on the difference $x - y$.

Then, the process Z is said to be **weakly stationary** (or, when no confusion is possible, **stationary**). Whenever Properties 1 and 2 are verified, we shall write, by a slight abuse of notation, $C(x, y) = C(x - y)$.

A random field Z as in Definition 5.19 is said "weakly" stationary, since Properties 1 and 2 only involve the covariance function of Z, but not its overall law. Plainly, if a process Z is weakly stationary *and* Gaussian, then Z is also strictly stationary, in the sense that, for every $h \in \mathbb{R}^m$, $Z(x) \overset{law}{=} Z(x + h)$, where the equality in law holds in the sense of finite-dimensional distributions.

One striking feature of stationary fields, is that they can be always represented in terms of stochastic integrals with respect to some complex-valued random noise.

Definition 5.20   Let μ be a positive finite measure on $\mathbb{R}^m$. A μ-**controlled complex noise** is a collection of complex-valued random variables of the type

$$M = \{M(B) : B \in \mathcal{B}(\mathbb{R}^m)\}, \tag{5.17}$$

($\mathcal{B}(\cdot)$ indicates the Borel σ-field) such that: (i) $E[M(B)] = 0$, (ii)

$$E\left[M(B)\overline{M(B)}\right] = \mu(B),$$

(iii) for every pair of disjoint sets (A, B), $M(A \cup B) = M(A) + M(B)$, a.s.-$P$, and $E\left[M(A)\overline{M(B)}\right] = 0$. The complex noise M in (5.17) is said to be **Gaussian** if W defines a centered complex Gaussian family of random variables.

Observe that Properties (i)–(iii) in the previous Definition yield that, for general sets A, B, $E\left[M(A)\overline{M(B)}\right] = \mu(A \cap B)$. It is a standard result that one can define stochastic integrals of the type $M(h) = \int_{\mathbb{R}^m} h(z)M(dz)$, where h is a complex-valued and bounded function, and also that, for any pair (g, h) of such functions

$$E\left[M(h)\overline{M(g)}\right] = \int_{\mathbb{R}^m} h(\lambda)\overline{g(\lambda)}\mu(d\lambda) \tag{5.18}$$

(note that (5.18) yields that the definition of $M(h)$ can be extended by isometry to arbitrary square-integrable functions h). The next (classic) result shows that such integrals may be used to describe arbitrary stationary fields.

Theorem 5.21 (Spectral representation theorem.)   *Let $Z = \{Z(x) : x \in \mathbb{R}^m\}$ be a (centered and square-integrable) stationary Gaussian process. Then, there exists a finite measure μ, as well as a μ-controlled complex noise M, such that, for every fixed $x \in \mathbb{R}^m$,*

$$Z(x) = \int_{\mathbb{R}^m} \exp(i\langle x, \lambda\rangle) M(d\lambda), \qquad (5.19)$$

with $\langle x, y\rangle = \sum_{i=1}^m x_i y_i$ the usual Euclidean product. Also

$$C(x - y) = E[Z(x)Z(y)] = \int_{\mathbb{R}^m} \exp(i\langle x - y, \lambda\rangle)\mu(d\lambda). \qquad (5.20)$$

*Moreover, if Z is Gaussian then M is also Gaussian. We call M and μ, respectively, the **spectral process** and the **spectral measure** of Z.*

Proof   We merely sketch the construction of the noise M, and we refer the reader e.g. to [3, pp. 110-111] for further details and references. First of all, we recall that (from Bochner Theorem – see [3, Th. 5.4.1]) since Z is stationary, then there exists a finite and positive measure μ on $\mathbb{R}^m$ such that relation (5.20) is verified. It follows that the mapping,

$$Z(x) \mapsto \exp(i\langle x, \cdot\rangle)$$

can be linearly extended to an isomorphism, say ϑ, between the $L^2(P)$ closed complex Hilbert space of random variables spanned by the $Z(x)$'s and $L^2_{\mathbb{C}}(\mu)$ (observe indeed that complex exponentials are dense in $L^2(\mu)$). The proof is concluded by verifying that the complex noise M is given by

$$M(A) = \vartheta^{-1}(\mathbf{1}_B), \quad B \in \mathcal{B}(\mathbb{R}^m). \qquad (5.21)$$

Note that (5.21) also implies that M is necessarily Gaussian whenever Z is Gaussian.   □

Now observe that, since Z is real-valued by assumption, the expression on the RHS of (5.19) must also define a real-valued random variable. This suggests that M and μ have some further structural properties, ensuring that the imaginary parts in (5.19) cancel out. We recall that a measure μ is called *symmetric* if, for every A, $\mu(A) = \mu(-A)$, where we use the notation $-A := \{x \in \mathbb{R}^m : -x \in A\}$.

Proposition 5.22 *Let M be the* μ*-controlled complex noise appearing in* *(5.19). Then, necessarily,* μ *is symmetric and M enjoys the property that, for* *every A,*

$$\overline{M(A)} = M(-A).\tag{5.22}$$

Proof Since $C(x-y)$ is real-valued, relation (5.20) yields that

$$\int_{\mathbb{R}^m} \exp(i\langle x,\lambda\rangle)\mu(d\lambda) = \int_{\mathbb{R}^m} \exp(-i\langle x,\lambda\rangle)\mu(d\lambda),$$

for every $x \in \mathbb{R}^m$, from which we deduce the symmetry of μ. Now observe that we can rewrite (5.22) as

$$\overline{\int_{\mathbb{R}^m} \mathbf{1}_A(\lambda) M(d\lambda)} = \int_{\mathbb{R}^m} \overline{\mathbf{1}_A(-\lambda)} M(d\lambda),$$

so that (5.22) is proved once it is shown that

$$\overline{\int_{\mathbb{R}^m} h(\lambda) M(d\lambda)} = \int_{\mathbb{R}^m} \overline{h(-\lambda)} M(d\lambda)\tag{5.23}$$

for every $h \in L^2_{\mathbb{C}}(\mu)$. However, since $Z(x)$ is real-valued, it is immediately seen from (5.19) that (5.23) is verified for every mapping h of the type $\lambda \mapsto h(\lambda) = e^{i\langle x,\lambda\rangle}$, $x \in \mathbb{R}^m$. The conclusion is achieved by a density argument. $\square$

Remark 5.23 (1) Suppose that the stationary field Z (and therefore M) is Gaussian. Then, μ is symmetric and, for every $x,y \in \mathbb{R}^m$,

$$E[Z(x)Z(y)] = \int_{\mathbb{R}^m} e^{i\langle x,\lambda\rangle} e^{-i\langle y,\lambda\rangle} \mu(d\lambda).$$

This last relation yields that the real Gaussian space generated by Z is indeed isomorphic to the Hilbert space $L^2_e(\mu)$, which is defined as the *real* Hilbert space spanned by complex even functions that are square-integrable with respect to μ. We recall that a complex-valued function f is said to be even, if $f(\lambda) = \overline{f(-\lambda)}$ for every λ.

(2) Let Ψ be any isomorphism from $L^2_e(\mu)$ onto $L^2([0,1])$. Then, Z has the same law as the process

$$x \mapsto \int_0^1 \Psi_s\left(e^{i\langle x,\cdot\rangle}\right) dW_s,$$

where $e^{i\langle x,\cdot\rangle}$ is shorthand for the mapping $\lambda \mapsto e^{i\langle x,\lambda\rangle}$, and

$$\left\{\Psi_s\left(e^{i\langle x,\cdot\rangle}\right) : s \in [0,1]\right\}$$

is the element of $L^2([0,1])$ defined as the image of $e^{i\langle x,\cdot\rangle}$ via Ψ.

(3) The reader is referred to Janson [105, p. 110] for an exhaustive analysis of complex Gaussian random measures.

5.4 Stationarity and weak isotropy in $\mathbb{R}^3$

We now focus on random fields defined on $\mathbb{R}^3$. We use the symbols $\|\cdot\|$ and $\langle \cdot, \cdot \rangle$, respectively, in order to denote the usual norm and scalar product on $\mathbb{R}^3$. To simplify the presentation, in this section we shall consider a real-valued field $Z = \{Z(x) : x \in \mathbb{R}^3\}$ verifying the following assumptions:

- Z is centered and *weakly isotropic*, in the specific sense that Z is weakly stationary in the sense of Definition 5.19 and moreover, for every $x, y \in \mathbb{R}^3$, the covariance function

$$E[Z(x)Z(y)], \quad x, y \in \mathbb{R}^3$$

depends uniquely on the norm $\|x - y\|$;
- the spectral measure of Z, noted μ, admits a density f with respect to the Lebesgue measure on $\mathbb{R}^3$.

Remark 5.24 (1) We shall write $E[Z(x)Z(y)] = C(\|x - y\|)$, in order to indicate the function $C : \mathbb{R}_+ \to \mathbb{R}$, that uniquely determines the covariance structure of Z.

(2) Thanks to isotropy, it is not difficult to prove that the density f uniquely depends on the Euclidean norm of its argument, that is, $f(x) = f(y)$ whenever $x, y \in \mathbb{R}^3$ are such that $\|x\| = \|y\|$.

Using spherical coordinates

$$x = (r_x \sin \vartheta_x \cos \varphi_x, r_x \sin \vartheta_x \sin \varphi_x, r_x \cos \vartheta_x),$$
$$\lambda = (r_\lambda \sin \vartheta_\lambda \cos \varphi_\lambda, r_\lambda \sin \vartheta_\lambda \sin \varphi_\lambda, r_\lambda \cos \vartheta_\lambda)$$

we have

$$C(\|(x_1, x_2, x_3)\|) = C(\|(r_x, 0, 0)\|) = C(r_x),$$

and the spectral representation of the covariance function becomes

$$C(r_x) = \int_{\mathbb{R}^3} \exp(i \langle x, \lambda \rangle) \mu(d\lambda) = \int_{\mathbb{R}^3} \exp(i r_x r_\lambda \cos \vartheta_\lambda) \mu(d\lambda).$$

Now define a measure $\widetilde{\mu}$ on $\mathbb{R}_+$ such that

$$\widetilde{\mu}([0, r_\lambda]) := \frac{\mu(B(r_\lambda))}{4\pi} ,$$

where we implicitly defined the ball $B(r_\lambda) := \{z : \|z\| \le r_\lambda\}$. We obtain (see [3, Section 5.4])

$$\int_{\mathbb{R}^3} \exp(ir_x r_\lambda \cos \vartheta_\lambda) \mu(d\lambda) = \int_0^\infty \int_{S^2} \exp(ir_x r_\lambda \cos \vartheta) \sin \vartheta d\vartheta d\varphi \widetilde{\mu}(dr_\lambda)$$

$$= 2\pi \int_0^\infty \int_0^\pi \exp(ir_x r_\lambda \cos \vartheta) \sin \vartheta d\vartheta \widetilde{\mu}(dr_\lambda) .$$

Now note that

$$\int_0^\pi \exp(ir_x r_\lambda \cos \vartheta) \sin \vartheta d\vartheta = - \int_0^\pi \exp(ir_x r_\lambda \cos \vartheta) d\cos \vartheta$$

$$= \int_{-1}^1 \exp(-ir_x r_\lambda u) du$$

$$= \frac{2\sin(r_x r_\lambda)}{r_x r_\lambda} ,$$

whence

$$C(r_x) = 4\pi \int_0^\infty \frac{\sin(r_x r_\lambda)}{r_x r_\lambda} \widetilde{\mu}(dr_\lambda) .$$

It is a trivial remark that the restriction to S^2 of Z is two-weakly isotropic. In what follows, we shall therefore investigate the relations between the two representations (5.10) and (5.19). Let us introduce the Bessel function of the first kind of order ν

$$J_\nu(x) = \sum_{k=0}^\infty (-1)^k \frac{(x/2)^{2k+\nu}}{k! \Gamma(k + \nu + 1)} ,$$

and recall that ([195, equation 5.17.3.14])

$$\exp(i \langle x, \lambda \rangle) = \sqrt{8\pi^3} \sum_{l=0}^\infty \frac{i^l J_{l+1/2}(r_x r_\lambda)}{\sqrt{r_x r_\lambda}} \sum_{m=-l}^l Y_{lm}(\vartheta_x, \varphi_x) \overline{Y}_{lm}(\vartheta_\lambda, \varphi_\lambda) .$$

Again, we denote by $\{Y_{lm} : l \ge 0, m = -l, ..., l\}$ the collection of the spherical harmonics, as defined in Section 3.4 of Chapter 3. Consider the stationary random field $\{Z(.)\}$, admitting the spectral representation (5.19)

$$Z(x) = \int_{\mathbb{R}^3} \exp(i \langle x, \lambda \rangle) M(d\lambda) .$$

We have the following statement (see [3, 125])

Proposition 5.25 *The following representation holds:*

$$Z(x) = \sum_{l=0}^{\infty} \sum_{m=-l}^{l} a_{lm}(r_x) Y_{lm}(\vartheta_x, \varphi_x), \tag{5.24}$$

where the series converges in $L^2(P)$,

$$a_{lm}(r_x) := \sqrt{8\pi^3} \int_0^{\infty} \frac{i^l J_{l+1/2}(r_x r_\lambda)}{(r_x r_\lambda)^{1/2}} W_{lm}(dr_\lambda), \tag{5.25}$$

where

$$E a_{l_1 m_1} \overline{a}_{l_2 m_2} = 0 \ for \ (l_1, m_1) \neq (l_2, m_2),$$

and for all subsets $A \subseteq \mathbb{R}$

$$W_{lm}(A) := \int_A \int_{S^2} \overline{Y}_{lm}(\vartheta_\lambda, \varphi_\lambda) \widetilde{M}(dr_\lambda, \sin \vartheta_\lambda d\vartheta_\lambda d\varphi_\lambda). \tag{5.26}$$

Here $\widetilde{M}(dr_\lambda, \sin \vartheta_\lambda d\vartheta_\lambda d\varphi_\lambda)$ is the image of the measure $M(.)$ in (5.19) under the spherical change of coordinates.

Loosely speaking, for $r_x = 1$, i.e. $x \in S^2$ equation (5.24) yields back (5.10). We stress that the results in this Section are not used later in the book, and hence the proofs are omitted.

Remark 5.26 To investigate the relationship between the *angular power spectrum* $\{C_l(r_x) = E|a_{lm}(r_x)|^2\}$ and the spectral density $f(.)$, introduce the radial spectral density

$$\widetilde{f}(r_\lambda) := f(r_\lambda, 0, 0)$$

and note that

$$E\left|\widetilde{M}(dr_\lambda, \sin \vartheta_\lambda d\vartheta_\lambda d\varphi)\right|^2 = \widetilde{f}(r_\lambda) r_\lambda^2 \sin \vartheta_\lambda d\vartheta_\lambda d\varphi dr_\lambda,$$

whence

$$C_l(r_x) = 8\pi^3 \int_0^{\infty} \left\{\frac{J_{l+1/2}(r_x r_\lambda)}{\sqrt{r_x r_\lambda}}\right\}^2 \widetilde{f}(r_\lambda) r_\lambda^2 dr_\lambda.$$

Indeed

$$\sum_l \frac{2l+1}{4\pi} C_l(r_x) = 8\pi^3 \int_0^{\infty} \sum_l \frac{2l+1}{4\pi} \left\{\frac{J_{l+1/2}(r_x r_\lambda)}{\sqrt{r_x r_\lambda}}\right\}^2 \widetilde{f}(r_\lambda) r_\lambda^2 dr_\lambda$$

$$= 4\pi \int_0^{\infty} \widetilde{f}(r_\lambda) r_\lambda^2 dr_\lambda = EZ^2(x),$$

as expected from (5.15), and where we have used the summation formula (see [86, Equation 8.532])

$$\sum_l \frac{2l+1}{4\pi} \left\{ \frac{J_{l+1/2}(r_x r_\lambda)}{\sqrt{r_x r_\lambda}} \right\}^2 = \frac{1}{2\pi^2} \, .$$

6

Characterizations of Isotropy

6.1 Introduction

One of the main results of the previous chapter (see Theorem 5.13) is that any square-integrable isotropic spherical field $\{T(x) : x \in S^2\}$ admits a decomposition of the type

$$T(x) = \sum_{l=0}^{\infty} \sum_{m=-l}^{l} a_{lm} Y_{lm}(x), \quad x \in S^2, \tag{6.1}$$

$$a_{lm} = \int_{S^2} T(x) \overline{Y_{lm}}(x) d\sigma(x), \tag{6.2}$$

where the harmonic coefficients a_{lm} are square-integrable complex-valued random variables, and the series converges in $L^2(P)$. This chapter focuses on the crucial question of characterizing the law of the array $\{a_{lm} : l \geq 0, \ m = -l, ..., l\}$ defined above. As we did earlier, for notational simplicity we shall consider zero-mean random fields, unless otherwise specified.

We start by reviewing some basic facts about the random spherical harmonic coefficients $\{a_{lm}\}$. Our first achievement will be a surprising characterization which has no analogue for standard Fourier expansions in the Abelian case: namely, following [10, 11], we shall prove that, while the $\{a_{lm}\}$ have always mean zero and are uncorrelated, they can be independent if and only if they are Gaussian. This fact will have deep consequences in the chapters to follow, in particular when we will study conditions for the asymptotic consistency of angular power spectra estimators. We then proceed with the analysis of the higher-order angular power spectra of an isotropic spherical random field.

As we did before, we find it useful to start from a simpler case, i.e. the classic situation of a stationary random field on the circle.

6.2 First example: the cyclic group

Fix $n \geq 1$. We consider a *strongly stationary* and *n-periodic* process on the integers, noted $X = \{X_t : t \geq 1\}$. By strong stationarity we mean that, for $k, \tau \in \mathbb{N}$, $(X_1, ..., X_k) \stackrel{law}{=} (X_{1+\tau}, ..., X_{k+\tau})$, whereas n-periodicity implies that $X_t = X_{t+n}$ for every integer t. We assume that X_1 (hence, by stationarity, every X_t) has finite moments of all orders, and also (for simplicity) that X has mean zero.

Since X can be reconstructed by using its first n coordinates $X_1, ..., X_n$, it follows that the law of X is completely determined by the law of its *discrete Fourier transform*, that is, by the random mapping

$$ j \mapsto w_j = \frac{1}{\sqrt{n}} \sum_{t=1}^{n} X_t \exp(-i\lambda_j t), \quad j = 1, 2, ..., n, $$

where $\lambda_j = \frac{2\pi j}{n}$ (see [30] for a classic reference on the results discussed in this introductory section). To ensure (notational) consistency with the discussion to follow, we will rather focus on the "harmonic" coefficients

$$ a_j = \frac{1}{\sqrt{n}} w_j = \frac{1}{n} \sum_{t=1}^{n} X_t \exp(-i\lambda_j t), \quad j = 1, 2, ..., n, $$

obtained by dividing the discrete Fourier transform by $1/\sqrt{n}$. We now ask the following (natural) question: *which conditions must be fulfilled by the joint moments of the vector $(a_1, ..., a_n)$ of the harmonic coefficients associated with a stationary and n-periodic X?*

Start by observing that, for any three frequencies $\lambda_{j_1}, \lambda_{j_2}, \lambda_{j_3} \in (0, 2\pi]$,

$$ E a_{j_1} a_{j_2} a_{j_3} = \frac{1}{n^3} \sum_{t_1, t_2, t_3 = 1}^{n} E[X_{t_1} X_{t_2} X_{t_3}] \exp\left(-i\{\lambda_{j_1} t_1 + \lambda_{j_2} t_2 + \lambda_{j_3} t_3\}\right) $$

which, due to the stationary property, for every $\tau \in \mathbb{N}$ must be equal to (assuming the moments exist)

$$ \frac{1}{n^3} \sum_{t_1, t_2, t_3 = 1}^{n} E X_{t_1 + \tau} X_{t_2 + \tau} X_{t_3 + \tau} \exp(-i\{\lambda_{j_1} t_1 + \lambda_{j_2} t_2 + \lambda_{j_3} t_3\}) $$

$$ \times \exp(-i\tau\{\lambda_{j_1} + \lambda_{j_2} + \lambda_{j_3}\}) $$

$$ = \frac{1}{n^3} \sum_{t_1, t_2, t_3 = 1}^{n} E X_{t_1} X_{t_2} X_{t_3} \exp(-i\{\lambda_{j_1} t_1 + \lambda_{j_2} t_2 + \lambda_{j_3} t_3\}) $$

$$ \times \exp(-i\tau\{\lambda_{j_1} + \lambda_{j_2} + \lambda_{j_3}\}) $$

$$ = \exp(-i\tau\{\lambda_{j_1} + \lambda_{j_2} + \lambda_{j_3}\}) E a_{j_1} a_{j_2} a_{j_3}. $$

This relation yields immediately the following well-known fact: a n-periodic process on the integers can be stationary only if $Ea_{j_1}a_{j_2}a_{j_3} = 0$ for every triple $\left(\lambda_{j_1}, \lambda_{j_2}, \lambda_{j_3}\right)$ such that $\lambda_{j_1} + \lambda_{j_2} + \lambda_{j_3} \neq 0$. Equivalently, we may say that the *bispectrum* of a stationary and n-periodic process X on the integers has necessarily the form

$$Ea_{j_1}a_{j_2}a_{j_3} = B(\lambda_{j_1}, \lambda_{j_2})\mathbb{I}(\lambda_{j_1} + \lambda_{j_2} + \lambda_{j_3} = 0), \qquad (6.3)$$

where $B(\cdot)$ is a deterministic function on $(0, 2\pi]^2$, and $\mathbb{I}(.)$ denotes an indicator. Analogous results could be established (by similar arguments) for *polyspectra* of arbitrary order (see [29]), that is: one can prove that, for X stationary and n-periodic, the joint moments of the harmonic coefficients take the form

$$Ea_{j_1}a_{j_2}...a_{j_p} = P(\lambda_{j_1}, ..., \lambda_{j_{p-1}})\mathbb{I}(\lambda_{j_1} + \cdots + \lambda_{j_p} = 0), \qquad (6.4)$$

for some deterministic function $P(\cdot)$ on $(0, 2\pi]^{p-1}$.

We now want to cast the above discussion into the framework of the representation theory for compact groups. From this standpoint, we can regard X as a process defined on the cyclic group Z_n, that is, on the set $\{1, ..., n\}$ endowed with the commutative group operation $s \circ t = (s + t) \bmod (n)$. As already discussed (see formulae (2.27)–(2.28) of Chapter 2), a complete family of irreducible matrix representations of this group (or equivalently, a complete family of characters) is provided by the mappings $t \mapsto \exp(i\lambda_j t)$, where $\lambda_j = 2\pi j/n$. In this framework, the stochastic Peter-Weyl Theorem 5.5 yields that

$$X_t = \sum_{j=1}^{n} a_j \exp(i\lambda_j t) .$$

Moreover, stationarity entails $X_{\tau \cdot t} = X_{t+\tau} \overset{law}{=} X_t$, which implies in turn $a_j \overset{law}{=} a_j \exp(i\lambda_j \tau)$ for all integers τ. This identity clearly remains unaffected if we take τ to be a discrete random variable, with values in $\{1, ..., n\}$ and independent of X. It follows that we can write

$$\begin{aligned} Ea_{j_1}a_{j_2}...a_{j_p} &= E\left\{a_{j_1}\exp(i\lambda_{j_1}\tau)a_{j_2}\exp(i\lambda_{j_2}\tau)...a_{j_p}\exp(i\lambda_{j_p}\tau)\right\} \\ &= Ea_{j_1}a_{j_2}\cdots a_{j_p}E\left\{\exp(i\lambda_{j_1}\tau)\exp(i\lambda_{j_2}\tau)\cdots\exp(i\lambda_{j_p}\tau)\right\} \\ &= Ea_{j_1}a_{j_2}\cdots a_{j_p} \times E\left\{\exp(i\left[\lambda_{j_1} + \cdots + \lambda_{j_p}\right]\tau)\right\}, \end{aligned}$$

from which we deduce once again the representation (6.4). Our point is that one can give some further algebraic flavour to this result, by rewriting the exponential inside the last expectation as follows

$$\exp\left(i\left[\lambda_{j_1} + \cdots + \lambda_{j_p}\right]\tau\right) = \exp(i\lambda_{j_1}\tau) \otimes \exp(i\lambda_{j_2}\tau) \otimes \cdots \otimes \exp(i\lambda_{j_p}\tau),$$

where we view the mappings

$$\tau \mapsto \exp(i\lambda_{j_1}\tau), \tau \mapsto \exp(i\lambda_{j_2}\tau), ..., \tau \mapsto \exp(i\lambda_{j_p}\tau)$$

as one-dimensional irreducible (matrix) representations of the cyclic group, for which pointwise product and tensor product coincide. As a consequence, we deduce that the joint moments of the harmonic coefficients of X must satisfy the tensor product relation

$$E a_{j_1} a_{j_2} \cdots a_{j_p} = E a_{j_1} a_{j_2} \cdots a_{j_p} \times E\left\{\exp(i\lambda_{j_1}\tau) \otimes \exp(i\lambda_{j_2}\tau) \otimes \cdots \otimes \exp(i\lambda_{j_p}\tau)\right\},$$
(6.5)

for any independent random integer τ. Note that (6.5) can be trivially rephrased by saying that the one-dimensional vector $E a_{j_1} a_{j_2} \cdots a_{j_p}$ is an eigenvector of the 1×1 matrix $E\{\exp(i\lambda_{j_1}\tau) \otimes \exp(i\lambda_{j_2}\tau) \otimes \cdots \otimes \exp(i\lambda_{j_p}\tau)\}$.

This simple remark serves as an ideal prelude for the rest of the chapter: indeed, a similar characterization of the polyspectra associated with isotropic spherical fields, involving eigenvalues of averages of tensor products of irreducible matrix representations, will be one of the principal achievements of the sections to follow.

Remark 6.1 Similar arguments can be used in order to characterize the joint moments of the Fourier coefficients associated with an isotropic random field on the circle.

6.3 The spherical harmonics coefficients

We shall now extend to an isotropic spherical field T the simple analysis on the cyclic group performed in the previous section, that is, we will deduce from the isotropic assumption several crucial properties of the higher-order power spectra associated with T (provided that the latter are well-defined). Despite some initial similarities, we shall soon discover that the structure of the angular polyspectra (that is, of the joint moments of harmonic coefficients – see Definition 6.18 below for a formal definition) of isotropic spherical fields is considerably richer (and much more complicated) than the one in the previous discrete-time example. Intuitively, the reason can be explained as follows. The fields we shall deal with are invariant under the action of the non-Abelian group $SO(3)$, rather than the Abelian group of translations we focussed on earlier. Group representations are then matrix-valued; it follows that the relationship between irreducible representations is provided by the rich and complicated Clebsch-Gordan matrices, rather than indicator functions.

We start by focussing on spherical fields of the type $T = \{T(x) : x \in S^2\}$, which are strongly isotropic and such that the mapping $x \mapsto T(x)$ belongs to $L^2(S^2, d\sigma)$ a.s.-P, where $d\sigma$ stands for the Lebesgue measure. We shall consider the harmonic decomposition (6.1) of T, and we shall often use the shorthand notation

$$a_{l.} = \{a_{lm} : m = -l, ..., l\},$$

to indicate the vector of harmonic coefficients associated with the frequency l. Note that, in view of the properties of spherical harmonics (see formula (3.40) of Chapter 3),

$$a_{lm} = (-1)^m \overline{a_{l-m}}, \quad l \geq 1, \quad m = 1, ..., l. \tag{6.6}$$

Remark 6.2 For the moment, we do not assume that T is square-integrable, so that we shall regard (6.1) as a pathwise ("omega by omega") decomposition. In particular, without further moment assumptions it is not possible to use the stochastic Peter-Weyl Theorem 5.5 directly, and the moments of the coefficients a_{lm} may as well not exist.

The following (easy) result entails a useful invariance property of the law of the vectors $a_{l.}$. As before, we use the notation $gx = g \cdot x$ to indicate the action of $g \in SO(3)$ on a point $x \in S^2$.

Lemma 6.3 *Let T be a strongly isotropic field on S^2, and let the harmonic coefficients $\{a_{lm}\}$ be defined according to (6.1). Then, for every $l \geq 0$ and every $g \in SO(3)$, we have*

$$D^l(g)a_{l.} \overset{law}{=} a_{l.}, \ l = 0, 1, 2, ... \tag{6.7}$$

The equality (6.7) must be understood in the sense of finite-dimensional distributions for sequences of random vectors, that is, (6.7) takes place if and only if, for every $k \geq 1$ and every $0 \leq l_1 < l_2 < \cdots < l_k$,

$$\{D^{l_1}(g)a_{l_1 .}, ..., D^{l_k}(g)a_{l_k .}\} \overset{law}{=} \{a_{l_1 .}, ..., a_{l_k .}\}. \tag{6.8}$$

Proof We provide the proof of (6.8) only when $k = 1$ and $l_1 = l \geq 1$. The general case is obtained analogously. By strong isotropy, we have that, for every $l \geq 1$, every $g \in SO(3)$ and every $x_1, ..., x_n \in S^2$, the equality (5.7) takes place. Now, (5.7) can be rewritten as follows:

$$\left\{\sum_l \sum_m a_{lm} Y_{lm}(x_1), ..., \sum_l \sum_m a_{lm} Y_{lm}(x_n)\right\} \tag{6.9}$$

$$\overset{law}{=} \left\{ \sum_l \sum_m a_{lm} Y_{lm}(g^{-1}x_1), ..., \sum_l \sum_m a_{lm} Y_{lm}(g^{-1}x_n) \right\}$$

$$= \left\{ \sum_l \sum_m a_{lm} \sum_{m'} D^l_{m'm}(g) Y_{lm'}(x_1), ..., \sum_l \sum_m a_{lm} \sum_{m'} D^l_{m'm}(g) Y_{lm'}(x_n) \right\}$$

$$= \left\{ \sum_l \sum_{m'} \widetilde{a}_{lm'} Y_{lm'}(x_1), ..., \sum_l \sum_{m'} \widetilde{a}_{lm'} Y_{lm'}(x_n) \right\},$$

where we write

$$\widetilde{a}_{lm'} := \sum_m a_{lm} D^l_{m'm}(g), \tag{6.10}$$

and we have used the fact that

$$\left\{ Y_{lm}(g^{-1}x_1), ..., Y_{lm}(g^{-1}x_n) \right\} \equiv \left\{ \sum_{m'} D^l_{m'm}(g) Y_{lm'}(x_1), ..., \sum_{m'} D^l_{m'm}(g) Y_{lm'}(x_n) \right\}, \tag{6.11}$$

which follows from the group representation property and the identity (3.44). To conclude, just observe that (6.9) implies that

$$\widetilde{a}_{lm'} = \int_{S^2} T(gx) \overline{Y_{lm'}(x)} dx, \quad m' = -l, ..., l,$$

yielding that, due to strong isotropy and with obvious notation, $\widetilde{a}_{l.} \overset{law}{=} a_{l.}$. The conclusion follows from the fact that, thanks to (6.10),

$$\widetilde{a}_{l.} = D^l(g) a_{l.}.$$

$\square$

Remark 6.4 As an application of Lemma 6.3, it can be proved that, if T is strongly isotropic, then, for all $l \geq 1$ such that $E\|a_{l.}\| < \infty$, one has $Ea_{lm} = 0$ for $m = -l, ..., l$. To see this, use (6.7), take the expected value on both sides, and exploit the fact that D^l defines an irreducible matrix representation.

A basic feature of the spectral representation coefficients for isotropic, centered and square-integrable random fields, is that the following orthogonality property takes place for $l, l_j \geq 1$

$$E[a_{l_im_i} \overline{a_{l_jm_j}}] = \delta^{l_j}_{l_i} \delta^{m_i}_{m_j} C_{l_i}, \tag{6.12}$$

where δ^a_b is the Kronecker symbol and the positive constant depends uniquely on l_i (and not on m_i). The sequence $\{C_l : l \geq 0\}$ is known as the *angular power spectrum* of the field. We now provide a proof of (6.12) (actually, a slightly stronger version).

Remark 6.5 The notation $\{C_l\}$ for the angular power spectrum is widely used in the literature. Observe that no confusion is possible with the Clebsch-Gordan matrices $C_{l_1 l_2}$, as they are always labeled with a double index of the type $l_1 l_2$.

The following statement, as most of those in this section, is taken from [10]. As usual we denote by A^* the complex conjugate of a given matrix A. Also, recall Remark 6.4.

Proposition 6.6 *Assume that T is strongly isotropic. Then, the following hold*

(1) For all l such that $E[\|a_{l.}\|^2] < \infty$,

$$Ea_{l.}a_{l.}^* = A_l I_{2l+1} , \qquad (6.13)$$

*where A_l is a non-negative constant depending uniquely on l, and I_{2l+1} denotes the $(2l + 1) \times (2l + 1)$ identity matrix. One usually adopts the notation: $C_0 = Var(a_{00}) = A_0 - E(a_{00})^2$, and $C_l = A_l$ and calls the collection $\{C_l : l \geq 0\}$ (whenever it is well defined) the **angular power spectrum** of the field T.*

(2) For all $l_1 \neq l_2$ such that $E[|a_{l_1.}|^2] < \infty$, $E[|a_{l_2.}|^2] < \infty$

$$Ea_{l_1.}a_{l_2.}^* = 0 \qquad (6.14)$$

(in the sense of the $(2l_1 + 1) \times (2l_2 + 1)$ zero matrix).

Proof 1. Let us denote by Γ_l the covariance matrix of the random vector $a_{l.}$. Since the vectors $a_{l.}$ and $D^l(g)a_{l.}$ have the same distribution (see Lemma 6.3), they have the same covariance matrix. This gives

$$\Gamma_l = D^l(g)\Gamma_l D^l(g)^* = D^l(g)\Gamma_l D^l(g)^{-1}$$

Since D^l is an irreducible matrix representation of $SO(3)$, by Schur Lemma (see Theorem 2.29), Γ_l is of the form $C_l I_{2l+1}$, where C_l is nonnegative (since Γ_l is nonnegative definite).

2. The representations D^{l_1} and D^{l_2} are not equivalent for $l_1 \neq l_2$ (they have different dimensions). Again by Schur Lemma, it follows that the identity

$$Ea_{l_1.}a_{l_2.}^* = D^{l_1}(g)Ea_{l_1.}a_{l_2.}^* D^{l_2}(g)^{-1}$$

can hold only if the left hand side is the zero matrix.   □

Remark 6.7 (1) In principle, Proposition 6.6 does not assume that T has finite moments.

(2) Relations (6.13) and (6.14) continue to hold if one assumes that T is two-weakly isotropic, that is, if T is square integrable, has constant mean and its covariance function is invariant with respect to rotations (note that, in this case, harmonic coefficients have finite second moments by construction). To see this, just observe that (6.13) and (6.14) only involve the covariance structure of the harmonic coefficients. Since harmonic coefficients are linear functionals of T, their covariances only depend on the covariance of T: in particular, the covariances of the a_{lm} associated with T coincide with those of an auxiliary Gaussian field with the same mean and covariance as T. Since a Gaussian two-weakly isotropic field is also strongly isotropic, Proposition 6.6 can be applied to deduce the conclusion.

(3) It is not difficult to find examples where square-integrability fails but the assumptions of Proposition 6.6 are fulfilled. Consider for instance a field of the type:

$$T(x) = \sum_{m=-l_1}^{l_1} a_{l_1 m_1} Y_{l_1 m_1}(x) + \sum_{m=-l_2}^{l_2} a_{l_2 m_2} Y_{l_2 m_2}(x) + \sum_{m=-l_3}^{l_3} b_{l_3 m_3} Y_{l_3 m_3}(x) \quad (6.15)$$

where $b_{l_3 m_3} = \eta \times a_{l_3 m_3}$, η is a random variable with infinite variance (e.g. a Cauchy random variable), and the three fields

$$\sum_{m=-l_1}^{l_1} a_{l_i m_i} Y_{l_i m_i}(\cdot), \quad i = 1, 2, 3,$$

are independent, square-integrable and strongly isotropic. It is easy to show that the field T is properly defined and strongly isotropic, although with infinite variance. Also, the conclusion of Proposition 6.6 holds for a_{l_1}. and a_{l_2}. (since they have finite variance), whereas it cannot be applied to b_{l_3}. (since its components have infinite variance). Albeit (6.15) is clearly an artificial model, some closely related fields may be of interest for practical applications: for instance, in CMB data analysis it is often the case that the observed field is a superposition of a signal plus foreground contamination, and the latter may be characterized by heavy tails at the highest multipoles (point sources). In such cases, it is of an obvious statistical interest to know that the standard properties of the spherical harmonics coefficients still hold at least for the multipoles where foreground contamination is absent.

The next result provides three further characterizations for the spherical harmonics coefficients of strongly isotropic fields. We recall that the (standard)

Cauchy distribution is defined as the probability measure on the real line having density

$$f(x) = \frac{1}{\pi(1 + x^2)}, \quad x \in \mathbb{R}.$$

It is easily checked that a random variable W has the Cauchy distribution if and only if arctan(W) is uniformly distributed on the interval $[-\pi/2, \pi/2]$.

Proposition 6.8 *Let T be a strongly isotropic random field. Then for all $l \geq 1$ such that $E\|a_{l.}\|^2 < \infty$, the following holds.*

(1) For all $m = 1, \ldots, l$,

$$\mathrm{Rea}_{lm} \overset{law}{=} \mathrm{Ima}_{lm} \quad and \quad \frac{\mathrm{Rea}_{lm}}{\mathrm{Ima}_{lm}} \sim Cauchy.$$

(2) For all $m = 1, 2, \ldots, l$, Rea_{lm} and Ima_{lm} are uncorrelated, with variance $E(\mathrm{Rea}_{lm})^2 = E(\mathrm{Ima}_{lm})^2 = C_l/2$.

(3) The marginal distribution of Rea_{lm}, Ima_{lm} is always symmetric, that is,

$$\mathrm{Rea}_{lm} \overset{law}{=} -\mathrm{Rea}_{lm}, \quad \mathrm{Ima}_{lm} \overset{law}{=} -\mathrm{Ima}_{lm}.$$

Proof (1) By selecting $g = (\alpha, \beta, \gamma)$ (in Euler angles) such that $\beta = \gamma = 0$, relation (6.7) entails that

$$a_{lm} \overset{law}{=} e^{-im\alpha} a_{lm},$$

for all $m = -l, \ldots, l$, $0 \leq \alpha < 2\pi$. This yields

$$\begin{pmatrix} \mathrm{Rea}_{lm} \\ \mathrm{Ima}_{lm} \end{pmatrix} \overset{law}{=} \begin{pmatrix} \cos\varphi & \sin\varphi \\ -\sin\varphi & \cos\varphi \end{pmatrix} \begin{pmatrix} \mathrm{Rea}_{\ell m} \\ \mathrm{Ima}_{\ell m} \end{pmatrix} \qquad (6.16)$$

for all $m = -l, \ldots, l$, $0 \leq \varphi < 2\pi$. Thus, the two-dimensional vector

$$(\mathrm{Rea}_{lm}, \mathrm{Ima}_{lm})'$$

has a distribution that is invariant with respect to rotations (in the plane). This implies that, in polar coordinates, this vector it can be written in the form

$$R(\cos(\Theta), \sin(\Theta)), \qquad (6.17)$$

where R is a random variable with values in $\mathbb{R}^+$, and Θ is uniform in $[-\pi, \pi]$. This entails immediately that arctan($\mathrm{Rea}_{\ell m}/\mathrm{Ima}_{\ell m}$) is uniform on $[-\frac{\pi}{2}, \frac{\pi}{2}]$, from which the result follows.

(2) This property can be deduced from the previous results of this section if T is square-integrable. From (6.17),

$$E[\mathrm{Rea}_{\ell m} \cdot \mathrm{Ima}_{\ell m}] = \int_0^{+\infty} r^2 d\mu_R(r) \int_{-\pi}^{\pi} \cos\vartheta \sin\vartheta d\vartheta = 0,$$

where μ_R stands for the law of R. 3. It suffices to take $\varphi_0 = \pi$ in (6.16). $\quad\square$

Remark 6.9 It is interesting to note how Proposition 6.8 implies that no information can be derived on the statistical distribution of an isotropic random field by the marginal distribution function of the ratios ($\mathrm{Re}a_{\ell m}/\mathrm{Im}a_{\ell m}$), a fact which has been apparently misunderstood in some cosmological papers. On the other hand, it may be possible to use these ratios to implement statistical tests on the assumption of isotropy, an issue which has gained a remarkable empirical relevance in the analysis of cosmological data.

It is clear that if the field T is Gaussian, then the random variables

$$\{a_{lm} : l \geq 0, \ m = -l, ..., l\}$$

define a complex-valued Gaussian family (each a_{lm} is indeed the a.s. limit of linear combinations of Gaussian random variables – i.e. Riemann sums). We now prove an independence result for this family of random variables. Note that, thanks to Proposition 6.6, these random variables are uncorrelated. We shall deduce independence by means of the following elementary result

Lemma 6.10 *Let Z_1, Z_2 be complex r.v.'s, centered and jointly Gaussian. Then they are independent if and only if*

$$E[Z_1\overline{Z_2}] = 0, \qquad E[Z_1Z_2] = 0 \tag{6.18}$$

Proof In one direction the statement is obvious. Let us assume that the two relations in (6.18) are satisfied. Then, if we set $Z_k = X_k + iY_k$, $k = 1, 2$, then

$$E[X_1X_2 + Y_1Y_2] + iE[-X_1Y_2 + Y_1X_2] = 0 \,,$$
$$E[X_1X_2 - Y_1Y_2] + iE[X_1Y_2 + Y_1X_2] = 0 \,.$$

From these one obtains $E[X_1X_2] = 0, E[Y_1Y_2] = 0, E[X_1Y_2] = 0$ and $E[Y_1X_2] = 0$. This means that each of the r.v.'s X_1, Y_1, X_2, Y_2 is uncorrelated with any of the other ones, so that, being jointly Gaussian, they are independent. $\quad\square$

Here is the announced independence characterization.

Proposition 6.11 *For a strongly isotropic Gaussian random field the r.v.'s a_{lm}, $l = 0, 1, \ldots$, $m = 0, \ldots, l$ are independent.*

Proof Let be $(l, m) \neq (l', m')$, $m > 0$, $m' > 0$. Then a_{lm} is uncorrelated with both $a_{l'm'}$ and $a_{l',-m'} = \overline{a_{l'm'}}$. Thus

$$E[a_{lm}\overline{a_{l'm'}}] = 0, \qquad E[a_{lm}a_{l'm'}] = E[a_{lm}\overline{a_{l',-m'}}] = 0$$

and the statement follows from Lemma 6.10. If one at least among m and m' is equal to 0, then either a_{lm} or $a_{l'm'}$ is real (or both are) and independence follows from absence of correlation as for the real case. □

We now present the main result of [10]: this finding is much less obvious and somewhat counterintuitive, and shows that the converse statement of Proposition 6.11 also holds.

Theorem 6.12 *For a strongly isotropic random field T, let l be such that $E|a_{l.}|^2 < \infty$. Then, the coefficients $(a_{l0}, a_{l1}, \ldots, a_{ll})$ are independent if and only if they are Gaussian.*

Proof This is an extended version (due to Paolo Baldi) of the original proof which was given in [10]. Up to replacing T with the "centered" field $T - \int_{S^2} T(x)dx$, we can assume, without loss of generality, that $a_{l0} = 0$. Isotropy implies that the two fields

$$T(x) = \sum_{l=1}^{\infty} \sum_{m=-l}^{l} a_{lm} Y_{lm}(x)$$

and

$$T(g^{-1}x) = \sum_{l=1}^{\infty} \sum_{m=-l}^{l} a_{lm} Y_{lm}(g^{-1}x) = \sum_{l=1}^{\infty} \sum_{m=-l}^{l} a_{lm} \sum_{m'=-l}^{l} D_{mm'}^{l}(g) Y_{lm'}(x)$$

have the same distribution. Therefore, for $m_1, m_2 > 0$ the two complex random variables

$$\widetilde{a}_{lm_1} = \sum_{m=-l}^{l} D_{mm_1}^{l}(g)a_{lm},$$

$$\widetilde{a}_{lm_2} = \sum_{m=-l}^{l} D_{mm_2}^{l}(g)a_{lm},$$

have the same joint distribution as a_{lm_1} and a_{lm_2}; in particular, they are independent for $m_1 \neq m_2$. We can hence write (exploiting $a_{l,-m} = (-1)^m \overline{a_{lm}}$)

$$\widetilde{a}_{lm_1} = D_{0m_1}^{l}(g)a_{l0} + \sum_{m=1}^{l} \left\{ D_{mm_1}^{l}(g)a_{lm} + (-1)^m D_{-mm_1}^{l}(g)\overline{a_{lm}} \right\},$$

$$\widetilde{a}_{lm_2} = D_{0m_2}^{l}(g)a_{l0} + \sum_{m=1}^{l} \left\{ D_{mm_2}^{l}(g)a_{lm} + (-1)^m D_{-mm_2}^{l}(g)\overline{a_{lm}} \right\}.$$

Now we view the coefficients a_{lm}, $m = 1, \ldots, l$ as bivariate real-valued vectors

$\{\mathrm{Re}a_{lm}, \mathrm{Im}a_{lm}\}$, and $D^l_{mm_i}(g)$, $i = 1, 2$ as 2×2 real matrices with elements

$$\begin{pmatrix} \mathrm{Re}(D^l_{mm_i}) & -\mathrm{Im}(D^l_{mm_i}) \\ \mathrm{Im}(D^l_{mm_i}) & \mathrm{Re}(D^l_{mm_i}) \end{pmatrix}.$$

On the other hand, a_{l0} is taken as the vector (a_{l0}, a_{l0}), and $D^l_{0m}(g)$ as a full rank 2×2 matrix ,

$$D^l_{0m}(g) = diag \left\{ \mathrm{Re}(D^l_{0m}(g)), \mathrm{Im}(D^l_{0m}(g)) \right\}.$$

By the Skitovich-Darmois Theorem (reported below), it follows that each one of the vectors $\{\mathrm{Re}a_{lm}, \mathrm{Im}a_{lm}\}$ is bivariate Gaussian, provided that we can prove there exist $g \in SO(3)$ such that the linear mapping

$$z \mapsto (D^l_{mm_i}(g) + (-1)^m D^l_{-mm_i}(g))z \;, m = 0, ..., l, \; i = 1, 2 \;,$$

are all nonsingular. This happens if and only if $\left| D^l_{mm_i}(g) \right| \neq \left| D^l_{-mm_i}(g) \right|$. For $l \geq 2$, take $m_1 = l, m_2 = l - 1$. Some simple algebra gives (with $\beta \in [-\pi/2, \pi/2]$ the second Euler angle of g)

$$\begin{aligned} \left| D^l_{ml}(\beta) \right| &= \left| d^l_{ml}(\beta) \right| \\ &= \left(\frac{(2l)!}{(l+m)!(l-m)!} \right)^{1/2} \left(\cos\frac{\beta}{2} \right)^{l+m} \left(\sin\frac{\beta}{2} \right)^{l-m} \\ &= \left(\frac{(2l)!}{(l+m)!(l-m)!} \right)^{1/2} \frac{\sin^l \beta}{2^l \sin^l \beta/2 \cos^l \beta/2} \left(\cos\frac{\beta}{2} \right)^{l+m} \left(\sin\frac{\beta}{2} \right)^{l-m} \\ &= \left(\frac{(2l)!}{(l+m)!(l-m)!} \right)^{1/2} \sin^l \beta \left(\tan\frac{\beta}{2} \right)^{-m} \\ &= \left(\frac{(2l)!}{(l+m)!(l-m)!} \right)^{1/2} \frac{1}{2^l} (1 + \cos\beta)^m \sin^{l-m} \beta \end{aligned}$$

where we have used

$$\tan\frac{\beta}{2} = \frac{\sin\beta}{1 + \cos\beta} \;.$$

By a similar argument, using $\tan\beta/2 = (1 - \cos\beta)/\sin\beta$ we obtain

$$\left| D^l_{-ml}(\beta) \right| = \left| d^l_{-ml}(\beta) \right| = \left(\frac{(2l)!}{(l+m)!(l-m)!} \right)^{1/2} \frac{1}{2^l} (1 - \cos\beta)^m \sin^{l-m} \beta \;,$$

whence the condition is satisfied, provided $\beta \neq \pi/2$. For $m = 0$, $D^l_{0m}(g)$ is trivially nonsingular. To conclude the argument, it suffices to note that

$$\begin{aligned} \left| D^l_{m,l-1}(\beta) \right| &= \left| d^l_{m,l-1}(\beta) \right| \\ &= \left(\frac{(2l-1)!}{(l+m)!(l-m)!} \right)^{1/2} \left(\cos\frac{\beta}{2} \right)^{l+m-1} \left(\sin\frac{\beta}{2} \right)^{l-m-1} (l\cos\beta - m) \end{aligned}$$

$$= \left(\frac{(2l-1)!}{(l+m)!(l-m)!} \right)^{1/2} (1 + \cos\beta)^m (\sin\beta)^{l-m-1} (l\cos\beta - m)$$

$$\left| d^l_{-m,l-1}(\beta) \right| = \left(\frac{(2l-1)!}{(l+m)!(l-m)!} \right)^{1/2} (1 - \cos\beta)^m (\sin\beta)^{l-m-1} (l\cos\beta + m) .$$

The case $l = 1$ can be dealt with by means of an *ad hoc* argument; note that in that case we obtain the identity representation of $SO(3)$.

□

Remark 6.13   It is important to note that Theorem 6.12 only applies to the frequency projections, but not to the random field as a whole. Indeed, Theorem 6.12 only implies that if, for a fixed l, the coefficients a_{lm}, $m = 0, 1, ..., l$, are independent, then the lth frequency projection

$$T_l(x) = \sum_{m=-l}^{l} a_{lm} Y_{lm}(x), \quad x \in S^2,$$

defines a Gaussian field. In particular, without further assumptions on the joint law of the T_l, it is not possible to deduce any information about the Gaussianity of the field T.

Theorem 6.14   *(Skitovich-Darmois) Let $X_1, \ldots, X_r$ be mutually independent random vectors in $\mathbb{R}^n$. If the linear statistics*

$$L_1 = \sum_{j=1}^{r} A_j X_j, \qquad L_2 = \sum_{j=1}^{r} B_j X_j,$$

are independent, for some real nonsingular $n \times n$ matrices A_j, B_j, $j = 1, \ldots, r$, then each of the vectors $X_1, \ldots, X_r$ is normally distributed.

As an application of Theorem 6.12, the following result formalizes the fact that, in general, one cannot deduce strong isotropy from weak isotropy. The proof makes use of Proposition 5.10.

Proposition 6.15   *For every $n \geq 2$, there exists a n-weakly isotropic spherical field T such that T is not strongly isotropic.*

Proof   Fix $l \geq 1$, and consider a vector

$$b_m, \quad m = -l, ..., l,$$

of centered complex-valued random variables such that: (i) b_0 is real, (ii) $b_{-m} = (-1)^m \overline{b_m}$ ($m = 1, ..., l$), (iii) the vector $\{b_0, ..., b_l\}$ is not Gaussian and is composed of independent random variables, (iv) for every $k = 1, ..., n$, the (possibly

mixed) moments of order k of the variables $\{b_0, ..., b_l\}$ coincide with those of a
vector $\{a_0, ..., a_l\}$ of independent, centered and complex-valued Gaussian random variables with common variance C_l and such that a_0 is real and, for every
$m = 1, ..., l$, the real and imaginary parts of a_m are independent and identically
distributed (the existence of a vector such as $\{b_0, ..., b_l\}$ is easily proved). Now
define the two fields

$$T(x) = \sum_{m=-l}^{l} b_m Y_{lm}(x) \quad \text{and} \quad T^*(x) = \sum_{m=-l}^{l} a_m Y_{lm}(x).$$

By Proposition 6.12, T^* is strongly isotropic, and also n-weakly isotropic by
Proposition 5.10. By construction, T is also n-weakly isotropic. However, T
cannot be strongly isotropic, since this would violate Proposition 6.12 (indeed,
if T was isotropic, we would have an example of an isotropic field whose
harmonic coefficients $\{b_0, ..., b_l\}$ are independent and non-Gaussian). $\quad\square$

Remark 6.16 As anticipated in the previous chapter, for finite variance, zero
mean isotropic random fields we have the important relationship

$$ET(x)T(y) = \sum_{l=1}^{\infty} \frac{2l+1}{4\pi} C_l P_l(\langle x, y \rangle), \qquad (6.19)$$

where $\langle x, y \rangle$ is the Euclidean inner product between x and y, and P_l is the lth
Legendre polynomial, as defined in Section 13.1.2. For $x = y$ we have thus

$$ET^2(x) = \sum_{l=1}^{\infty} \frac{2l+1}{4\pi} C_l. \qquad (6.20)$$

(6.19) is an easy consequence of (3.42), i.e.

$$\sum_{m=-l}^{l} Y_{lm}(x)\overline{Y_{lm}(y)} = \frac{2l+1}{4\pi} P_l(\langle x, y \rangle), \quad x, y \in S^2. \qquad (6.21)$$

Remark 6.17 Most of the results of this section can be extended to isotropic
random fields on general homogeneous spaces of compact groups, see [11].

6.4 Group representations and polyspectra

In the previous section, we proved that spherical harmonic coefficients for
isotropic random fields can be independent (at a fixed frequency) only in the
Gaussian case. In this section, we investigate the constraints that are imposed

on spectral moments by isotropy, under very general non-Gaussian circumstances.

As before, we use the symbol $A \otimes B$ to indicate the *Kronecker product* between two matrices A and B. For future reference, we recall some further standard properties of Kronecker (tensor) products and direct sums of matrices (see Magnus and Neudecker [128]): we have

$$\oplus_{i=1}^{n}(A_i B_i) = \left(\oplus_{i=1}^{n} A_i\right)\left(\oplus_{i=1}^{n} B_i\right), \qquad (6.22)$$

$$\left(\oplus_{i=1}^{n} A_i\right) \otimes B = \oplus_{i=1}^{n}(A_i \otimes B) \qquad (6.23)$$

and, provided all matrix products are well-defined,

$$(AB \otimes C) = (A \otimes I_n)(B \otimes C). \qquad (6.24)$$

Here, $\oplus_{i=1}^{n} A_i$ is defined as the block diagonal matrix $diag\{A_1, ..., A_n\}$ if A_i is a set of square matrices of order $r_i \times r_i$, whereas it is defined as the stacked column vector of order $\left(\sum_{i=1}^{n} r_i\right) \times 1$ if the A_i are $r_i \times 1$ column vectors.

Definition 6.18 (1) Let the field T admit the representation (6.1), and suppose that, for some $n \geq 2$, $E |a_{lm}|^n < \infty$ for every l, m. Then, T is said to have **finite spectral moments** of order n.

(2) Suppose that T has finite spectral moments of order $n \geq 2$. The **(angular) polyspectrum of order** $n - 1$, associated with T, is given by the collection of vectors

$$S_{l_1...l_n} = E\left[a_{l_1.} \otimes a_{l_2.} \otimes \cdots \otimes a_{l_n.}\right], \qquad (6.25)$$

where $0 \leq l_1, l_2, ..., l_n$. Note that the vector $S_{l_1...l_n}$ appearing in (6.25) has dimension $(2l_1 + 1) \times \cdots \times (2l_n + 1)$.

The next theorem connects the invariance properties of the vectors $\{a_{l.}\}$ to the representations of the non-commutative group $SO(3)$, whose Haar measure with unit mass is denoted once again by dg. We also denote by $\{D^l : 0, 1, ...\}$ the sequence of Wigner D-matrices (with integer indices), as defined in Section 3.3. We need to establish some further notation. For every $0 \leq l_1, l_2, ..., l_n$, we shall write

$$\Delta_{l_1...l_n} := \int_{SO(3)} \left\{D^{l_1}(g) \otimes D^{l_2}(g) \otimes \cdots \otimes D^{l_n}(g)\right\} dg, \qquad (6.26)$$

$$\Delta_{l_1...l_n}(g) := D^{l_1}(g) \otimes D^{l_2}(g) \otimes \cdots \otimes D^{l_n}(g), \quad g \in SO(3), \qquad (6.27)$$

and use the symbol $S_{l_1...l_n}$ (whenever is well-defined), as given in formula

(6.25). We stress that $\Delta_{l_1...l_n}$ and $\Delta_{l_1...l_n}(g)$ are square matrices with $(2l_1 + 1) \times \cdots \times (2l_n + 1)$ rows and $S_{l_1...l_n}$ is a column vector with $(2l_1 + 1) \times \cdots \times (2l_n + 1)$ elements. The following result applies to an arbitrary $n \geq 2$: see [102] for some related results in the case $n = 3, 4$.

Proposition 6.19 *Let T be a strongly isotropic field with moments of order $n \geq 2$. Then, for every $0 \leq l_1, l_2, ..., l_n$ and every fixed $g^* \in SO(3)$,*

$$\Delta_{l_1...l_n} S_{l_1...l_n} = S_{l_1...l_n} \qquad (6.28)$$

$$\Delta_{l_1...l_n}(g^*) S_{l_1...l_n} = S_{l_1...l_n}. \qquad (6.29)$$

On the other hand, fix $n \geq 2$ and assume that $T(x)$ is a (not necessarily isotropic) random field on the sphere s.t. $\sup_x (E|T(x)|^n) < \infty$. Then $T(.)$ is P-almost surely Lebesgue square integrable and the n-th order spectral moments of T exist and are finite. If moreover (6.28) holds for every $0 \leq l_1 \leq \cdots \leq l_n$, then for every $g \in SO(3)$,

$$E\left[D^{l_1}(g)a_{l_1.} \otimes \cdots \otimes D^{l_n}(g)a_{l_n.} \right] = E\left[a_{l_1.} \otimes \cdots \otimes a_{l_n.} \right], \qquad (6.30)$$

and T is n-weakly isotropic.

Proof By strong isotropy and Lemma 6.3, one has

$$E\left\{ D^{l_1}(g)a_{l_1.} \otimes \cdots \otimes D^{l_n}(g)a_{l_n.} \right\} = E\left\{ a_{l_1.} \otimes \cdots \otimes a_{l_n.} \right\} \forall g \in SO(3), \, l_1, ..., l_n \in \mathbb{N}^n.$$

Now assume that g is sampled randomly (and independently of the $\{a_{l.}\}$) according to some probability measure, say P_0, on $SO(3)$. From the property (6.24) of tensor products and trivial manipulations, we obtain (with obvious notation and by independence)

$$E\left\{ D^{l_1}(\cdot)a_{l_1.} \otimes \cdots \otimes D^{l_n}(\cdot)a_{l_n.} \right\} = E\left\{ \left[D^{l_1}(\cdot) \otimes \cdots \otimes D^{l_n}(\cdot) \right] \left[a_{l_1.} \otimes \cdots \otimes a_{l_n.} \right] \right\}$$

$$= E_0\left\{ D^{l_1}(\cdot) \otimes \cdots \otimes D^{l_n}(\cdot) \right\} E\left\{ a_{l_1.} \otimes \cdots \otimes a_{l_n.} \right\}.$$

Now, if one chooses P_0 to be equal to the Haar (uniform) measure on $SO(3)$,

$$E_0\left\{ D^{l_1}(\cdot) \otimes \cdots \otimes D^{l_n}(\cdot) \right\} = \Delta_{l_1...l_n},$$

thus giving (6.28). On the other hand, if one chooses P_0 to be equal to the Dirac mass at some $g^* \in SO(3)$, one has that

$$E_0\left\{ D^{l_1}(\cdot) \otimes \cdots \otimes D^{l_n}(\cdot) \right\} = \Delta_{l_1...l_n}(g^*),$$

which shows that (6.29) is satisfied.

Now let T satisfy the assumptions of the second part of the statement for some

$n \geq 2$. We recall first that the representation (6.1) continues to hold, in a path-wise sense. To see that the nth order joint moments of the harmonic coefficients a_{lm} are finite it is enough to use Jensen's inequality, along with a standard version of the Fubini Theorem, to obtain that

$$E\,|a_{lm}|^n = E\left|\int_{S^2} T(x)\overline{Y_{lm}(x)}dx\right|^n \leq E\int_{S^2} |T(x)|^n |Y_{lm}(x)|^n dx$$

$$\leq \left\{\sup_{x\in S^2} |Y_{lm}(x)|^n\right\}\left\{\sup_{x\in S^2} E|T(x)|^n\right\}$$

$$\leq \left(\frac{2l+1}{4\pi}\right)^{n/2}\left\{\sup_{x\in S^2} E|T(x)|^n\right\} < \infty\,.$$

It is then straightforward that, if $S_{l_1\ldots l_n}$ satisfies (6.28), for any fixed $\overline{g} \in SO(3)$

$$E\left\{\left[D^{l_1}(\overline{g})\otimes\ldots\otimes D^{l_n}(\overline{g})\right][a_{l_1.}\otimes\cdots\otimes a_{l_n.}]\right\}$$

$$= \left[D^{l_1}(\overline{g})\otimes\cdots\otimes D^{l_n}(\overline{g})\right]E\,[a_{l_1.}\otimes\cdots\otimes a_{l_n.}]$$

$$= \left[D^{l_1}(\overline{g})\otimes\cdots\otimes D^{l_n}(\overline{g})\right]\Delta_{l_1\ldots l_n}S_{l_1\ldots l_n}$$

$$= \left\{\left[D^{l_1}(\overline{g})\otimes\cdots\otimes D^{l_n}(\overline{g})\right]\int_{SO(3)}\left\{D^{l_1}(g)\otimes\cdots\otimes D^{l_n}(g)\right\}dg\right\}S_{l_1\ldots l_n}$$

$$= \left\{\int_{SO(3)}\left\{D^{l_1}(\overline{g}g)\otimes D^{l_2}(\overline{g}g)\otimes\cdots\otimes D^{l_n}(\overline{g}g)\right\}dg\right\}S_{l_1\ldots l_n}$$

$$= \Delta_{l_1\ldots l_n}S_{l_1\ldots l_n} = E\,\{a_{l_1.}\otimes\cdots\otimes a_{l_n.}\}\,,$$

which proves the n-th spectral moment is invariant to rotations. The fact that T is n-weakly isotropic is a consequence of the spectral representation (6.1). $\square$

Note that relation (6.28) can be rephrased by saying that, for a strongly isotropic field, the joint moment vector $E\,\{a_{l_1.}\otimes a_{l_2.}\otimes\cdots\otimes a_{l_n.}\}$ must be an *eigenvector* of the matrix (6.26) for every $n \geq 2$ and every $0 \leq l_1 \leq \cdots \leq l_n$.

Remark 6.20 Note that (6.28) is the "non-commutative" analogous of relation (6.5) for random sequences defined on the cyclic group. As anticipated, the fact that $SO\,(3)$ is not commutative (and therefore that its irreducible representations have dimensions ≥ 1) explains the presence of averages of non-trivial matrix products.

6.5 Angular polyspectra and the structure of $\Delta_{l_1...l_n}$

Our aim in this section is to investigate more deeply the structure of the matrix $\Delta_{l_1...l_n}$ appearing in (6.26), in order to derive an explicit characterization for the angular polyspectra. Our analysis involves the use of the Clebsch-Gordan coefficients $\{C^{lm}_{l_1 m_1 l_2 m_2}\}$, as defined in Section 3.5.

6.5.1 Spectra of strongly isotropic fields

As a preliminary example, we deal with the case $n = 2$.

Proposition 6.21   *For integers* $l_1, l_2 \geq 0$,

$$\Delta_{l_1 l_2} = \int_{SO(3)} \left\{ D^{l_1}(g) \otimes D^{l_2}(g) \right\} dg = \delta^{l_2}_{l_1} C^{00}_{l_1.l_2.} (C^{00}_{l_1.l_2.})', \tag{6.31}$$

that is: if $l_1 \neq l_2$, *then* $\Delta_{l_1 l_2}$ *is a* $(2l_1 + 1)(2l_2 + 1) \times (2l_1 + 1)(2l_2 + 1)$ *zero matrix; if* $l_1 = l_2$, *then* $\Delta_{l_1 l_2} = \Delta_{l_1 l_1}$ *is given by* $C^{00}_{l_1.l_1.} (C^{00}_{l_1.l_1.})'$.

Proof   Using the equivalence of the two representations $D^{l_1}(g) \otimes D^{l_2}(g)$ and $\oplus^{l_2+l_1}_{\lambda=|l_2-l_1|} D^{\lambda}(g)$, as well as the definition of the Clebsch-Gordan matrices, we obtain that

$$\int_{SO(3)} \left\{ D^{l_1}(g) \otimes D^{l_2}(g) \right\} dg = C_{l_1 l_2} \left[\int_{SO(3)} \left\{ \oplus^{l_2+l_1}_{\lambda=|l_2-l_1|} D^{\lambda}(g) \right\} dg \right] C^{*}_{l_1 l_2}. \tag{6.32}$$

Now, if $l_1 \neq l_2$, then the right-hand side of (6.32) is equal to the zero matrix since, as a consequence of the Peter-Weyl Theorem and for $\lambda \neq 0$, the entries of $D^{\lambda}(\cdot)$ are orthogonal to the constants. If $l_1 = l_2$, then the integrated matrix on the right-hand side of (6.32) becomes $\int_{SO(3)} \left\{ \oplus^{2l_1}_{\lambda=0} D^{\lambda}(g) \right\} dg$, that is, a $(2l_1 + 1)^2 \times (2l_1 + 1)^2$ matrix which is zero everywhere, except for the entry in the top-left corner, which is equal to one (since $\int_{SO(3)} dg = 1$). The proof is concluded by checking that

$$C_{l_1 l_1} \left[\int_{SO(3)} \left\{ \oplus^{2l_1}_{\lambda=0} D^{\lambda}(g) \right\} dg \right] C^{*}_{l_1 l_1} = C^{00}_{l_1.l_1.} (C^{00}_{l_1.l_1.})'.$$

$\square$

Remark 6.22   Recall that $C^{00}_{l_1.l_2.}$ is a *column* vector of dimension

$$(2l_1 + 1)(2l_2 + 1),$$

corresponding to the first column of the matrix $C_{l_1 l_2}$. Also, according e.g. to [195, formula 8.5.1.1], we have that

$$C_{l_1 . l_2.}^{00} = \left\{ \frac{(-1)^{m_1}}{\sqrt{2l_1 + 1}} \delta_{l_1}^{l_2} \delta_{m_1}^{-m_2} \right\}_{m_1 = -l_1, \ldots, l_1; m_2 = -l_2, \ldots, l_2}. \tag{6.33}$$

Proposition 6.21 provides a characterization of the spectrum of a strongly isotropic field. As indicated in the statement, this result should be compared with Proposition 6.6 of this chapter.

Corollary 6.23 *Let T be a strongly isotropic field with second moments, and let the vectors of the harmonic coefficients $\{a_{l.}\}$ be defined according to (6.1). Then, for any integers $l_1, l_2 \geq 0$, we have that*

$$E\{a_{l_1.} \otimes a_{l_2.}\} = C_{l_1}^{\#} \times \left\{ \frac{(-1)^{m_1}}{\sqrt{2l_1 + 1}} \delta_{l_1}^{l_2} \delta_{m_1}^{-m_2} \right\} \tag{6.34}$$

for some $C_{l_1}^{\#} \geq 0$ depending uniquely on l_1, and where the expression between brackets denotes the $(2l_1 + 1)(2l_2 + 1)$-dimensional vector appearing in (6.33). The relation between $C_{l_1}^{\#}$ and the constant C_{l_1} appearing in (6.13) is $C_{l_1}^{\#} = \left(\sqrt{2l_1 + 1} \right) \times C_{l_1}$.

Proof According to (6.28), we have that

$$E\{a_{l_1.} \otimes a_{l_2.}\} = \delta_{l_1}^{l_2} C_{l_1 . l_2.}^{00} (C_{l_1 . l_2.}^{00})' E\{a_{l_1.} \otimes a_{l_2.}\},$$

implying that $E\{a_{l_1.} \otimes a_{l_2.}\}$ is (a) equal to the zero vector for $l_1 \neq l_2$, and (b) of the form $C_{l_1}^{\#} \times C_{l_1 . l_2.}^{00}$, for some constant $C_{l_1}^{\#}$, when $l_1 = l_2$. To see that $C_{l_1}^{\#}$ cannot be negative, just observe that $a_{l_1 0}$ is real-valued for every $l_1 \geq 0$, so that (6.34) yields that

$$C_{l_1}^{\#} = \left(\sqrt{2l_1 + 1} \right) \times E\left(a_{l_1 0}^2 \right),$$

which automatically proves the last claim in the statement.   □

In the subsequent two subsections, we shall obtain, for every $n \geq 3$, a characterization of $\Delta_{l_1 \ldots l_n}$ and $E\{a_{l_1.} \otimes \cdots \otimes a_{l_n.}\}$, respectively analogous to (6.31) and (6.34).

6.5.2 The structure of $\Delta_{l_1 \ldots l_n}$

We first need to establish some further notation.

Definition 6.24 Fix $n \geq 3$. For integers $l_1, \ldots, l_n \geq 0$, we define $C_{l_1\ldots l_n}$ to be the unitary matrix, of dimension

$$\prod_{j=1}^{n}(2l_j + 1) \times \prod_{j=1}^{n}(2l_j + 1),$$

connecting the following two equivalent representations of $SO(3)$

$$D^{l_1}(.) \otimes D^{l_2}(.) \otimes \cdots \otimes D^{l_n}(.) \tag{6.35}$$

and

$$\bigoplus_{\lambda_1=|l_2-l_1|}^{l_2+l_1} \bigoplus_{\lambda_2=|l_3-\lambda_1|}^{l_3+\lambda_1} \cdots \bigoplus_{\lambda_{n-1}=|l_n-\lambda_{n-2}|}^{l_n+\lambda_{n-2}} D^{\lambda_{n-1}}(.). \tag{6.36}$$

Remark 6.25 (1) Fix $l_1, \ldots, l_n \geq 0$, as well as $g \in SO(3)$. Then, the matrix

$$\bigoplus_{\lambda_1=|l_2-l_1|}^{l_2+l_1} \bigoplus_{\lambda_2=|l_3-\lambda_1|}^{l_3+\lambda_1} \cdots \bigoplus_{\lambda_{n-1}=|l_n-\lambda_{n-2}|}^{l_n+\lambda_{n-2}} D^{\lambda_{n-1}}(g) \tag{6.37}$$

is a block-diagonal matrix, obtained as follows. (a) Consider vectors of integers $(\lambda_1, \ldots, \lambda_{n-1})$ satisfying the relations $|l_2 - l_1| \leq \lambda_1 \leq l_1 + l_2$, and $|l_{k+1} - \lambda_{k-1}| \leq \lambda_k \leq l_{k+1} + \lambda_{k-1}$, for $k = 2, \ldots, n-1$. (b) Introduce a (total) order $\prec_0$ on the collection of these vectors by saying that

$$(\lambda_1, \ldots, \lambda_{n-1}) \prec_0 \left(\lambda'_1, \ldots, \lambda'_{n-1}\right), \tag{6.38}$$

whenever either $\lambda_1 < \lambda'_1$, or there exists $k = 2, \ldots, n-2$ such that $\lambda_j = \lambda'_j$ for every $j = 1, \ldots, k$, and $\lambda_{k+1} < \lambda'_{k+1}$. (c) Associate to each vector $(\lambda_1, \ldots, \lambda_{n-1})$ the matrix $D^{\lambda_{n-1}}(g)$. (d) Construct a block-diagonal matrix by disposing the matrices $D^{\lambda_{n-1}}(g)$ from the top-left corner to the bottom-right corner, in increasing order with respect to $\prec_0$. As an example, consider the case where $n = 3$ and $l_1 = l_2 = l_3 = 1$. Here, the vectors (λ_1, λ_2) involved in the direct sum (6.36) are (in increasing order with respect to $\prec_0$)

$$(0, 1), \quad (1, 0), \quad (1, 1), \quad (1, 2), \quad (2, 1), \quad (2, 2) \quad \text{and} \quad (2, 3),$$

and the matrix (6.37) is therefore given by

$$\begin{pmatrix} D^1(g) & \ldots & \ldots & \ldots & \ldots & \ldots & \ldots \\ \ldots & 1 & \ldots & \ldots & \ldots & \ldots & \ldots \\ \ldots & \ldots & D^1(g) & \ldots & \ldots & \ldots & \ldots \\ \ldots & \ldots & \ldots & D^2(g) & \ldots & \ldots & \ldots \\ \ldots & \ldots & \ldots & \ldots & D^1(g) & \ldots & \ldots \\ \ldots & \ldots & \ldots & \ldots & \ldots & D^2(g) & \ldots \\ \ldots & \ldots & \ldots & \ldots & \ldots & \ldots & D^3(g) \end{pmatrix} \tag{6.39}$$

where the dots indicate *zero* entries, and we have used the fact that $D^0(g) \equiv 1$.

(2) The fact that the representation (6.36) has dimension $\prod_{j=1}^{n}\left(2l_j + 1\right)$ is a direct consequence of formula (3.59).

(3) The fact that the two representations (6.35) and (6.36) are equivalent can be proved by iteration. Indeed, it can be checked that (6.35) is equivalent to

$$\oplus_{\lambda_1=|l_2-l_1|}^{l_2+l_1} D^{\lambda_1}(.) \otimes D^{l_3}(\cdot) \otimes \cdots \otimes D^{l_n}(\cdot),$$

which is in turn equivalent to

$$\oplus_{\lambda_1=|l_2-l_1|}^{l_2+l_1} \oplus_{\lambda_2=|l_3-\lambda_1|}^{l_3+\lambda_1} D^{\lambda_2}(.) \otimes D^{l_4}(\cdot) \otimes \cdots \otimes D^{l_n}(\cdot).$$

By iterating the same procedure until all tensor products have disappeared (that is, by successively replacing the tensor product $D^{\lambda_k}(.) \otimes D^{l_{k+2}}(\cdot)$ with $\oplus_{\lambda_{k+1}=|l_{k+2}-\lambda_k|}^{l_{k+2}+\lambda_k} D^{\lambda_2}(.)$ for $k = 2, ..., n-1$), one obtains the desired conclusion.

For every $n \geq 3$ and every $l_1, ..., l_n \geq 0$, the elements of the matrix $C_{l_1...l_n}$, introduced in Definition 6.24, can be written in the form $C_{l_1m_1...l_nm_n}^{\lambda_1...\lambda_{n-1}\mu_{n-1}}$. The indices $(m_1, ..., m_n)$ are such that $-l_i \leq m_i \leq l_i$ $(i = 1, ..., n)$ and label rows; on the other hand, the indices $(\lambda_1...\lambda_{n-1}, \mu_{n-1})$ label columns, and verify the relations $|l_2 - l_1| \leq \lambda_1 \leq l_1 + l_2$, $|l_{k+1} - \lambda_{k-1}| \leq \lambda_k \leq l_{k+1} + \lambda_{k-1}$ $(k = 2, ..., n-1)$ and $-\lambda_{n-1} \leq \mu_{n-1} \leq \lambda_{n-1}$. It is well known (see e.g. [195]) that the quantity $C_{l_1m_1...l_nm_n}^{\lambda_1...\lambda_{n-1}\mu_{n-1}}$ can be represented as a *convolution* of the Clebsch-Gordan coefficients introduced in Section 3.5, namely:

$$C_{l_1m_1...l_nm_n}^{\lambda_1,...,\lambda_{n-1}\mu_{n-1}} = C_{l_1m_1...l_{n-1}m_{n-1}}^{\lambda_1,...,\lambda_{n-2}} \cdot C_{\lambda_{n-2}\cdot l_n m_n}^{\lambda_{n-1}\mu_{n-1}} \tag{6.40}$$

$$= \sum_{\mu_{n-2}}\left\{ \sum_{\mu_1,...,\mu_{n-3}} C_{l_1m_1l_2m_2}^{\lambda_1\mu_1} C_{\lambda_1\mu_1l_3m_3}^{\lambda_2\mu_2}...C_{\lambda_{n-3}\mu_{n-3}l_{n-1}m_{n-1}}^{\lambda_{n-2}\mu_{n-2}} \right\} C_{\lambda_{n-2}\mu_{n-2}l_nm_n}^{\lambda_{n-1}\mu_{n-1}}$$

$$= \sum_{\mu_1,...,\mu_{n-2}} C_{l_1m_1l_2m_2}^{\lambda_1\mu_1} C_{\lambda_1\mu_1l_3m_3}^{\lambda_2\mu_2}...C_{\lambda_{n-3}\mu_{n-3}l_{n-1}m_{n-1}}^{\lambda_{n-2}\mu_{n-2}} C_{\lambda_{n-2}\mu_{n-2}l_nm_n}^{\lambda_{n-1}\mu_{n-1}}. \tag{6.41}$$

Remark 6.26 Given an enumeration of the coefficients $C_{l_1m_1...l_nm_n}^{\lambda_1...\lambda_{n-1}\mu_{n-1}}$, the matrix $C_{l_1...l_n}$ can be built (analogously to the case of the Clebsch-Gordan matrices of Section 3.5) by disposing rows (from top to bottom) and columns (from left to right) increasingly according to two separate total orders. The order $<_r$ on the symbols $(m_1, ..., m_n)$ is obtained by setting that $(m_1, ..., m_n) <_r \left(m'_1, ..., m'_n\right)$ whenever either $m_1 < m'_1$, or there exists $k = 2, ..., n-1$ such that $m_j = m'_j$ for every $j = 1, ..., k$, and $m_{k+1} < m'_{k+1}$. The order $<_c$ on the symbols

$(\lambda_1...\lambda_{n-1}, \mu_{n-1})$ is obtained by setting that $(\lambda_1...\lambda_{n-1}, \mu_{n-1}) \prec_c \left(\lambda'_1...\lambda'_{n-1}, \mu'_{n-1}\right)$ whenever either $(\lambda_1,...,\lambda_{n-1}) \prec_0 \left(\lambda'_1,...,\lambda'_{n-1}\right)$, as defined in (6.38), or $\lambda_i = \lambda'_i$ for every $i = 1,...,n-1$ and $\mu_{n-1} < \mu'_{n-1}$.

We have also the following (useful) alternative representation of generalized Clebsch-Gordan matrices.

Proposition 6.27 *For every $n \geq 3$ and every $l_1,...,l_n \geq 0$, the matrix $C_{l_1...l_n}$ can be represented as follows*

$$C_{l_1...l_n} = \{C_{l_1 l_2 l_3...l_{n-1}} \otimes I_{2l_n+1}\} \left\{\oplus_{\lambda_1=|l_2-l_1|}^{l_2+l_1} \cdots \oplus_{\lambda_{n-2}=|l_n-\lambda_{n-3}|}^{l_n+\lambda_{n-3}} C_{\lambda_{n-2} l_n}\right\},$$

where I_m indicates a $m \times m$ identity matrix. Also, one has that

$$C_{l_1...l_n} = (C_{l_1 l_2} \otimes I_{2l_3+1} \otimes \cdots \otimes I_{2l_n+1}) \times \left[(\oplus_{\lambda=|l_2-l_1|}^{l_2+l_1} C_{\lambda l_3}) \otimes \cdots \otimes I_{2l_n+1}\right]$$

$$\times \cdots \times \left[\oplus_{\lambda_1=|l_2-l_1|}^{l_2+l_1} \cdots \oplus_{\lambda_{n-2}=|l_n-\lambda_{n-3}|}^{l_n+\lambda_{n-3}} C_{\lambda_{n-2} l_n}\right],$$

where $\times$ stands for the usual product between matrices.

Remark 6.28 For every $n \geq 3$ and every $l_1,...,l_n \geq 0$, we define $E_{l_1...l_n}$ to be the $\Pi_{j=1}^n \left(2l_j + 1\right) \times \Pi_{j=1}^n \left(2l_j + 1\right)$ square matrix

$$E_{l_1...l_n} := \oplus_{\lambda_1=|l_2-l_1|}^{l_2+l_1} \cdots \oplus_{\lambda_{n-1}=|l_n-\lambda_{n-2}|}^{l_n+\lambda_{n-2}} \delta_{\lambda_{n-1}}^0 I_{2\lambda_{n-1}+1}. \tag{6.42}$$

In other words, $E_{l_1...l_n}$ is the diagonal matrix built from the matrix (6.37), by replacing every block of the type $D^{\lambda_{n-1}}(g)$, with $\lambda_{n-1} > 0$, with a $(2\lambda_{n-1} + 1) \times (2\lambda_{n-1} + 1)$ zero matrix, and by letting the 1×1 blocks $D^0(g) = 1$ unchanged. For instance, by setting $n = 3$ and $l_1 = l_2 = l_3 = 1$ (and by using (6.39)) we obtain a 27×27 matrix E_{111} whose entries are all zero, except for the fourth element (starting from the top-left corner) of the main diagonal.

The following result states that the matrix $\Delta_{l_1...l_n}$ can be diagonalized in terms of $C_{l_1...l_n}$ and $E_{l_1...l_n}$.

Proposition 6.29 *The matrix $\Delta_{l_1...l_n}$ can be diagonalized as*

$$\Delta_{l_1...l_n} = C_{l_1...l_n} E_{l_1...l_n} C^*_{l_1...l_n}, \tag{6.43}$$

where $E_{l_1...l_n}$ is the matrix introduced in (6.42).

Proof   We have that

$$\Delta_{l_1\ldots l_n} \tag{6.44}$$

$$= \int_{SO(3)} D^{l_1}(g) \otimes D^{l_2}(g) \otimes \cdots \otimes D^{l_n}(g)\, dg \tag{6.45}$$

$$= \int_{SO(3)} \left[C_{l_1\ldots l_n} \oplus_{\lambda_1=|l_2-l_1|}^{l_2+l_1} \oplus_{\lambda_2=|l_3-\lambda_1|}^{l_3+\lambda_1} \cdots \oplus_{\lambda_{n-1}=|l_n-\lambda_{n-2}|}^{l_n+\lambda_{n-2}} D^{\lambda_{n-1}}(g) C^*_{l_1\ldots l_n} \right] dg.$$

By linearity and by the definition of the integral of a matrix-valued function, the last line of (6.44) equals

$$C_{l_1\ldots l_n} \left[\oplus_{\lambda_1=|l_2-l_1|}^{l_2+l_1} \oplus_{\lambda_2=|l_3-\lambda_1|}^{l_3+\lambda_1} \cdots \oplus_{\lambda_{n-1}=|l_n-\lambda_{n-2}|}^{l_n+\lambda_{n-2}} \int_{SO(3)} D^{\lambda_{n-1}}(g)\, dg \right] C^*_{l_1\ldots l_n}.$$

Now observe that, if $\lambda_{n-1} > 0$, then $\int_{SO(3)} D^{\lambda_{n-1}}(g)\, dg$ equals a $(2\lambda_{n-1}+1) \times (2\lambda_{n-1}+1)$ zero matrix, whereas $\int_{SO(3)} D^0(g)\, dg = \int_{SO(3)} 1\, dg = 1$. The conclusion is obtained by resorting to the definition of $E_{l_1\ldots l_n}$ given in (6.42).   □

Remark 6.30   As mentioned in Chapter 3, expression (6.45) is known as a generalized Gaunt integral. By using the identity (3.35), we see that Proposition 6.29 yields a powerful tool for the evaluation of integrals of multiple products in spherical harmonics, i.e.

$$\mathcal{G}\{l_1,m_1;\ldots;l_r,m_r\} := \int_{S^2} Y_{l_1,m_1}(x) \cdots Y_{l_r,m_r}(x)\, d\sigma(x).$$

By iterating e.g. the arguments in the proof of Proposition 3.43 (or, equivalently, by integrating out both sides of identity 5.6.2.12 in [195]), one has the general expression

$$\mathcal{G}\{l_1,m_1;\ldots;l_r,m_r\} = (-1)^{m_r} \sqrt{\frac{(2l_1+1)\cdots(2l_{r-1}+1)}{(4\pi)^{r-2}(2l_r+1)}} \sum_{\lambda_1,\ldots,\lambda_{r-3}} C^{\lambda_1 0}_{l_1 0 l_2 0} \cdots C^{l_r 0}_{\lambda_{r-3} 0 l_{r-1} 0}$$

$$\times \sum_{\mu_1,\ldots,\mu_{r-3}} C^{\lambda_1 \mu_1}_{l_1 m_1 l_2 m_2} \cdots C^{l_r -m_r}_{\lambda_{r-3}\mu_{r-3} l_{r-1} m_{r-1}}. \tag{6.46}$$

The following special case already appears in Section 3.5.2:

$$\mathcal{G}\{l_1,m_1;l_2,m_2;l,-m\} = (-1)^m \sqrt{\frac{(2l_1+1)(2l_2+1)}{4\pi(2l+1)}} C^{lm}_{l_1 m_1 l_2 m_2} C^{l0}_{l_1 0 l_2 0}.$$

6.6 Reduced polyspectra of arbitrary orders

Combining the previous Proposition with (6.19), we obtain one of the main results of this chapter (compare with the results of [136]), that is, the existence of *reduced polyspectra* of arbitrary orders for isotropic spherical fields. This result should be compared with the existence of the functions $B(\cdot)$ and $P(\cdot)$, appearing in formulae (6.3) and (6.4), for periodic random sequences on the integers.

Theorem 6.31 *If a random field is strongly isotropic with finite moments of order $n \geq 3$, then for every $l_1, ..., l_n$ there exists an array $P_{l_1...l_n}(\lambda_1, ..., \lambda_{n-3})$, with*
$$|l_2 - l_1| \leq \lambda_1 \leq l_2 + l_1, |l_3 - \lambda_1| \leq \lambda_2 \leq l_3 + \lambda_1, ..., |l_{n-2} - \lambda_{n-4}| \leq \lambda_{n-3} \leq l_{n-2} + \lambda_{n-4},$$
such that

$$Ea_{l_1 m_1}...a_{l_n m_n} = (-1)^{m_n} \sum_{\lambda_1 = |l_2 - l_1|}^{l_2 + l_1} \cdots \sum_{\lambda_{n-3}} C^{\lambda_1...\lambda_{n-3}l_n, -m_n}_{l_1 m_1...l_{n-1}m_{n-1}} P_{l_1...l_n}(\lambda_1, ..., \lambda_{n-3}) \quad (6.47)$$

$$C^{\lambda_1...\lambda_{n-3}l_n, -m_n}_{l_1 m_1...l_{n-1}m_{n-1}} = \sum_{\mu_1} \cdots \sum_{\mu_{n-3}} C^{\lambda_1 \mu_1}_{l_1 m_1 l_2 m_2} C^{\lambda_2 \mu_2}_{\lambda_1 \mu_1 l_3 m_3} \cdots C^{l_n, -m_n}_{\lambda_{n-3}\mu_{n-3}l_{n-1}m_{n-1}} . \quad (6.48)$$

Remark 6.32 For a fixed $n \geq 2$, the real-valued array $\{P_{l_1...l_n}(\cdot) : l_1, ..., l_n \geq 0\}$ is called the *reduced polyspectrum of order $n-1$* associated with the underlying strongly isotropic random field.

Proof of Theorem 6.31 By Proposition 6.19 and Proposition 6.29, if the random field is isotropic, then

$$S_{l_1...l_n} = C_{l_1...l_n} E_{l_1...l_n} C^*_{l_1...l_n} S_{l_1...l_n} ,$$

that is, because $C_{l_1...l_n}$ is unitary

$$C^*_{l_1...l_n} S_{l_1...l_n} = E_{l_1...l_n} C^*_{l_1...l_n} S_{l_1...l_n} .$$

It follows that $S_{l_1...l_n}$ is a solution if and only if the column vector $C^*_{l_1...l_n} S_{l_1...l_n}$ has zeroes corresponding to the zeroes of $E_{l_1...l_n}$, whereas the elements corresponding to unity can be arbitrary. In view of the orthonormality properties of $C^*_{l_1...l_n}$, this condition is met if, and only if, $S_{l_1...l_n}$ is a linear combination of the columns in the matrix $C^*_{l_1...l_n}$ corresponding to non-zero elements of the diagonal $E_{l_1...l_n}$. We can write these linear combinations explicitly as

$$\sum_{\lambda_1 = |l_2 - l_1|}^{l_2 + l_1} \sum_{\lambda_2 = |l_3 - \lambda_1|}^{l_3 + \lambda_1} \cdots \sum_{l = |l_n - \lambda_{n-2}|}^{l_n + \lambda_{n-2}} \sum_{m=-l}^{l} C^{\lambda_1...\lambda_{n-2}l, m}_{l_1 m_1...l_n m_n} \widetilde{P}_{l_1...l_n}(\lambda_1, ..., \lambda_{n-2}) \delta^0_l$$

$$= \sum_{\lambda_1 = |l_2 - l_1|}^{l_2 + l_1} \sum_{\lambda_2 = |l_3 - \lambda_1|}^{l_3 + \lambda_1} \cdots \sum_{l = |l_n - \lambda_{n-2}|}^{l_n + \lambda_{n-2}} \sum_{m=-l}^{l}$$

$$\times \left\{ \sum_{\mu_1,\dots,\mu_{n-2}} C^{\lambda_1\mu_1}_{l_1 m_1 l_2 m_2} C^{\lambda_2\mu_2}_{\lambda_1\mu_1 l_3 m_3} \cdots C^{lm}_{\lambda_{n-2}\mu_{n-2} l_n m_n} \delta^0_l \right\} \widetilde{P}_{l_1\dots l_n}(\lambda_1, \dots, \lambda_{n-3}, \lambda_{n-2})$$

$$= \sum_{\lambda_1=|l_2-l_1|}^{l_2+l_1} \sum_{\lambda_2=|l_3-\lambda_1|}^{l_3+\lambda_1} \cdots \sum_{l=|l_n-\lambda_{n-2}|}^{l_n+\lambda_{n-2}}$$

$$\left\{ \sum_{\mu_1,\dots,\mu_{n-2}} C^{\lambda_1\mu_1}_{l_1 m_1 l_2 m_2} C^{\lambda_2\mu_2}_{\lambda_1\mu_1 l_3 m_3} \cdots C^{00}_{\lambda_{n-2}\mu_{n-2} l_n m_n} \right\} \widetilde{P}_{l_1\dots l_n}(\lambda_1, \dots, \lambda_{n-3}, \lambda_{n-2}).$$

Recalling again that ([195], formula 8.5.1.1)

$$C^{0m}_{l_1 m_1 l_2 m_2} = \frac{(-1)^{m_1}}{\sqrt{2l_1+1}} \delta^{l_2}_{l_1} \delta^{-m_2}_{m_1} \delta^0_m ,$$

we obtain that

$$= \sum_{\lambda_1=|l_2-l_1|}^{l_2+l_1} \sum_{\lambda_2=|l_3-\lambda_1|}^{l_3+\lambda_1} \cdots \sum_{l=|l_n-\lambda_{n-2}|}^{l_n+\lambda_{n-2}}$$

$$\left\{ \sum_{\mu_1,\dots,\mu_{n-2}} C^{\lambda_1\mu_1}_{l_1 m_1 l_2 m_2} C^{\lambda_2\mu_2}_{\lambda_1\mu_1 l_3 m_3} \cdots \frac{(-1)^{m_n}}{\sqrt{2l_n+1}} \delta^{l_n}_{\lambda_{n-2}} \delta^{-m_n}_{\mu_{n-2}} \right\} \widetilde{P}_{l_1\dots l_n}(\lambda_1, \dots, \lambda_{n-3}, \lambda_{n-2})$$

$$= (-1)^{m_n} \sum_{\lambda_1=|l_2-l_1|}^{l_2+l_1} \sum_{\lambda_2=|l_3-\lambda_1|}^{l_3+\lambda_1} \cdots \sum_{\lambda_{n-3}=|l_{n-1}-l_n|}^{l_{n-1}+l_n}$$

$$\left\{ \sum_{\mu_1,\dots,\mu_{n-2}} C^{\lambda_1\mu_1}_{l_1 m_1 l_2 m_2} C^{\lambda_2\mu_2}_{\lambda_1\mu_1 l_3 m_3} \cdots C^{l_n-m_n}_{\lambda_{n-3}\mu_{n-3}.l_{n-1} m_{n-1}} \right\} P_{l_1\dots l_n}(\lambda_1, \dots, \lambda_{n-3})$$

$$= (-1)^{m_n} \sum_{\lambda_1=|l_2-l_1|}^{l_2+l_1} \sum_{\lambda_2=|l_3-\lambda_1|}^{l_3+\lambda_1} \cdots \sum_{\lambda_{n-3}=|l_{n-1}-l_n|}^{l_{n-1}+l_n} C^{\lambda_1\dots\lambda_{n-3}l_n,-m_n}_{l_1 m_1\dots l_{n-1} m_{n-1}} P_{l_1\dots l_n}(\lambda_1, \dots, \lambda_{n-3}),$$

where we have set

$$P_{l_1\dots l_n}(\lambda_1, \dots, \lambda_{n-3}) := \frac{1}{\sqrt{2l_n+1}} \widetilde{P}_{l_1\dots l_n}(\lambda_1, \dots, \lambda_{n-3}, l_n) .$$

All there is left to show is that the coefficients of this linear combination are necessarily real. To see this, it is sufficient to specialize the previous discussion to the case where $m_1 = m_2 = \dots = m_n = 0,$, and to observe that, in this case

$$E a_{l_1 0} \cdots a_{l_n 0} = \sum_{\lambda_1} \cdots \sum_{\lambda_{n-3}} C^{\lambda_1\dots\lambda_{n-3}l_n,0}_{l_1 0\dots l_{n-1} 0} P_{l_1\dots l_n}(\lambda_1, \dots, \lambda_{n-3})$$

is real by definition (note indeed that the columns of $C_{l_1\dots l_n}$ are linearly independent). $\qquad\square$

Let us illustrate the previous results by some more examples.

Example 6.33 For $n = 3$, Theorem 6.31 implies that, under isotropy the *bispectrum* ([102]) satisfies

$$B_{l_1 m_1 l_2 m_2 l_3 m_3} := E a_{l_1 m_1} a_{l_2 m_2} a_{l_3 m_3} = (-1)^{m_3} C^{l_3 -m_3}_{l_1 m_1 l_2 m_2} P_{l_1 l_2 l_3} \ .$$

From this last relation, we can recover the so-called *reduced bispectrum*, noted $b_{l_1 l_2 l_3}$, (see for instance [102], [131] and [132]), which satisfies indeed the relationship

$$P_{l_1 l_2 l_3} = b_{l_1 l_2 l_3} C^{l_3 0}_{l_1 0 l_2 0} \sqrt{\frac{(2l_1 + 1)(2l_2 + 1)}{(2l_3 + 1) 4\pi}} \ .$$

For $n = 4$ (i.e. the trispectrum, see again [102]) we obtain the expression

$$E a_{l_1 m_1} a_{l_2 m_2} a_{l_3 m_3} a_{l_4 m_4} = (-1)^{m_4} \sum_{\lambda = |l_2 - l_1|}^{l_2 + l_1} C^{\lambda l_4 - m_4}_{l_1 m_1 l_2 m_2 l_3 m_3} P_{l_1 l_2 l_3 l_4}(\lambda)$$

$$= \sum_{\lambda = |l_2 - l_1|}^{l_2 + l_1} \sum_{\mu = -\lambda}^{\lambda} C^{\lambda \mu}_{l_1 m_1 l_2 m_2} C^{l_4, -m_4}_{\lambda \mu l_3 m_3} P_{l_1 l_2 l_3 l_4}(\lambda) \ .$$

The next result gives a further probabilistic characterization of the reduced bispectrum.

Proposition 6.34 *Fix $n \geq 2$. A real-valued array $\{A_{l_1 \dots l_n}(\cdot) : l_1, \dots, l_n \geq 0\}$ is the reduced polyspectrum of order $n - 1$ (resp. the reduced cumulant polyspectrum of order $n-1$) of some strongly isotropic random field if, and only if, there exists a sequence $\{X_l : l \geq 0\}$ of zero-mean real-valued random variables such that*

$$\sum_{l \geq 0} (2l + 1) E \left[X_l^2 \right] < +\infty$$

and, for every $l_1, \dots, l_n \geq 0$

$$E (X_{l_1} \cdots X_{l_n}) = \sum_{\lambda_1 = l_2 - l_1}^{l_2 + l_1} \cdots \sum_{\lambda_{n-3}} C^{\lambda_1 \dots \lambda_{n-3} l_n, 0}_{l_1 0 \dots l_{n-1} 0} A_{l_1 \dots l_n}(\lambda_1, \dots, \lambda_{n-3}). \qquad (6.49)$$

Proof For the necessity it is enough to take $X_l = a_{l0}$, where a_{l0} is the harmonic coefficient of index $(l, 0)$ associated with a strongly isotropic field with moments of all orders. For the sufficiency, we consider first the (anisotropic) random field

$$Z(x) = \sum_{l \geq 0} X_l Y_{l0}(x).$$

Then, by taking $T(x) = Z(gx)$, where g is sampled randomly with the uniform

Haar measure on $SO(3)$, one obtains a random field with the desired charac-
teristics.   □

6.7 Some examples

In this section we provide explicit computations for the reduced polyspectra
$P_{l_1...l_n}$ ($n \geq 2$), for some models of physical interest. In what follows, we
shall be concerned in particular with polyspectra of *Gaussian subordinated*
isotropic fields, that is, random fields that can be written as a deterministic and
non-linear function of some collection of Gaussian isotropic fields. In general,
this class of random fields allow for a clear-cut mathematical treatment, whilst
covering a great array of empirically relevant circumstances.

6.7.1 A simple physical model

The general Gaussian subordinated model considered in this section has the
form

$$T = \sum_{j=1}^{q} f_j H_j \left(T_G / \sqrt{E\left(T_G^2\right)}\right) = f_1 T_G + f_2(T_G^2/E\left(T_G^2\right) - 1) + ..., \quad (6.50)$$

where f_j is a real constant, $H_j(.)$ denotes the jth Hermite polynomial (as introduced in Definition 4.8), and T_G is a Gaussian, zero-mean isotropic random
field. Recall that our definition of Hermite polynomials is such that $H_1(x) = x$,
$H_2(x) = x^2 - 1$, $H_3(x) = x^3 - 3x$, and so on. In this section, when no further
specification is needed, the spectral decomposition of the underlying Gaussian
field T_G as before is written

$$T_G = \sum_{lm} a_{lm} Y_{lm}.$$

We shall sometimes use the following notation, that will be further exploited
in the next chapter,

$$T = \sum_{lm} \widetilde{a}_{lm} Y_{lm} = \sum_{j=1}^{q} f_j a_{lm;j} Y_{lm}, \quad (6.51)$$

$$a_{lm;j} = \int_{S^2} H_j \left(T_G(x) / \sqrt{E\left(T_G^2\right)}\right) \overline{Y_{lm}}(x)\, dx, \quad (6.52)$$

$$\widetilde{a}_{lm} = \sum_{j=1}^{q} a_{lm;j}. \quad (6.53)$$

For instance, models of Cosmic Microwave Background radiation are currently dominated by assumptions such as the Sachs-Wolfe model with the so-called *Bardeen's potential* (see e.g. [18] or [51]). In its simplest version, the latter can be written down explicitly as

$$T = T_G + f_{NL}(T_G^2 - ET_G^2),\tag{6.54}$$

where f_{NL} is a nonlinearity parameters which depends upon physical constants in the associated *"slow-roll" inflationary model* (see e.g. [18]). Note that (6.54) has can be written in the form (6.50), by setting $f_1 = 1$, $f_2 = f_{NL} \times E\left(T_G^2\right)$ and $f_j = 0$, for $j \geq 3$. The value of the constant $f_{NL} \times E\left(T_G^2\right)$ is expected to be very small, namely of the order $10^{-3}/10^{-4}$ (see [18]). To simplify the discussion, we now assume that $ET_G^2 = 1$. In this case, by using (6.51)–(6.53), we have that

$$\widetilde{a}_{lm} = a_{lm} + f_{NL}a_{lm;2},$$

$$a_{lm}(2) = \int_{S^2} T^2 \overline{Y}_{lm} dx = \int_{S^2} \sum_{\ell_1 \ell_2} \sum_{m_1 m_2} a_{\ell_1 m_1} a_{\ell_2 m_2} Y_{\ell_1 m_1} Y_{\ell_2 m_2} \overline{Y}_{lm} dx$$

$$= \sum_{\ell_1 \ell_2} \sum_{m_1 m_2} a_{\ell_1 m_1} a_{\ell_2 m_2} \sqrt{\frac{(2\ell_1 + 1)(2\ell_2 + 1)}{(2l + 1)4\pi}} C_{\ell_1 0 \ell_2 0}^{l0} C_{\ell_1 m_1 \ell_2 m_2}^{lm}.$$

It follows that

$$\widetilde{C}_l := E|\widetilde{a}_{lm}|^2 = C_l + 2f_{NL}^2 \sum_{l_1 l_2} C_{l_1} C_{l_2} \frac{(2l_1 + 1)(2l_2 + 1)}{4\pi(2l + 1)} \left(C_{l_1 0 l_2 0}^{l0}\right)^2,$$

so that

$$Var(T) = \sum_l \frac{2l + 1}{4\pi} \widetilde{C}_l$$

$$= \sum_l \frac{2l + 1}{4\pi} C_l + 2f_{NL}^2 \sum_{l_1 l_2} C_{l_1} C_{l_2} \frac{(2l_1 + 1)(2l_2 + 1)}{(4\pi)^2} \sum_l \left(C_{l_1 0 l_2 0}^{l0}\right)^2$$

$$= \sum_l \frac{2l + 1}{4\pi} C_l + 2f_{NL}^2 \left\{ \sum_{l_1} C_{l_1} \frac{(2l_1 + 1)}{4\pi} \right\}^2$$

$$= Var(T_G) + f_{NL}^2 Var(H_2(T_G)),$$

as expected, due to the orthogonality properties of Hermite polynomials. For the bispectrum, we obtain therefore

$$E\widetilde{a}_{l_1 m_1} \widetilde{a}_{l_2 m_2} \widetilde{a}_{l_3 m_3}$$

$$= E\left\{ (a_{l_1 m_1} + f_2 a_{l_1 m_1;2})(a_{l_2 m_2} + f_2 a_{l_2 m_2;2})(a_{l_3 m_3} + f_2 a_{l_3 m_3;2}) \right\}$$

$$= f_2 E a_{l_1 m_1; 2} a_{l_2 m_2} a_{l_3 m_3} + f_2 E a_{l_1 m_1} a_{l_2 m_2; 2} a_{l_3 m_3}$$
$$+ f_2 E a_{l_1 m_1} a_{l_2 m_2} a_{l_3 m_3; 2} + f_2^3 E a_{l_1 m_1; 2} a_{l_2 m_2; 2} a_{l_3 m_3; 2}$$
$$= (-1)^{m_3} C_{l_1 m_1 l_2 m_2}^{l_3 - m_3} P_{l_1 l_2 l_3},$$

where

$$P_{l_1 l_2 l_3} = 6 f_2 \sqrt{\frac{(2l_1 + 1)(2l_2 + 1)}{(2l_3 + 1)4\pi}} C_{l_1 0 l_2 0}^{l_3 0} \{C_{l_1} C_{l_2} + C_{l_1} C_{l_3} + C_{l_2} C_{l_3}\} \quad (6.55)$$

$$+ f_2^3 \sum_{\ell_1 \ell_2 \ell_3} C_{\ell_1 0 \ell_2 0}^{l_1 0} C_{\ell_1 0 \ell_3 0}^{l_2 0} C_{\ell_2 0 \ell_3 0}^{l_3 0} \frac{(2\ell_1 + 1)(2\ell_2 + 1)(2\ell_3 + 1)}{\sqrt{(4\pi)^3}} \quad (6.56)$$

$$\times \frac{8(-1)^{l_3}}{\sqrt{2l_3 + 1}} \left\{ \begin{matrix} \ell_1 & \ell_2 & \ell_3 \\ l_3 & l_2 & l_1 \end{matrix} \right\} \{C_{\ell_1} C_{\ell_2} C_{\ell_3}\} .$$

The lack of symmetry with respect to the l_3 term is only apparent and can be easily dispensed with by permuting the multipoles in $C_{l_1 m_1 l_2 m_2}^{l_3 m_3}$ or using expression (3.68). Formula (6.55) is consistent with the cosmological literature, where (6.56) is considered a higher order term and hence neglected (see again [102]).

6.7.2 The connection with higher order moments

We now provide a simple result, connecting the reduced polyspectrum with the higher order moments of the associated spherical random field.

Proposition 6.35 *The following identity holds for every isotropic field with finite moments of order p and with a reduced polyspectrum*

$$\left\{ P_{l_1 \dots l_p} (\cdot) : l_1, \dots, l_p \geq 0 \right\} :$$

for every $x \in S^2$,

$$ET(x)^p \equiv \sum_{l_1, \dots, l_p} \sqrt{\frac{(2l_1 + 1) \cdots (2l_p + 1)}{(4\pi)^p}} \sum_{\lambda_1, \dots, \lambda_{p-3}} P_{l_1 \dots l_p}(\lambda_1, \dots, \lambda_{p-3}) C_{l_1 0 \dots l_p - 2 0}^{\lambda_1 \dots \lambda_{p-3} l_p, 0} .$$

Proof We exploit the trivial fact that

$$T(x) \overset{law}{=} T(\mathbf{e}_3) = \sum_l a_{l0} Y_{l0}(\mathbf{e}_3) = \sum_l a_{l0} \sqrt{\frac{2l + 1}{4\pi}},$$

where $\mathbf{e}_3 = (0, 0, 1)$ is the North Pole and we used the fact that, for $m \neq 0$,

$Y_{lm}(\mathbf{e}_3) = 0$ and $Y_{l0}(0) = \sqrt{\frac{2l+1}{4\pi}}$. Hence,

$$
ET^p = \sum_{l_1 \dots l_p} \sqrt{\frac{(2l_1 + 1) \cdots (2l_p + 1)}{(4\pi)^p}} E\left\{a_{l_10} \dots a_{l_p0}\right\}
$$

$$
= \sum_{l_1 \dots l_p} \sqrt{\frac{(2l_1 + 1) \cdots (2l_p + 1)}{(4\pi)^p}} \sum_{\lambda_1 \dots \lambda_{p-3}} P_{l_1 \dots l_p}(\lambda_1, \dots, \lambda_{p-3}) C_{l_10 \dots l_{p-2}0}^{\lambda_1 \dots \lambda_{p-3} l_p, 0}.
$$

$\square$

Example 6.36 Take $T = H_q(T_G)$, where H_q is the qth Hermite polynomial. Then $ET^p = c_{pq}\left\{ET^2\right\}^{qp/2}$, where the integer c_{pq} denotes the cardinality of the set $\Gamma_{\overline{F}}(p, q)$, that is, c_{pq} counts the number of diagrams without flat edges associated with a table with p rows and q columns (just use the graphical methods discussed Section 4.3, in particular Proposition 4.15). Therefore, we have the identity

$$
\sum_{l_1 \dots l_p} \sqrt{\frac{(2l_1 + 1) \cdots (2l_p + 1)}{(4\pi)^p}} \sum_{\lambda_1 \dots \lambda_{p-3}} P_{l_1 \dots l_p}(\lambda_1, \dots, \lambda_{p-3}) C_{l_10 \dots l_{p-2}0}^{\lambda_1 \dots \lambda_{p-3} l_p, 0}
$$

$$
= c_{pq} \left\{\sum_l \frac{(2l + 1)}{4\pi} C_l\right\}^{pq/2}.
$$

6.7.3 The χ_ν^2 polyspectrum

Previously in (6.56), we have implicitly derived the "χ_1^2 bispectrum", that is, the bispectrum associated with a field of the type $T = H_2(T_G)$, where T_G is Gaussian, centered, isotropic and with unit variance. More precisely, with the notation (6.51)–(6.53), we deduce from (6.56) that

$$
Ea_{l_1m_1;2}a_{l_2m_2;2}a_{l_3m_3;2}
$$

$$
= \sum_{\ell_1\ell_2\ell_3} \sum_{\ell_4\ell_5\ell_6} \sum_{\mu_1 \dots \mu_6} C_{\ell_10\ell_20}^{l_10} C_{\ell_1\mu_1\ell_2\mu_2}^{l_1m_1} C_{\ell_30\ell_40}^{l_20} C_{\ell_3\mu_3\ell_4\mu_4}^{l_2m_2} C_{\ell_50\ell_60}^{l_30} C_{\ell_5\mu_5\ell_6\mu_6}^{l_3m_3} \times
$$

$$
\sqrt{\frac{(2\ell_1 + 1) \times \cdots \times (2\ell_6 + 1)}{(2l_1 + 1)(2l_2 + 1)(2l_3 + 1)(4\pi)^3}} E\left\{a_{\ell_1\mu_1}a_{\ell_2\mu_2}a_{\ell_3\mu_3}a_{\ell_4\mu_4}a_{\ell_5\mu_5}a_{\ell_6\mu_6}\right\}
$$

$$= 8(-1)^{l_3-m_3} \sum_{\ell_1\ell_2\ell_3} C^{l_10}_{\ell_10\ell_20} C^{l_20}_{\ell_10\ell_30} C^{l_30}_{\ell_20\ell_30} \frac{(2\ell_1+1)(2\ell_2+1)(2\ell_3+1)}{\sqrt{(4\pi)^3}} \quad (6.57)$$

$$\times \frac{C^{l_3-m_3}_{l_1m_1l_2m_2}}{\sqrt{2l_3+1}} \begin{Bmatrix} \ell_1 & \ell_2 & \ell_3 \\ l_3 & l_2 & l_1 \end{Bmatrix} \{C_{\ell_1}C_{\ell_2}C_{\ell_3}\}, \quad (6.58)$$

see [195, p. 260 ; p. 454]. We now wish to extend these results to polyspectra of order $p = 4, 5, 6$ for random fields of the type $T = H_2(T_G)$, where (as above) T_G is Gaussian, centered, isotropic and with unit variance . Here we find it more convenient to focus on the cumulant polyspectrum (compare [102]), defined as follows.

Definition 6.37 The **cumulant polyspectrum** of an isotropic spherical random field with random spherical harmonic coefficients $\{a_{lm}\}$ is defined by

$$Cum\left(a_{l_1m_1},...,a_{l_pm_p}\right) =: (-1)^{l_p-m_p} \sum_{\lambda_1...\lambda_{p-3}} C^{\lambda_1...\lambda_{p-3}l_p,-m_p}_{l_1m_1...l_{p-1}m_{p-1}} \times P^C_{l_1...l_p}\left(\lambda_1,...,\lambda_{p-3}\right).$$

The cumulant polyspectrum is easily shown to exist and may be simpler to analyze in some circumstances: for instance, in the Gaussian case it is trivially zero for all $p \geq 3$. For quadratic fields, we have the following result.

Proposition 6.38 *The cumulant polyspectrum associated with the harmonic coefficients of an isotropic random field of the type $H_2(T_G)$ (where T_G is Gaussian and isotropic, with angular power spectrum $\{C_l : l \geq 0\}$) is given by*

$$P^{C;1}_{l_1l_2l_3l_4}(\lambda) = 48 \sqrt{\frac{(2\lambda+1)}{(4\pi)^4(2l_4+1)}} \sum_{\ell_1...\ell_4} C_{\ell_1}...C_{\ell_4} C^{l_10}_{\ell_10\ell_20} C^{l_30}_{\ell_20\ell_30} C^{l_40}_{\ell_30\ell_40} C^{l_20}_{\ell_40\ell_10}$$

$$\times (2\ell_1+1)\cdots(2\ell_4+1)(-1)^{l_1+l_2+\ell_2+\ell_4} \begin{Bmatrix} l_1 & l_2 & \lambda \\ \ell_4 & \ell_2 & \ell_1 \end{Bmatrix} \begin{Bmatrix} \lambda & l_3 & l_4 \\ \ell_3 & \ell_4 & \ell_2 \end{Bmatrix},$$

for $p = 4$,

$$P^{C;1}_{l_1...l_5}(\lambda_1,\lambda_2) = 384 \sqrt{\frac{(2\lambda_1+1)(2\lambda_2+1)}{(4\pi)^5(2l_5+1)}}$$

$$\times \sum_{\ell_1,...,\ell_5} C_{\ell_1}\cdots C_{\ell_5} C^{l_10}_{\ell_10\ell_20} C^{l_20}_{\ell_20\ell_30} C^{l_40}_{\ell_30\ell_40} C^{l_50}_{\ell_40\ell_50} C^{l_30}_{\ell_50\ell_10}$$

$$\times (2\ell_1+1)\cdots(2\ell_5+1)(-1)^{\ell_1+\ell_5+l_3} \begin{Bmatrix} l_1 & l_2 & \lambda_1 \\ l_3 & \ell_1 & \ell_2 \end{Bmatrix}$$

$$\times \begin{Bmatrix} \lambda_1 & l_3 & \lambda_2 \\ \ell_5 & \ell_3 & \ell_1 \end{Bmatrix} \begin{Bmatrix} \lambda_2 & l_4 & l_5 \\ \ell_4 & \ell_5 & \ell_3 \end{Bmatrix},$$

for p = 5, and

$$P^{C;1}_{l_1...l_6}(\lambda_1,\lambda_2,\lambda_3) = 3840\sqrt{\frac{(2\lambda_1+1)(2\lambda_2+1)(2\lambda_3+1)}{(4\pi)^6(2l_5+1)}}$$

$$\times \sum_{\ell_1,...,\ell_5} C_{\ell_1}...C_{\ell_6}C^{l_10}_{\ell_10\ell_20}C^{l_20}_{\ell_20\ell_30}C^{l_30}_{\ell_30\ell_40}C^{l_50}_{\ell_40\ell_50}C^{l_60}_{\ell_50\ell_60}C^{l_40}_{\ell_60\ell_10}$$

$$\times(2\ell_1+1)\cdots(2\ell_6+1)(-1)^{\lambda_1+\ell_3+\ell_6+l_4}\begin{Bmatrix} l_1 & l_2 & \lambda_1 \\ \ell_3 & \ell_1 & \ell_2 \end{Bmatrix}$$

$$\times\begin{Bmatrix} \lambda_1 & l_5 & \lambda_2 \\ \ell_5 & \ell_3 & \ell_1 \end{Bmatrix}\begin{Bmatrix} \lambda_2 & l_3 & l_4 \\ \ell_4 & \ell_5 & \ell_3 \end{Bmatrix},$$

for p = 6.

Proof The result can be proved by means of slightly more advanced graphical techniques (with respect to the ones presented in Section 4.5) for convolutions of Clebsch-Gordan coefficients, as described in [195, Chapters 11 and 12]. Here, we only provide the complete proof for the case $p = 6$. Let $\{a_{\ell m}\}$ be the random harmonic coefficients associated with the underlying Gaussian field T_G. By definition, the field $H_2(T_G)$ admits the expansion

$$H_2(T_G) = \sum_{l\geq 0}\sum_{m=-l}^{l} a_{lm;2}Y_{lm},$$

where

$$a_{lm;2} = \sum_{\ell_1 m_1 \ell_2 m_2} a_{\ell_1 m_1}a_{\ell_2 m_2}\int_{S^2} Y_{\ell_1 m_1}(x)Y_{\ell_2 m_2}(x)\overline{Y_{lm}(x)}dx$$

$$= \sum_{\ell_1 m_1 \ell_2 m_2} a_{\ell_1 m_1}a_{\ell_2 m_2}\begin{pmatrix} \ell_1 & \ell_2 & l \\ m_1 & m_2 & -m \end{pmatrix}\times(-1)^m$$

$$\times\begin{pmatrix} \ell_1 & \ell_2 & l \\ 0 & 0 & 0 \end{pmatrix}\sqrt{\frac{(2\ell_1+1)(2\ell_2+1)(2l+1)}{4\pi}}$$

$$= \sum_{\ell_1 m_1 \ell_2 m_2} a_{\ell_1 m_1}a_{\ell_2 m_2}C^{lm}_{\ell_1 m_1 \ell_2 m_2}C^{lm}_{\ell_10\ell_20}\sqrt{\frac{(2\ell_1+1)(2\ell_2+1)}{4\pi(2l+1)}}.$$

By using once again the multilinearity of cumulants, one obtains that

$$Cum\{a_{l_1 m_1;2},...,a_{l_6 m_6;2}\}$$
$$= \sum_{\ell_{11}m_{11}\ell_{12}m_{12}}\cdots\sum_{\ell_{61}m_{61}\ell_{61}m_{61}} Cum\{a_{\ell_{11}m_{11}}a_{\ell_{12}m_{12}},...,a_{\ell_{61}m_{61}}a_{\ell_{62}m_{62}}\}$$

$$\times \prod_{j=1}^{6} \left\{ C^{l_j m_j}_{\ell_{j1} m_{j1} \ell_{j2} m_{j2}} C^{l_j m_j}_{\ell_{j1} 0 \ell_{j2} 0} \sqrt{\frac{(2\ell_{j1} + 1)(2\ell_{j2} + 1)}{4\pi(2l_j + 1)}} \right\}.$$

For a given $\mathbf{lm} = (\ell_{11} m_{11}, \ell_{12} m_{12}; ...; \ell_{61} m_{61}, \ell_{62} m_{62})$, the quantity

$$Cum\{a_{\ell_{11} m_{11}} a_{\ell_{12} m_{12}}, ..., a_{\ell_{61} m_{61}} a_{\ell_{62} m_{62}}\}$$

is computed as follows:

- Build the 6×2 table

$$\Lambda\,(\mathbf{lm}) = \begin{bmatrix} \ell_{11} m_{11} & \ell_{12} m_{12} \\ \ell_{21} m_{21} & \ell_{22} m_{22} \\ \ell_{31} m_{31} & \ell_{32} m_{32} \\ \ell_{41} m_{41} & \ell_{42} m_{42} \\ \ell_{51} m_{51} & \ell_{52} m_{52} \\ \ell_{61} m_{61} & \ell_{62} m_{62} \end{bmatrix}.$$

- Define the class $M = \Gamma_{\overline{F}}(\Lambda\,(\mathbf{lm})) \cap \Gamma_C(\Lambda\,(\mathbf{lm}))$ of connected, Gaussian non-flat diagrams over $\Lambda\,(\mathbf{lm})$ (see Section 4.3.1).
- For every $\gamma \in M$, write

$$\delta(\gamma) = \prod_{\{\ell_{ab} m_{ab}, \ell_{cd} m_{cd}\} \in \gamma} \delta^{\ell_{ab}}_{\ell_{cd}} \delta^{-m_{cd}}_{m_{ab}} (-1)^{m_{ab}} C_{\ell_{ab}}$$

(where δ^b_a is the usual Kronecker symbol)
- Use the diagram formulae of Section 4.3.1 to obtain that

$$Cum\{a_{\ell_{11} m_{11}} a_{\ell_{12} m_{12}}, ..., a_{\ell_{61} m_{61}} a_{\ell_{62} m_{62}}\} = \sum_{\gamma \in M} \delta(\gamma).$$

It follows that

$$Cum\{a_{l_1 m_1;2}, ..., a_{l_6 m_6;2}\}$$

$$= \sum_{\mathbf{lm}} \sum_{\gamma \in M} \delta(\gamma) \prod_{j=1}^{6} \left\{ C^{l_j m_j}_{\ell_{j1} m_{j1} \ell_{j2} m_{j2}} C^{l_j m_j}_{\ell_{j1} 0 \ell_{j2} 0} \sqrt{\frac{(2\ell_{j1} + 1)(2\ell_{j2} + 1)}{4\pi(2l_j + 1)}} \right\},$$

where the first sum runs over all vectors of the type

$$\mathbf{lm} = (\ell_{11} m_{11}, \ell_{12} m_{12}; ...; \ell_{61} m_{61}, \ell_{62} m_{62}) .$$

The proof now follows directly from the announced (advanced) graphical techniques. In particular, the previous term can be associated with an hexagon, having in each vertex an outward line corresponding to a "free" (i.e. not summed up) index $l_i m_i$, $i = 1, ..., 6$. An expression for convolutions of Clebsch-Gordan coefficients corresponding to such a configuration can be found in [195, p. 461,

Eq. 12.1.6.30]. From this, standard combinatorial arguments and a convenient relabeling of the indexes, we obtain the result for $p = 6$. Note that $3840 = 2^5 5!$ is the number of automorphisms between graphs belonging to the class M defined above. □

We recall that the Clebsch-Gordan coefficients $\left\{C^{c0}_{a0b0}\right\}$ are identically zero unless $a + b + c$ is even; it is hence easy to see that the previous polyspectra are non-zero only if the sum $\left\{l_1 + \cdots + l_p\right\}$ is even as well.

From the previous Proposition, we can derive the corresponding expressions for the cumulant polyspectra for χ^2_ν random field.

Definition 6.39   We say the random field $T_{\chi^2_\nu}$ has a chi-square law with $\nu \geq 1$ degrees of freedom if there exist ν independent and identically distributed Gaussian random fields T_i such that

$$T_{\chi^2_\nu} \overset{law}{=} T^2_1 + \cdots + T^2_\nu .$$

It is trivial to show that $T_{\chi^2_\nu}$ is mean-square continuous and isotropic if T_i is. We have the following

Proposition 6.40   *The cumulant polyspectra of $T_{\chi^2_\nu}$ (for $p \geq 2$) are given by*

$$P^{C;\nu}_{l_1...l_p}(\lambda_1, ..., \lambda_{p-3}) = \nu P^{C;1}_{l_1...l_p}(\lambda_1, ..., \lambda_{p-3}).$$

Proof   Note that the cumulant polyspectra of order $p \geq 2$ of $T_{\chi^2_\nu}$ coincide with those of the centered field $T_{\chi^2_\nu} - ET_{\chi^2_\nu}$ (due to the translation-invariance properties of cumulants). Then, the proof is an immediate consequence of Proposition 6.38 and the of the standard multilinearity properties of cumulants. □

Remark 6.41   The previous results can be exploited to develop a probabilistic algorithm to compress information on Clebsch-Gordan coefficients (see [136]). A simple computation on the cardinality of these coefficients shows that

$$\#\left\{C^{l_3m_3}_{l_1m_1l_2m_2} : l_1, l_2, l_3 \leq L, \left|C^{l_3m_3}_{l_1m_1l_2m_2}\right| \neq 0\right\} \approx O(L^6) .$$

It is therefore clear how for most applications the storage of Clebsch-Gordan coefficients for future usage is simply unfeasible, whatever the supercomputing facilities (for instance, for CMB data analysis, $L \approx 3\times10^3$ is currently required, so that the number of Clebsch-Gordan coefficients to be saved would exceed 10^{20}). Let us consider again a chi-square field as defined before, i.e.

$$T_{\chi^2}(x) = H_2(T_G(x)) = \sum_{lm} a_{lm;2} Y_{lm}(x) ;$$

we have proved earlier in (6.57) that

$$E a_{l_1 m_1;2} a_{l_2 m_2;2} a_{l_3 m_3;2} = (-1)^{m_3} C^{l_3 m_3}_{l_1 m_1 l_2 m_2} h_{l_1 l_2 l_3}$$

where

$$h_{l_1 l_2 l_3} := 8 \sum_{\ell_1 \ell_2 \ell_3} C^{l_1 0}_{\ell_1 0 \ell_2 0} C^{l_2 0}_{\ell_1 0 \ell_3 0} C^{l_3 0}_{\ell_2 0 \ell_3 0} \frac{(2\ell_1 + 1)(2\ell_2 + 1)(2\ell_3 + 1)}{\sqrt{(4\pi)^3}}$$

$$\times \frac{1}{\sqrt{2l_3 + 1}} \left\{ \begin{array}{ccc} \ell_1 & \ell_2 & \ell_3 \\ l_1 & l_2 & l_3 \end{array} \right\} \{ C_{\ell_1} C_{\ell_2} C_{\ell_3} \} ,$$

which can be calculated analytically and stored, with cardinality

$$\# \left\{ h_{l_1 l_2 l_3} : l_1, l_2, l_3 \le L, \left| C^{l_3 0}_{l_1 0 l_2 0} \right| \neq 0 \right\} \approx O(L^3) .$$

Let us assume we simulate B times $T_{\chi^2}(x)$, a task which can be trivially accomplished by simply squaring a Gaussian field. We store the triangular arrays $\{a^i_{lm}\}_{l=1,\dots,L;m=-l,\dots,l}$, $i = 1, \dots, B$; here the dimension is of order $B \times L^2$. We can then recover any value $C^{l_3 m_3}_{l_1 m_1 l_2 m_2}$ by means of the Monte Carlo estimate

$$\widehat{C}^{l_3 m_3}_{l_1 m_1 l_2 m_2} = h^{-1}_{l_1 l_2 l_3} \sum_{i=1}^{B} \frac{a^{(i)}_{l_1 m_1} a^{(i)}_{l_2 m_2} a^{(i)}_{l_3 m_3}}{B} ,$$

which requires B steps and $B \times L^2 + L^3$ storage capacity, as opposed to L^6 storage capacity by the direct method. We leave for further research a more thorough investigation on the convergence properties of this algorithm; we stress, however, that the procedure we advocate is completely general, i.e. it does not depend on peculiar features of the group $SO(3)$ we are currently considering. We believe, hence, that similar ideas can be implemented for the numerical estimation of Clebsch-Gordan coefficients for other compact groups of interest for theoretical physicists.

7

Limit Theorems for Gaussian Subordinated Random Fields

7.1 Introduction

In the previous Chapter 6, we provided several characterizations of random spherical harmonic coefficients, associated with Gaussian and non-Gaussian isotropic fields. We will now use limit theorems in order to investigate some relations between Gaussianity and non-Gaussianity for spherical fields.

The purpose of this chapter is indeed to discuss a class of central limit results, involving high-frequency (or high-resolution) asymptotics. As already discussed in the Introduction, this type of limit theorems represents the mathematical counterpart of a situation which is often encountered when dealing with cosmological data, namely: one is faced with a single observation of some infinite-dimensional object (e.g. the spherical CMB radiation), that is detected at higher and higher degrees of accuracy. In the particular case of CMB data, the increasing accuracy corresponds to higher and higher observed frequencies, in such a way that usual asymptotic statistical procedures (such as e.g. asymptotic tests of Gaussianity) can only be implemented in the high-frequency sense.

In short, our aim here can be summarized as follows. Consider a Gaussian subordinated field $F(T(x))$, and write again the spectral representation

$$F(T(x)) = \sum_{lm} a_{lm} Y_{lm}(x) = \sum_{l} F_l(T(x)) \, ,$$

$$a_{lm} = \int_{S^2} F_l(T(x)) \, \overline{Y}_{lm}(x) d\sigma(x) \, , \, F_l(T(x)) = \sum_{m} a_{lm} Y_{lm}(x) \, .$$

It is obvious that, for Gaussian random fields (for instance, when $F(T(x)) = T(x)$), the Fourier components $F_l(T(x))$ are themselves Gaussian (by linearity). Our question is: *are there circumstances such that the (normalized) components $F_l(T(x))$ are asymptotically Gaussian, even if $F(T(x))$ is not?*

We stress the relevance of this investigation for the statistical analysis of the field $F(T)$. For instance, if asymptotic Gaussianity holds, then testing for non-Gaussianity at high frequencies may become meaningless; on the other hand, such a central limit result may justify the use of Gaussian likelihoods at high frequencies.

Formally, in what follows we shall look for sufficient (and sometimes, also necessary) conditions on F and on the law of T to have that the following two phenomena take place: (**I**) as $l \to +\infty$, for a fixed m and for an appropriate sequence $\tau_1(l)$ ($l \geq |m|$), the sequence

$$\tau_1(l) \times a_{lm} = \tau_1(l) \int_{S^2} F(T(z)) \overline{Y_{lm}(z)} dz, \quad l \geq |m|$$

converges in law to a Gaussian random variable (real-valued for $m = 0$, and complex-valued for $m \neq 0$); (**II**) for a suitable real-valued sequence $\tau_2(l)$ ($l \geq 0$) and for l sufficiently large, the finite-dimensional distributions of the field

$$\tau_2(l) \times \widetilde{T}_l(\cdot) = \tau_2(l) \sum_{m=-l}^{l} a_{lm} Y_{lm}(\cdot),$$

are close (for instance, in the sense of twice differentiable test functions – see Theorem 4.21) to those of a real spherical Gaussian field.

In order to keep the presentation as simple as possible, and to better illustrate our methods, we shall mainly focus on the case where F is equal to a Hermite polynomial of degree 2 or 3. This Chapter is largely based on our paper [135], to which the reader is referred for more general statements.

In the following proofs, one of our main tools is the "simplified method of moments", as described in Section 4.4. These techniques, combined with the use of group representation theory, lead to one of the main themes of this chapter: the derivation of sufficient (or necessary and sufficient) conditions for (**I**) and (**II**), expressed in terms of convolutions of Clebsch-Gordan coefficients. As discussed earlier, these coefficients are widely used in quantum mechanics, and admit a well-known interpretation in terms of probability amplitudes related to the coupling of angular momenta in a quantum mechanical system (see Liboff [127] , Varshalovich, Moskalev and Khersonskii [195] or Section 7.6 below). It follows that most of our results can be given an alternative interpretation in terms of random couplings of quantum particles (see below).

Furthermore, we will also show that many of our conditions can be alternatively restated in terms of 'bridges' of random walks on $\widehat{SO(3)}$ (the dual of $SO(3)$). The definition of such random walks differs from the one given e.g. in

[89], although we will show that the two approaches can be related through the notion of mixed quantum state (see Section 7.6).

An analogous connection with random walks on $\mathbb{Z}^d$ and stationary random fields on a d-dimensional torus was pointed out in [134]: the case $d = 1$ (the circle) is the object of the next (introductory) section.

7.2 First example: the circle

Consider a centered real-valued Gaussian field $V = \{V(\theta) : \theta \in \mathbb{T}\}$ defined on the torus $\mathbb{T} = [0, 2\pi)$ (that we regard as an Abelian compact group with group operation given by $xy = (x + y) \mod(2\pi)$). We suppose that the law of V is isotropic, i.e. that $V(\theta) \overset{law}{=} V(x\theta)$ (in the sense of stochastic processes) for every $x \in \mathbb{T}$, and also $EV(\theta)^2 = 1$.

As explained in Section 5.2, V can be decomposed (in the mean-square sense) as $V(\theta) = \sum_{l \in \mathbb{Z}} a_l e^{il\theta}$. We write $\Gamma_l^V = \mathbb{E}|a_l|^2$ (note that $\Gamma_l^V = \Gamma_{-l}^V$ and $1 = EV(\theta)^2 = \sum_l \Gamma_l^V$). Fix $q \geq 2$, and consider the Hermite-subordinated field $H_q[V](\theta) = H_q(V(\theta))$, where q is the qth Hermite polynomial defined in formula (4.14) of Chapter 4. The field $H_q[V]$ is of course isotropic, and also admits a decomposition of the type $H_q[V](\theta) = \sum_{l \in \mathbb{Z}} a_l^{(q)} e^{il\theta}$.

Write N, N' to indicate a pair of independent centered Gaussian random variables with common variance equal to $1/2$. In [134], the following natural question has been studied: *for a fixed $q \geq 2$, which conditions must be satisfied by the sequence $\{\Gamma_l^V : l \in \mathbb{Z}\}$ in order to have that the high-frequency CLT*

$$\frac{a_l^{(q)}}{Var\left(a_l^{(q)}\right)^{1/2}} = \frac{\int_{\mathbb{T}} H_q[V](\theta) e^{-il\theta} d\theta}{Var\left(a_l^{(q)}\right)^{1/2}} \overset{law}{\underset{l \to \infty}{\longrightarrow}} N + iN'. \tag{7.1}$$

holds? Note that (7.1) is genuinely a high-frequency result, since the limit is taken by letting the frequency index l diverge to infinity.

One of the main findings of [134] is that necessary and sufficient conditions for (7.1) to take place can be neatly expressed in terms of convolutions of the coefficients Γ_l^V. From a probabilistic standpoint, a very convenient way of expressing this result involves the use of random walks on $\mathbb{Z}$. Indeed, we have that (7.1) holds if and only if , for every $p = 1, ..., q - 1$,

$$\limsup_{\substack{l \to \infty \\ j \in \mathbb{Z}}} P\left[U_p = j \mid U_q = l\right] = 0, \tag{7.2}$$

where $\{U_n : n \geq 0\}$ is the random walk on $\mathbb{Z}$ whose law is given by $U_0 = 0$ and

$$P\left[U_{n+1} = j \mid U_n = k\right] = \Gamma_{j-k}^V. \tag{7.3}$$

Remark 7.1   Condition (7.2) can be interpreted as follows. For every l, define a "bridge" of length q, from 0 to l, by conditioning U to equal l at time q. Then, (7.2) is verified if and only if, the probability that the bridge hits j at time p converges to zero, uniformly on j, as l diverges to infinity. Plainly, when (7.2) is verified for every one also has that

$$\sup_{j_1,\ldots,j_{q-1}\in\mathbb{Z}} P\left[U_1 = j_1, \ldots, U_{q-1} = j_{q-1} \mid U_q = l\right] \to 0,$$

when $l \to \infty$, meaning that, asymptotically, there is no "privileged path" of length q linking 0 and l.

Fig. 7.1 provides an illustration of three such paths for $q = 4$ and $l = 5$.

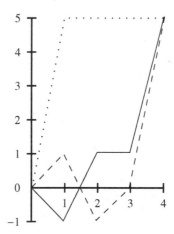

Figure 7.1   Three paths of length 4, linking 0 and 5.

The correspondence between (7.1) and the "random walk bridge" (7.2) has been used in [134] to establish explicit conditions on the power spectrum $\{\Gamma_l^V\}$ to have that (7.1) holds. More precisely,

Proposition 7.2 (Exponential/algebraic duality on the circle.)   *Fix $q \geq 2$.*

(1) The CLT (7.1) holds if Γ_l^V is asymptotically equivalent to a sequence of the type $l^p \exp(-Bl)$, where $B > 0$ and $p \in \mathbb{R}$.
(2) The CLT (7.1) does not hold if Γ_l^V is asymptotically equivalent to a sequence of the type $l^{-\alpha}$ where $\alpha > 1$.

A striking phenomenon described in the following sections is that an analogous duality can be deduced for subordinated random spherical fields. As anticipated, the underlying combinatorial structure is that of random walks on the dual of $SO(3)$.

7.3 Preliminaries on Gaussian-subordinated fields

As everywhere in this book, we denote by S^2 the unit sphere and by $d\sigma(x) = \sin\vartheta d\vartheta d\phi$ the non-normalized Lebesgue measure on S^2, where $x = (\varphi, \vartheta)$ in spherical coordinates. From now on, we shall denote by $T = \{T(x) : x \in S^2\}$ a centered, real-valued and Gaussian random field on S^2. We also suppose that T is strongly isotropic, that is, for every $g \in SO(3)$ we have that $T(x) \overset{law}{=} T(gx)$, where the equality holds in the sense of finite-dimensional distributions (see Chapter 6). We assume that the process is normalized in such a way that $ET(x)^2 = 1$. According to Theorem 5.13, we know that under isotropy T admits the spectral decomposition

$$T(x) = \sum_{l=0}^{\infty} \sum_{m=-l}^{l} a_{lm;1} Y_{lm}(x) = \sum_{l=0}^{\infty} T_l(x), \quad x \in S^2, \qquad (7.4)$$

where $a_{lm;1} := \int_{S^2} T(x) \overline{Y_{lm}(x)} dx$ (the role of the subscript "$lm; 1$", that we already encountered in the previous chapter, will be clarified in the following discussion), $T_l(x) := \sum_{m=-l}^{l} a_{lm;1} Y_{lm}(x)$, and the convergence takes place in $L^2(P)$ for every fixed x, as well as in $L^2(P \otimes d\sigma(x))$.

Assumption. The angular power spectrum $\{C_l : l \geq 0\}$, as defined in (6.12) is such that $C_l > 0$ for every l. Note that the results of this chapter could be extended without difficulties (but at the cost of a heavier notation) to the case of a power spectrum such that $C_l \neq 0$ for infinitely many l's.

In the subsequent sections, we shall obtain high-frequency CLTs for centered isotropic spherical fields that are subordinated to the Gaussian field T defined above.

Definition 7.3 (Subordinated fields)   Let $L_0^2(\mathbb{R}, e^{-z^2/2}dz)$ indicate the class of real-valued functions $F(z)$ on $\mathbb{R}$, which are square-integrable with respect to the measure $e^{-z^2/2}dz$ and such that $\int F(z) e^{-z^2/2}dz = 0$. A (centered) random field $\widetilde{T} = \{\widetilde{T}(x) : x \in S^2\}$ is said to be **subordinated** to the Gaussian field T appearing in (6.1) if there exists $F \in L_0^2(\mathbb{R}, e^{-z^2/2}dz)$ such that $\widetilde{T}(x) = F[T](x)$, $\forall x \in S^2$, where the symbol $F[T](x)$ stands for $F(T(x))$. Whenever $\widetilde{T}$ is subordinated, we will rather use the notation $F[T](x)$ instead of $\widetilde{T}(x)$, in order to emphasize the role of the function F. Of course, if $F(z) = z$, then $F[T](x) = \widetilde{T}(x) = T(x)$.

It is immediate to check that, since T is isotropic, a subordinated field $F[T](\cdot)$ as in Definition 7.3 is necessarily isotropic. As a consequence, following again Chapter 5 we deduce that $F[T]$ admits the spectral representation

$$F[T](x) = \sum_{l=0}^{\infty} \sum_{m=-l}^{l} a_{lm}(F) Y_{lm}(x) = \sum_{l=0}^{\infty} F[T]_l(x), \quad x \in S^2, \qquad (7.5)$$

with convergence in $L^2(P)$ (for fixed x) and in $L^2\left(\Omega \times S^2, dP \otimes d\sigma(x)\right)$. Here,

$$a_{lm}(F) := \int_{S^2} F[T](y) \overline{Y_{lm}(y)} d\sigma(y), \text{ and} \qquad (7.6)$$

$$F[T]_l(x) := \sum_{m=-l}^{l} a_{lm}(F) Y_{lm}(x). \qquad (7.7)$$

The complex-valued array $\{a_{lm}(F) : l \geq 0, \ m = -l, ..., l\}$ always enjoys the following properties (**a**)-(**c**), that are basically a rewriting of Proposition 6.6 and Proposition 6.8: (**a**) for every $l \geq 0$, the random variable $a_{l0}(F)$ is real-valued and centered; (**b**) for every $l \geq 1$, and every $m = 1, ..., l$, the random variable $a_{lm}(F)$ is complex-valued, centered and such that

$$a_{lm}(F) = (-1)^m \overline{a_{l-m}(F)} \ ; \ E(\mathrm{Re}\,(a_{lm}(F)) \, \mathrm{Im}\,(a_{lm}(F))) = 0$$
$$E(\mathrm{Re}\,(a_{lm}(F))^2) \ = E(\mathrm{Im}\,(a_{lm}(F))^2) = E(a_{l0}(F)^2)/2 = C_l(F)/2,$$

where the finite constant $C_l(F) \geq 0$ depends solely on F and l; (**c**) $E(a_{lm}(F) \times \overline{a_{l'm'}(F)}) = 0, \ \forall\,(l, m) \neq (l', m')$. We stress that, in general, it is no longer true that $\mathrm{Re}\,(a_{lm}(F))$ and $\mathrm{Im}\,(a_{lm}(F))$ are independent random variables.

Let H_q be the qth Hermite polynomial – see Definition 4.8. When a subordinated field has the form (for $q \geq 2$) $H_q[T](x), x \in S^2$ (that is, when $F = H_q$ in Definition 7.3), we will use the shorthand notation:

$$T^{(q)}(x) := H_q[T](x), \quad x \in S^2, \qquad (7.8)$$

$$a_{lm;q} := a_{lm}\left(H_q\right), \qquad (7.9)$$

$$T_l^{(q)}(x) := H_q[T]_l(x), \quad l \geq 1, x \in S^2, \qquad (7.10)$$

$$\overline{T}_l^{(q)}(x) := Var\left(T_l^{(q)}(x)\right)^{-1/2} T_l^{(q)}(x), \quad l \geq 1, x \in S^2, \qquad (7.11)$$

$$\widetilde{C}_l^{(q)} := C_l\left(H_q\right) = E|a_{lm;q}|^2, l \geq 1, m = -l, ..., l. \qquad (7.12)$$

We conclude the section with an easy Lemma.

Lemma 7.4 *Let $F[T](x)$, $x \in S^2$, be an (isotropic) subordinated field as in Definition A. Then, for every $l \geq 1$ we have the following:*

(1) The random field $x \mapsto F[T]_l(x)$ defined in (7.7) is real-valued and isotropic;

(2) For every fixed $x \in S^2$, $F[T]_l(x) \overset{law}{=} \sqrt{\frac{2l+1}{4\pi}} a_{l0}(F)$, where the coefficient $a_{l0}(F)$ is defined according to (7.6), and consequently $E(F[T]_l(x)^2) = \frac{2l+1}{4\pi} C_l(F)$;

(3) The normalized random field

$$\overline{F[T]}_l(x) = \left[\frac{(2l+1) C_l(F)}{4\pi}\right]^{-1/2} F[T]_l(x) \tag{7.13}$$

has a covariance structure given by: for every $x, y \in S^2$,

$$E\left(\overline{F[T]}_l(x) \times \overline{F[T]}_l(y)\right) = P_l(\langle x, y \rangle), \tag{7.14}$$

where $P_l(\cdot)$ is the lth Legendre polynomial defined in the Appendix and, as before, $\langle x, y \rangle$ is the Euclidean inner product between x and y.

Proof The fact that $F[T]_l$ is isotropic is a consequence of Proposition 5.16. Now set as before $\mathbf{e}_3 = (0,0)$ (in spherical coordinates). By exploiting Point (1) and the definition of the Y_{lm}'s, we now deduce that

$$F[T]_l(x) \overset{law}{=} F[T]_l(\mathbf{e}_3) = \sum_{m=-l}^{l} a_{lm}(F) Y_{lm}(\mathbf{e}_3) = \sqrt{\frac{2l+1}{4\pi}} a_{l0}(F),$$

giving Point (2) in the statement. Finally, to prove relation (7.14) we use (6.21) to deduce that, for every $x, y \in S^2$,

$$E(F[T]_l(x) F[T]_l(y)) = C_l(F) \frac{2l+1}{4\pi} P_l(\langle x, y \rangle),$$

thus giving the desired conclusion (recall that $P_l(1) = 1$). $\qquad\square$

For instance, a consequence of Lemma 7.4 is that, for every $q \geq 2$,

$$E(T_l^{(q)}(x)^2) = (2l+1) \widetilde{C}_l^{(q)}/4\pi \tag{7.15}$$

where we used the notation introduced at (7.8)-(7.12), so that

$$\overline{T}_l^{(q)}(x) = [(2l+1) \widetilde{C}_l^{(q)}/4\pi]^{-1/2} T_l^{(q)}(x) .$$

The main aim of the subsequent sections is to provide a solution to the following problems.

(P-I) For a fixed $q \geq 2$, find conditions on the power spectrum $\{C_l : l \geq 0\}$ of T, to have that the subordinated process $T^{(q)} = \{T^{(q)}(x) : x \in S^2\}$ defined in (7.8) is such that, for every $x \in S^2$,

$$\sqrt{(2l+1)\,\widetilde{C}_l^{(q)}/4\pi} \times T_l^{(q)}(x) \xrightarrow{law} N, \tag{7.16}$$

as $l \to +\infty$, where $N = N(0,1)$ is a centered standard Gaussian random variable.

(P-II) Under the conditions found at **(P-I)**, study the asymptotic behaviour, as $l \to +\infty$, of the vector

$$\sqrt{(2l+1)\,\widetilde{C}_l^{(q)}/4\pi} \times \left(T_l^{(q)}(x_1), ..., T_l^{(q)}(x_k)\right), \tag{7.17}$$

for every $x_1, ..., x_k \in S^2$.

7.4 High-frequency CLTs

The aim of this section is to obtain conditions for high-frequency CLTs in terms of the Gaunt integrals defined in Section 3.5.2 (see also Chapter 6).

7.4.1 Hermite subordination

We focus on the spherical field $T^{(q)}$ ($q \geq 2$) defined in (7.8), which is obtained by composing the Gaussian field T in (6.1) with the qth Hermite polynomial H_q. Our first purpose is to characterize the asymptotic Gaussianity (when $l \to \infty$) of the spherical harmonic coefficients $\{a_{lm;q}\}$ defined in (7.9).

As anticipated, in the subsequent statements, a crucial role will be played by the Gaunt integrals $\mathcal{G}\{l_1, m_1; ...; l_r, m_r\}$. We recall that these symbols are defined as the integral of the product of several spherical harmonics, namely:

$$\mathcal{G}\{l_1, m_1; ...; l_r, m_r\} := \int_{S^2} Y_{l_1, m_1}(x) \cdots Y_{l_r, m_r}(x)\, d\sigma(x). \tag{7.18}$$

Some special explicit values of Gaunt integrals have been given in Section 3.5.2, whereas a general expression appears in formula (6.46).

Theorem 7.5 *Fix $q \geq 2$.*

(1) For every $l \geq 1$, the positive constant $\widetilde{C}_l^{(q)}$ in (7.12) (which does not depend on m) equals the quantity

$$q! \sum_{l_1, m_1} \cdots \sum_{l_q, m_q} C_{l_1} C_{l_2} \cdots C_{l_q} \left| \mathcal{G}\{l_1, m_1; ...; l_q, m_q; l, -m\} \right|^2 \tag{7.19}$$

$$= q! \sum_{l_1,\ldots,l_q=0}^{\infty} C_{l_1} \cdots C_{l_q} \frac{4\pi}{2l+1} \left\{ \prod_{i=1}^{q} \frac{2l_i+1}{4\pi} \right\} \sum_{L_1 \ldots L_{q-2}} \left\{ C_{l_1 0 \ldots l_q 0}^{L_1 L_2 \ldots L_{q-2} l,0} \right\}^2 \quad (7.20)$$

for every $m = -l,\ldots,l$, where the (generalized) Gaunt integral $\mathcal{G}\{\cdot\}$ is defined via (7.18), and we used the notation for convolutions of Clebsch-Gordan coefficients introduced in (6.40).

(2) Fix $m \neq 0$. As $l \to +\infty$, the following two conditions (A) and (B) are equivalent: (A)

$$(\widetilde{C}_l^{(q)})^{-1/2} \times a_{lm;q} \overset{law}{\to} N + iN', \quad (7.21)$$

where $N, N' \sim N(0, 1/2)$ are independent; (B) for every $p = \frac{q-1}{2}+1,\ldots,q-1$, if $q-1$ is even, and every $p = q/2,\ldots,q-1$ if $q-1$ is odd

$$(\widetilde{C}_l^{(q)})^{-2} \sum_{n_1,j_1} \cdots \sum_{n_{2(q-p)},j_{2(q-p)}} C_{j_1} \cdots C_{j_{2(q-p)}} \left| \sum_{l_1,m_1} \cdots \sum_{l_p,m_p} C_{l_1} \cdots C_{l_p} \right.$$

$$\times \mathcal{G}\{l_1,m_1;\ldots;l_p,m_p;j_1,n_1;\ldots;j_{q-p},n_{q-p};l,-m\} \times \quad (7.22)$$

$$\left. \times \mathcal{G}\{l_1,m_1;\ldots;l_p,m_p;j_{q-p+1},n_{q-p+1};\ldots;j_{2(q-p)},n_{2(q-p)};l,-m\} \right|^2 \to 0$$

(3) Let N be a centered Gaussian random variable with unitary variance. As $l \to \infty$, the CLT

$$(\widetilde{C}_l^{(q)})^{-1/2} \times a_{l0;q} \overset{law}{\to} N \quad (7.23)$$

takes place if and only if the asymptotic condition (7.22) holds for $m = 0$ and for every $p = \frac{q-1}{2}+1,\ldots,q-1$, if $q-1$ is even, and every $p = q/2,\ldots,q-1$ if $q-1$ is odd.

Proof The proof is based on the simplified method of moments discussed in Section 4.4. Our first task is therefore to build a Gaussian field, having the same law as T, in terms of a standard Brownian motion $W = \{W_t : t \in [0, 1]\}$. To this end, we denote (as before) by $L^2_{\mathbb{C}}([0, 1]) = L^2_{\mathbb{C}}([0, 1], d\lambda)$ the class of complex-valued functions that are square-integrable with respect to the restriction of the Lebesgue measure $d\lambda$ to $[0, 1]$. Select a complex-valued family

$$\{g_{lm} : l \geq 0, \ -l \leq m \leq l\} \subseteq L^2_{\mathbb{C}}([0, 1])$$

with the following five properties: (1) g_{l0} is real for every $l \geq 0$, (2) $g_{lm} = (-1)^m \overline{g_{l-m}}$, (3) $\int g_{lm} \overline{g_{l'm'}} d\lambda = 0$, $\forall (l, m) \neq (l', m')$, (4) $\int \text{Re}(g_{lm}) \text{Im}(g_{lm}) d\lambda = 0$, (5) $\int \text{Re}(g_{lm})^2 d\lambda = \int \text{Im}(g_{lm})^2 d\lambda = \int g_{l0}^2 d\lambda/2 = C_l/2$, where $\{C_l : l \geq 0\}$ is the power spectrum of the Gaussian field T. A simple covariance computation shows that the following identity in law holds:

$$\{a_{lm;1} : l \geq 0, \ -l \leq m \leq l\} \stackrel{law}{=} \{I_1(g_{lm}) : l \geq 0, \ -l \leq m \leq l\},$$

where $I_1(g_{lm}) = \int_0^1 g_{lm}dW = \int_0^1 \mathrm{Re}(g_{lm})dW + i\int_0^1 \mathrm{Im}(g_{lm})dW$ is the usual (complex-valued) Wiener-Itô integral of g_{lm} with respect to W. From this last relation, it also follows that, in the sense of stochastic processes, $T(x) \stackrel{law}{=} I_1\left(\sum_{l=0}^{\infty} \sum_{m=-l}^{l} g_{lm}Y_{lm}(x)\right)$ (note that the function $z \mapsto \sum_{l,m} g_{lm}(z)Y_{lm}(x)$ is real-valued for every fixed $x \in S^2$ and with norm equal to 1). Now define

$$L_{s,\mathbb{C}}^2([0,1]^q)$$

to be the class of complex-valued and symmetric functions on $[0,1]^q$, that are square-integrable with respect to Lebesgue measure. For every

$$f \in L_{s,\mathbb{C}}^2([0,1]^q),$$

we define $I_q(f) = I_q(\mathrm{Re}(f)) + iI_q(\mathrm{Im}(f))$ to be the multiple Wiener-Itô integral, of order q, of f with respect to the Brownian motion W. From formula (4.18) in Chapter 4 and the previous discussion it follows that, for every $q \geq 2$,

$$T^{(q)}(x) = H_q(T(x)) \stackrel{law}{=} I_q\left[\left\{\sum_{l=0}^{\infty} \sum_{m=-l}^{l} g_{lm}Y_{lm}(x)\right\}^{\otimes q}\right], \qquad (7.24)$$

where the equality in law holds in the sense of finite dimensional distributions and, for every $f \in L_{\mathbb{C}}^2([0,1])$, we use the notation $f^{\otimes q}(a_1,...,a_q) = f(a_1) \times \cdots \times f(a_q)$. Now set $h_{l,m}^{(q)} = (-1)^m \sum_{l_1,m_1} \cdots \sum_{l_q,m_q} g_{l_1 m_1} \cdots g_{l_q m_q} G\{l_1,m_1;...; l_q,m_q; l,-m\}$, so that

$$a_{lm;q} \stackrel{law}{=} \int_{S^2} I_q\left[\left\{\sum_{l=0}^{\infty} \sum_{m=-l}^{l} g_{lm}Y_{lm}(x)\right\}^{\otimes q}\right] \overline{Y_{lm}(x)}dx = I_q\left[h_{l,m}^{(q)}\right] \qquad (7.25)$$

and (7.19) follows immediately from the isometry relation

$$E\left[\left|I_q\left[h_{l,m}^{(q)}\right]\right|^2\right] = q!\left\|h_{l,m}^{(q)}\right\|_{L^2([0,1]^q)}^2,$$

which is a consequence of (4.7) (to obtain (7.25) we interchanged stochastic and deterministic integration, by means of a standard stochastic Fubini argument). To prove that (7.20) is equal to (7.19), observe first that from the unitary properties of higher-order Clebsch-Gordan matrices we have

$$\sum_{m_1=-l_1}^{l_1} \cdots \sum_{m_q=-l_q}^{l_q} C_{l_1 m_1 ... l_q m_q}^{L_1 L_2 ... L_{q-2} l, m} C_{l_1 m_1 ... l_q m_q}^{L_1' L_2' ... L_{q-2}' l, m} = \delta_{L_1}^{L_1'} ... \delta_{L_{q-2}}^{L_{q-2}'}$$

(the right-hand side of the previous expression does not depend on m). Then, using the unitary properties of Clebsch-Gordan coefficients and exploiting (6.46), we deduce that

$$
\sum_{m_1=-l_1}^{l_1} \cdots \sum_{m_q=-l_q}^{l_q} G\{l_1, m_1; \ldots; l_q, m_q; l, -m\}^2
$$

$$
= \frac{4\pi}{2l+1} \left\{ \prod_{i=1}^{q} \frac{2l_i+1}{4\pi} \right\} \sum_{L_1,\ldots,L_{q-2}} \left\{ C_{l_0\ldots l_q 0}^{L_1 L_2 \ldots L_{q-2} l, 0} \right\}^2.
$$

This proves Point (1) in the statement. To prove Point (2), recall that, according to Proposition 4.24 , relation (7.21) holds if and only if

$$
(\widetilde{C}_l^{(q)})^{-2} \left\| h_{l,m}^{(q)} \otimes_p \overline{h_{l,m}^{(q)}} \right\|_{L^2([0,1]^{2(q-p)})}^2 \to 0,
$$

for every $p = 1, \ldots, q-1$, where the complex-valued (and not necessarily symmetric) function $h_{l,m}^{(q)} \otimes_p \overline{h_{l,m}^{(q)}}$ (which is an element of $L^2([0,1]^{2(q-p)})$) is defined as the *contraction*

$$
h_{l,m}^{(q)} \otimes_p \overline{h_{l,m}^{(q)}} \left(a_1, \ldots, a_{2(q-p)}\right) \tag{7.26}
$$

$$
= \int_{[0,1]^p} h_{l,m}^{(q)}\left(\mathbf{x}_p, a_1, \ldots, a_{q-p}\right) \overline{h_{l,m}^{(q)}\left(\mathbf{x}_p, a_{q-p+1}, \ldots, a_{2(q-p)}\right)} d\mathbf{x}_p,
$$

for every $(a_1, \ldots, a_{2(q-p)}) \in [0,1]^{2(q-p)}$, where $d\mathbf{x}_p$ is the Lebesgue measure on $[0,1]^p$. Since, trivially, $\|h_{l,m}^{(q)} \otimes_p \overline{h_{l,m}^{(q)}}\|^2 = \|h_{l,m}^{(q)} \otimes_{q-p} \overline{h_{l,m}^{(q)}}\|^2$ (we stress that, in the last equality, the first norm is taken in $L^2([0,1]^{2(q-p)})$, whereas the second is in $L^2([0,1]^{2p})$), it is sufficient to check that the norm of $h_{l,m}^{(q)} \otimes_p \overline{h_{l,m}^{(q)}}$ is asymptotically negligible for every $p = \frac{q-1}{2}+1, \ldots, q-1$, if $q-1$ is even, and every $p = q/2, \ldots, q-1$ if $q-1$ is odd. It follows that the result is proved once it is shown that, for every p in such range, the norm $\|h_{l,m}^{(q)} \otimes_p \overline{h_{l,m}^{(q)}}\|^2$ equals the multiple sum appearing in (7.22). To see this, use (7.26) to deduce that (recall that Gaunt integrals are real-valued)

$$
h_{l,m}^{(q)} \otimes_p \overline{h_{l,m}^{(q)}} \left(a_1, \ldots, a_{2(q-p)}\right)
$$

$$
= \sum_{n_1,j_1} \cdots \sum_{n_{2(q-p)},j_{2(q-p)}} g_{j_1 n_1} \cdots g_{j_{q-p} n_{q-p}} \overline{g_{j_{q-p+1} n_{q-p+1}}} \cdots \overline{g_{j_{2(q-p)} n_{2(q-p)}}}
$$

$$
\sum_{l_1,m_1} \cdots \sum_{l_p,m_p} C_{l_1} \cdots C_{l_p} G\{l_1, m_1; \ldots; l_p, m_p; j_1, n_1; \ldots; j_{q-p}, n_{q-p}; l, -m\}
$$

$$
G\{l_1, m_1; \ldots; l_p, m_p; j_{q-p+1}, n_{q-p+1}; \ldots; j_{2(q-p)}, n_{2(q-p)}; l, -m\},
$$

and the result is obtained by using the orthogonality properties of the g_{jn}'s.

Point (3) in the statement is proved in exactly the same way, by first observing that $a_{l0;q}$ is a real-valued random variable, and then by applying Theorem 4.18.

$\square$

Remark 7.6   We have the relation $E\left[T^{(q)}(x)^2\right] = q!\left[E\left\{T(x)^2\right\}\right]^q$.

Now recall that, according to Part 2 of Lemma 7.4, $T_l^{(q)}(x) \overset{law}{=} \sqrt{\frac{2l+1}{4\pi}} a_{l0;q}$, so that relation (7.15) holds. This gives immediately a first (exhaustive) solution to Problem (**P-I**), as stated in Section 7.3.

Corollary 7.7   *For every $q \geq 2$ the following conditions are equivalent:*

(1) The CLT (7.16) holds for every $x \in S^2$;
(2) The asymptotic relation (7.22) holds for $m = 0$ and for every $p = \frac{q-1}{2} + 1, ..., q - 1$, if $q - 1$ is even, and every $p = q/2, ..., q - 1$ if $q - 1$ is odd.

To deal with Problem (**P-II**) of Section 7.3, we recall the notation $\overline{T}_l^{(q)}$ (indicating the lth normalized frequency component of $T^{(q)}$) introduced in (7.11). We also introduce (for every $l \geq 1$) the *normalized lth frequency component* of the Gaussian field T, which is defined as

$$\overline{T}_l(x) = \frac{T_l(x)}{Var(T_l(x))^{1/2}} = \frac{T_l(x)}{(\frac{2l+1}{4\pi}C_l)^{1/2}}, \quad x \in S^2. \qquad (7.27)$$

According to Lemma 7.4 (in the special case $F(z) = z$), $\overline{T}_l$ is a real-valued, isotropic, centered and Gaussian field. Moreover, we have that $\mathbb{E}[\overline{T}_l(x)\overline{T}_l(y)] = \mathbb{E}[\overline{T}_l^{(q)}(x)\overline{T}_l^{(q)}(y)] = P_l(\langle x, y\rangle)$, for every $q \geq 2$ and every $l \geq 1$. The next result – which gives an exhaustive solution to Problem (**P-II**) – states that, whenever Condition 1 (or, equivalently, Condition 2) in the statement of Corollary 7.7 is verified (and without *any* additional assumption), the "distance" (which is defined in terms of the supremum over a suitable class of test functions) between the finite dimensional distributions of the normalized field $\overline{T}_l^{(q)}$ and those of $\overline{T}_l$ converge to zero. The proof (left to the reader) is an immediate consequence of Theorem 4.21.

Theorem 7.8   *Let $q \geq 2$ be fixed, and suppose that Condition 1 (or 2) of Corollary 7.7 is satisfied. Then, for every $k \geq 1$, every $x_1, ..., x_k \in S^2$, and every twice differentiable function $\phi : \mathbb{R}^k \to \mathbb{R}$, such that $\|\phi''\|_\infty < \infty$ (see (4.34)),*

$$\lim_{k \to +\infty} \left| E\left[\phi\left(\overline{T}_l^{(q)}(x_1), ..., \overline{T}_l^{(q)}(x_k)\right)\right] - E\left[\phi\left(\overline{T}_l(x_1), ..., \overline{T}_l(x_k)\right)\right] \right| = 0.$$

For every real B > 0, the above convergence takes place uniformly on the class of those test functions ϕ such that $\|\phi''\|_\infty \leq B$.

7.5 Convolutions and random walks

In this section, we study in more detail the conditions for the CLTs proved in Section 7.4 for the (Hermite) frequency components $T_l^{(q)}$, $l \geq 0$. In particular, we shall establish sufficient conditions that are expressed in terms of the asymptotic behaviour of the angular power spectrum $\{C_l : l \geq 0\}$. The results of Section 7.5.2 cover the case $q = 2$ and $q = 3$. Section 7.5.3 contains some partial findings for the case of a general q, as well as several conjectures. These results are used in Section 7.7 to deduce explicit conditions on the rate of decay of the angular power spectrum $\{C_l : l \geq 0\}$.

7.5.1 Convolutions on $\widehat{SO}(3)$

In the light of Part 3 of Theorem 7.5 and by Corollary 7.7, we will focus on the sequence $\{a_{l0;q} : l \geq 0\}$ (see (7.9)), whose behaviour as $l \to +\infty$ yields an asymptotic characterization of the fields $T_l^{(q)}(\cdot)$ defined in (7.10). A crucial point is the simple fact that the numerator of (7.22), for $m = 0$, can be developed as a multiple sum involving products of four generalized Gaunt integrals, so that, by (7.18), the asymptotic expressions appearing in Theorem 7.5 can be studied by means of the properties of linear combinations of products of Clebsch-Gordan coefficients. As anticipated, a very efficient tool for our analysis will be the use of convolutions on $\mathbb{N}$, that we endow with an hypergroup structure isomorphic to $\widehat{SO}(3)$, i.e. the dual of $SO(3)$. This will be the object of the subsequent discussion.

From now on, and for the rest of the section, we shall fix a sequence

$$\{C_l : l \geq 0\},$$

representing the angular power spectrum of an isotropic centered, normalized Gaussian field T over S^2, as in Section 7.3. Whenever convenient we shall write

$$\Gamma_l := (2l+1)C_l, \quad l \geq 0, \tag{7.28}$$

so that, for $l \geq 1$ and up to the constant $1/4\pi$, the parameter Γ_l represents the variance of the projection of the Gaussian field T in (6.1) on the frequency l:

indeed, according to Lemma 7.4, $Var(T_l) = \Gamma_l/4\pi$. Also, we define the following convolutions of the coefficients Γ_l (in the following expressions, the sums over indices l_i, L_i ... range implicitly from 0 to $+\infty$):

$$\widehat{\Gamma}_{2,l} = \sum_{l_1,l_2} \Gamma_{l_1}\Gamma_{l_2}(C^{l0}_{l_10l_20})^2 , \tag{7.29}$$

$$\widehat{\Gamma}_{3,l} = \sum_{L_1,l_3} \widehat{\Gamma}_{2,L_1}\Gamma_{l_3}(C^{l0}_{L_10l_30})^2 \tag{7.30}$$

$$= \sum_{l_1,l_2,l_3} \Gamma_{l_1}\Gamma_{l_2}\Gamma_{l_3} \sum_{L_1}(C^{L_1l,0}_{l_10l_20l_30})^2, \dots \tag{7.31}$$

$$\widehat{\Gamma}_{q,l} = \sum_{L_1,l_q} \widehat{\Gamma}_{q-1,L_{q-1}}\Gamma_{l_q}(C^{l0}_{L_{q-1}0l_q0})^2 \tag{7.32}$$

$$= \sum_{l_1\dots l_q} \Gamma_{l_1}..\Gamma_{l_q} \sum_{L_1..L_{q-2}}(C^{L_1\dots L_{q-2}l,0}_{l_10\dots l_q0})^2 \tag{7.33}$$

(see (6.48) for the notation). The equalities in formulae (7.31) and (7.33) follow from induction and the properties of Clebsch-Gordan coefficients. It will be also convenient to define a *-convolution of order $p \geq 2$ as:

$$\widehat{\Gamma}^*_{p,l;l_1} = \sum_{l_2}\cdots\sum_{l_p}\Gamma_{l_2}\cdots\Gamma_{l_p}\sum_{L_1,\dots,L_{p-2}}\left\{C^{L_10}_{l_10l_20}C^{L_20}_{L_10l_30}\cdots C^{l0}_{L_{p-2}0l_p0}\right\}^2$$

$$= \sum_{l_2}\cdots\sum_{l_p}\Gamma_{l_2}\cdots\Gamma_{l_p}\sum_{L_1,\dots,L_{p-2}}\left\{C^{L_1\dots l,0}_{l_10l_20\dots l_p0}\right\}^2. \tag{7.34}$$

Note that the number of sums following the equalities in formula (7.34) is $p-1$: however, we choose to keep the symbol p to denote *-convolutions, since it is consistent with the probabilistic representations given in formulae (7.38) and (7.39) below. The above *-convolution has the following property: for every $p = 2, \dots, q$

$$\sum_{l_1}\widehat{\Gamma}_{q+1-p,l_1}\widehat{\Gamma}^*_{p,l;l_1} = \widehat{\Gamma}_{q,l} \text{ , and, in particular, } \sum_{l_1}\Gamma_{l_1}\widehat{\Gamma}^*_{q,l;l_1} = \widehat{\Gamma}_{q,l} \text{ .}$$

The *-convolution of order 2 can be written more explicitly as

$$\widehat{\Gamma}^*_{2,l;l_1} = \sum_{l_2}\Gamma_{l_2}(C^{l0}_{l_10l_20})^2. \tag{7.35}$$

Remark 7.9 (1) (*Probabilistic interpretation of the convolutions*) Write first $\Gamma_* \triangleq \sum_l \Gamma_l$ (plainly, in our framework $\Gamma_* = 4\pi$, but the following discussion applies to coefficients $\{\Gamma_l\}$ such that $\Gamma_* > 0$ is arbitrary) so that $l \longmapsto \Gamma_l/\Gamma_*$ defines a probability on $\mathbb{N}$. Recall that, for fixed l_1, l_2, the application $l \longmapsto$

$(C^{l0}_{l_1 0 l_2 0})^2$ is a probability on $\mathbb{N}$. Now define the law of a (homogeneous) Markov chain $\{Z_n : n \geq 1\}$ as follows:

$$P\{Z_1 = l\} = \Gamma_l / \Gamma_* \tag{7.36}$$

$$P\{Z_{n+1} = l \mid Z_n = L\} = \sum_{l_0} \frac{\Gamma_{l_0}}{\Gamma_*} \left(C^{l0}_{l_0 0 L 0}\right)^2. \tag{7.37}$$

It is clear that $P\{Z_q = l\} = \widehat{\Gamma}_{q,l} / (\Gamma_*)^q$, and also, for $p \geq 2$,

$$\frac{\widehat{\Gamma}^*_{p,l;l_1}}{(\Gamma_*)^{p-1}} = P\{Z_p = l \mid Z_1 = l_1\} \tag{7.38}$$

$$\frac{\widehat{\Gamma}^*_{p,l;l_1} \widehat{\Gamma}_{q+1-p,l_1}}{(\Gamma_*)^q} = P\{(Z_q = l) \cap (Z_{q+1-p} = l_1)\} \quad (q > p - 1). \tag{7.39}$$

The following quantity will be crucial in the subsequent sections:

$$\frac{\widehat{\Gamma}^*_{q+1-p,l;\lambda} \widehat{\Gamma}_{p,\lambda}}{\sum_L \widehat{\Gamma}_{p,L} \widehat{\Gamma}^*_{q+1-p,l;L}} = \frac{\widehat{\Gamma}^*_{q+1-p,l;\lambda} \widehat{\Gamma}_{p,\lambda}}{\widehat{\Gamma}_{q,l}} = P\{Z_p = \lambda \mid Z_q = l\} \quad (q > p); \tag{7.40}$$

observe that the last relation in (7.40) derives from

$$\widehat{\Gamma}^*_{q+1-p,l;\lambda} / (\Gamma_*)^{q-p} = P\{(Z_{q+1-p} = l) \mid (Z_1 = \lambda)\}$$

$$= P\{(Z_q = \lambda) \mid (Z_p = l)\},$$

where the last equality is a consequence of the homogeneity of Z. Note also that we can identify each natural number $l \geq 0$ with an irreducible representation of $SO(3)$. It follows that the formal addition $l_1 + l_2 \triangleq \sum_l l(C^{l0}_{l_1 0 l_2 0})^2$ may be used to endow $\widehat{SO(3)}$ with an hypergroup structure. In this sense, we can interpret the chain $\{Z_n : n \geq 1\}$ as a random walk on the hypergroup $\widehat{SO(3)}$, in a spirit similar to [89]. In Section 7.6, we will discuss a physical interpretation of these convolutions and establish a precise connection between the objects introduced in this section and the notion of convolution appearing in [89].

(2) (*A comparison with the Abelian case*) As already discussed in Section 7.2, in [134], where similar problems are addressed in the case of homogenous spaces of Abelian groups, convolutions over $\mathbb{Z}$ are used extensively. This kind of convolutions, that we note $_A\widehat{\Gamma}_{q,l}$ ($q \geq 2$, $l \in \mathbb{Z}$) are obtained as in (7.29)-(7.35), by taking sums over $\mathbb{Z}$ (instead than over $\mathbb{N}$) and by replacing the Clebsch-Gordan symbols $(C^{l0}_{l_1 0 l_2 0})^2$ with the indicator $\mathbf{1}_{l_1 + l_2 = l}$. As we noted earlier, these indicator functions do indeed provide the Clebsch-Gordan coefficients associated with the irreducible representations of the

1-dimensional torus $\mathbb{T} = [0, 2\pi)$, regarded as a compact Abelian group with group operation $xy = (x + y) (\mathrm{mod}(2\pi))$ (this is equivalent to the trivial relation $e^{il_1 x} e^{il_2 x} = \sum_l \mathbf{1}_{l_1 + l_2 = l} e^{ilx} = e^{i(l_1 + l_2)x}$). Note also that in the Abelian case one has $_A \widehat{\Gamma}_{p,l;l_1}^* = _A \widehat{\Gamma}_{p,l-l_1}$. Also, if $\Gamma_l = \Gamma_l^V$, where $\{\Gamma_l^V\}$ is the power spectrum of the Gaussian field V on $\mathbb{T}$, we have that $_A \widehat{\Gamma}_{q,l}^V = P\left[U_q = l\right]$, where $\{U_n\}$ is the random walk given in (7.3).

7.5.2 The cases $q = 2$ and $q = 3$

In this subsection, we provide a sufficient condition on the spectrum $\{C_l : l \geq 0\}$ (or, equivalently, on $\{\Gamma_l : l \geq 0\}$, as defined in (7.28)) to have the CLT (7.23) in the quadratic and cubic cases $q = 2, 3$. The proofs are very technical and require very careful manipulations of Clebsch-Gordan coefficients. For brevity's sake, they are not reported here.[1]

Proposition 7.10   *For $q = 2$, a sufficient condition for the CLT (7.23) is the following asymptotic relation*

$$\lim_{l \to +\infty} \sup_{\lambda} \frac{\Gamma_\lambda \sum_{l_2} \Gamma_{l_2} \left\{ C_{\lambda 0 l_2 0}^{l0} \right\}^2}{\sum_{l_1, l_2} \Gamma_{l_1} \Gamma_{l_2} (C_{l_1 0 l_2 0}^{l0})^2} = \lim_{l \to +\infty} \sup_{\lambda} P\{Z_1 = \lambda | Z_2 = l\} = 0, \qquad (7.41)$$

where the $\{\Gamma_l\}$ are given by (7.28) and $\{Z_l\}$ is the Markov chain defined in formulae (7.36) and (7.37).

Remark 7.11   Note that, using (7.33) and (7.35), condition (7.41) becomes

$$\lim_{l \to \infty} \sup_{\lambda} \frac{\Gamma_\lambda \widehat{\Gamma}_{2,l;\lambda}^*}{\sum_{l_1} \Gamma_{l_1} \widehat{\Gamma}_{2,l;l_1}} = 0 \ . \qquad (7.42)$$

Note also that if, in the convolutions (7.33), we replace each squared Clebsch-Gordan coefficient $\left(C_{l_1 0 l_2 0}^{l0} \right)^2$ by the indicator $\mathbf{1}_{l_1 + l_2 = l}$ and extends the sums over $\mathbb{Z}$, we obtain the relation

$$\lim_{l \to \infty} \sup_{l_1} \frac{\Gamma_{l_1} \Gamma_{l-l_1}}{\sum_{l_1} \Gamma_{l_1} \Gamma_{l-l_1}} = 0. \qquad (7.43)$$

In particular, when $\{\Gamma_l\} = \{\Gamma_l^V\}$ (the power spectrum of the field V on $\mathbb{T}$) it is not difficult to show that formula (7.43) gives exactly the asymptotic (necessary and sufficient) condition (7.2).

---

[1] Full details are given in the ArXiv preprint 0706.2851v1.

Proposition 7.12 *A sufficient condition for the CLT (7.23) when $q = 3$ is*

$$\limsup_{l \to \infty} \frac{\sum_{l_1 l_2 j_1} \Gamma_{l_1} \Gamma_{l_2} \Gamma_{j_1} \left\{ C^{L_1 l, 0}_{l_1 0 l_2 0 j_1 0} \right\}^2}{\sum_{L_1} \sum_{l_1, l_2, l_3} \Gamma_{l_1} \Gamma_{l_2} \Gamma_{l_3} \left\{ C^{L_1 l, 0}_{l_1 0 l_2 0 l_3 0} \right\}^2} = 0, \ and \tag{7.44}$$

$$\limsup_{l \to \infty} \frac{\sum_{l_1 l_2 L_1} \Gamma_{l_1} \Gamma_{l_2} \Gamma_{j_1} \left\{ C^{L_1 l, 0}_{l_1 0 l_2 0 j_1 0} \right\}^2}{\sum_{L_1} \sum_{l_1, l_2, l_3} \Gamma_{l_1} \Gamma_{l_2} \Gamma_{l_3} \left\{ C^{L_1 l, 0}_{l_1 0 l_2 0 l_3 0} \right\}^2} = 0. \tag{7.45}$$

Remark 7.13 In the light of (7.33)-(7.35) and of the definition of the random walk Z given in (7.36) and (7.37), it is not difficult to see that (7.44) can be rewritten as

$$\limsup_{l \to \infty} \frac{\widehat{\Gamma}_{2,\lambda} \sum_{j_1} \Gamma_{j_1} \left\{ C^{l 0}_{\lambda 0 j_1 0} \right\}^2}{\widehat{\Gamma}_{3,l}} = \limsup_{l \to \infty} \frac{\widehat{\Gamma}_{2,\lambda} \widehat{\Gamma}^*_{2,l;\lambda}}{\sum_{L_1} \left[\widehat{\Gamma}_{2,L_1} \widehat{\Gamma}^*_{1,l;L_1} \right]} \tag{7.46}$$

$$= \limsup_{l \to \infty} P \left[Z_2 = \lambda \mid Z_3 = l \right] = 0. \tag{7.47}$$

Likewise, we obtain that (7.45) is equivalent to

$$\limsup_{l \to \infty} \frac{\Gamma_{j_1} \widehat{\Gamma}^*_{3,l;j_1}}{\sum_{L_1} \sum_{l_1, l_2, l_3} \Gamma_{l_1} \Gamma_{l_2} \Gamma_{l_3} \left\{ C^{L_1 l, 0}_{l_1 0 l_2 0 l_3 0} \right\}^2} \tag{7.48}$$

$$= \limsup_{l \to \infty} P \left[Z_1 = j_1 \mid Z_3 = l \right] = 0.$$

It should be noted that the two conditions (7.46) and (7.48) can be written compactly as

$$\lim_{l \to \infty} \max_{q=1,2} \sup_{j_1} \frac{\widehat{\Gamma}_{q,j_1} \widehat{\Gamma}^*_{3-q,l;j_1}}{\sum_{L_1} \sum_{l_1, l_2, l_3} \Gamma_{l_1} \Gamma_{l_2} \Gamma_{l_3} \left\{ C^{L_1 l 0}_{l_1 0 l_2 0 l_3 0} \right\}^2} = 0. \tag{7.49}$$

Relation (7.49) parallels once again analogous conditions established for stationary fields on a torus – see [134].

7.5.3 The case of a general q: results and conjectures

The following proposition gives a general version of the results discussed in the previous Section. The proof (omitted) is rather long, and can be obtained with an extensive use of graphical techniques for the manipulations of multiple products of Clebsch-Gordan coefficients.

Proposition 7.14 *Fix $q \geq 4$. Then, a sufficient condition to have the asymptotic relation (7.22) in the case $p = q - 1$ is the following;*

$$\lim_{l \to \infty} \left\{ \sup_{\lambda} \frac{\widehat{\Gamma}_{q-1,\lambda} \widehat{\Gamma}^*_{2,l;\lambda}}{\sum_L \widehat{\Gamma}_{q-1,L} \widehat{\Gamma}^*_{1,l;L}} + \sup_{\lambda} \frac{\widehat{\Gamma}^*_{q,l;\lambda} \Gamma_{\lambda}}{\sum_L \widehat{\Gamma}_{q-1,L} \widehat{\Gamma}^*_{1,l;L}} \right\}$$

$$= \lim_{l \to \infty} \left\{ \sup_{\lambda} \frac{\widehat{\Gamma}_{q-1,\lambda} \widehat{\Gamma}^*_{2,l;\lambda}}{\widehat{\Gamma}_{q,l}} + \sup_{\lambda} \frac{\widehat{\Gamma}^*_{q,l;\lambda} \Gamma_{\lambda}}{\widehat{\Gamma}_{q,l}} \right\} = 0. \tag{7.50}$$

Remark 7.15 (1) As in the proofs of Proposition 7.10 and Proposition 7.12, a crucial technique in proving Proposition 7.14 consists in the simplification of sums of products of Clebsch-Gordan coefficients by means of the general relation

$$\sum_{m_1 m_2} C^{L_1 M_1}_{l_1 m_1 l_2 m_2} C^{L_3 M_3}_{l_1 m_1 l_2 m_2} = \delta^{L_3}_{L_1} \delta^{M_3}_{M_1}. \tag{7.51}$$

More generally, such convolutions can be dealt with by means of the graphical techniques we discussed at the end of Chapter 5.

(2) Note that, since $q \geq 4$ and according to Part C of Theorem 7.5, condition (7.22) *is only necessary* to have the CLT (7.23), so that (7.50) cannot be used to deduce the asymptotic Gaussianity of the frequency components of Hermite-subordinated fields of the type $H_q[T]$. Some conjectures concerning the case $q \geq 4$, $p \neq q - 1$ are presented at the end of the section.

(3) Observe that, in terms of the random walk $\{Z_n\}$ defined in (7.36)-(7.37),

$$\frac{\widehat{\Gamma}_{q-1,\lambda} \widehat{\Gamma}^*_{2,l;\lambda}}{\widehat{\Gamma}_{q,l}} = P\{Z_{q-1} = \lambda \mid Z_q = l\}$$

$$\frac{\widehat{\Gamma}^*_{q,l;\lambda} \Gamma_{\lambda}}{\widehat{\Gamma}_{q,l}} = P\{Z_1 = \lambda \mid Z_q = l\}.$$

As mentioned before, the relation (7.22) (which implies (7.23)), in the general case where $q \geq 4$ and $p \neq q - 1$, is still being investigated, as it requires a hard analysis of higher order Clebsch-Gordan coefficients by means of graphical techniques (see for instance [195, Ch. 11]). As in [135], it is however natural to propose the following conjecture. Recall that we focus on the CLT (7.23) because of the equality in law $T_l^{(q)}(x) = \sqrt{\frac{2l+1}{4\pi}} a_{l0;q}$, and Corollary 7.7.

Conjecture A *(Weak) A sufficient condition for the CLT (7.23) is*

$$\lim_{l \to \infty} \max_{1 \le p \le q-1} \sup_{\lambda} \frac{\widehat{\Gamma}_{p,\lambda} \widehat{\Gamma}^*_{q+1-p,l;\lambda}}{\sum_L \widehat{\Gamma}_{p,L_{q-2}} \widehat{\Gamma}^*_{q+1-p,l;L}} \tag{7.52}$$

$$= \lim_{l \to \infty} \max_{1 \le p \le q-1} \sup_{\lambda} P \left\{ Z_p = \lambda \mid Z_q = l \right\} = 0 .$$

It is worth emphasizing how condition (7.52) is the exact analogous of the necessary and sufficient condition (7.3), established in [134] for the high-frequency CLT on the torus $\mathbb{T} = [0, 2\pi)$. This remarkable circumstance led the authors in [135] to suggest the following (much more general and somewhat imprecise) extension.

Conjecture B *(Strong) Let T be an isotropic Gaussian field defined on the homogeneous space of a compact group G, and set $T^{(q)} = H_q(T)$ ($q \ge 2$). Then, the high-frequency components of $T^{(q)}$ are asymptotically Gaussian if, and only if, it holds a condition of the type*

$$\lim_{l \to l_0} \max_{1 \le p \le q-1} \sup_{\lambda \in \widehat{G}} \frac{\widehat{\Gamma}^*_{p,\lambda} \widehat{\Gamma}_{q+1-p,l;\lambda}}{\sum_{L \in \widehat{G}} \widehat{\Gamma}^*_{p,L} \widehat{\Gamma}_{q+1-p,l;L}} = 0 , \tag{7.53}$$

where $\widehat{G}$ is the dual of G, l_0 is some point at the boundary of $\widehat{G}$, and the convolutions $\widehat{\Gamma}$ and $\widehat{\Gamma}^*$ are defined (analogously to (7.29)-(7.34)) on the power spectrum of T, by means of the appropriate Clebsch-Gordan coefficients of the group.

Remark 7.16 In terms of Z, condition (7.52) can be further interpreted as follows: for every l, define a "bridge" of length q, by conditioning Z to equal l at time q. Then, (7.52) is verified if, and only if, the probability that the bridge hits λ at time q converges to zero, uniformly on λ, as $l \to +\infty$. It is also evident that, when (7.52) is verified for every $p = 1, ..., q - 1$, one also has that

$$\lim_{l \to +\infty} \sup_{\lambda_1, ..., \lambda_{q-1} \in \mathbb{N}} P \left[Z_1 = \lambda_1, ..., Z_{m-1} = \lambda_{q-1} \mid Z_q = l \right] = 0, \tag{7.54}$$

meaning that, asymptotically, the law of Z does not charge any "privileged path" of length q leading to l. The interpretation of condition (7.54) in terms of bridges can be reinforced by putting by convention $Z_0 = 0$, so that the probability in (7.54) is that of the particular path $0 \to \lambda_1 \to ... \to \lambda_{q-1} \to l$, associated with a random bridge linking 0 and l.

7.6 Further remarks

7.6.1 Convolutions as mixed states

We recall that, in quantum mechanics, it is customary to consider two possible initial states for a particle, i.e. those provided by the so-called *pure states*, where the state of a particle is given, and those provided by the so-called *mixed states*, where the state of the particle is given by a mixture (in the usual probabilistic sense) over different quantum states. We refer the reader to [127] for an introduction to these ideas. From this standpoint, the quantity $\widehat{\Gamma}_{q,l}$ defined in (7.33) is the probability associated to a mixed state, where the mixing is performed over all possible values of the total angular momentum. To illustrate this point, we use the standard bra-ket notation $|l0\rangle$ to indicate the state of a particle having total angular momentum equal to l and projection 0 on the z-axis. By using this formalism, the quantity $\widehat{\Gamma}_{q,l}$ can be obtained as follows:

(i)   consider a system of q particles $\alpha_1, ..., \alpha_q$ such that each α_j is in the mixed state Ξ according to which a particle is in the state $|k0\rangle$ with probability Γ_k/Γ_* ($k \geq 0$);

(ii) obtain $\widehat{\Gamma}_{q,l}$ as the probability that the elements of this system are coupled pairwise to form a particle in the state $|l0\rangle$.

Now denote by $\mathbf{A}_{p,|\lambda 0\rangle}$ the event that the first p particles $\alpha_1, ..., \alpha_p$ have coupled pairwise to generate the state $|\lambda 0\rangle$. Then,

$$\frac{\widehat{\Gamma}_{p+1,\lambda}\widehat{\Gamma}^*_{q-p,l;\lambda}}{\widehat{\Gamma}_{q,l}} = \Pr\left\{\text{the } q \text{ particles generate } |l0\rangle \mid \mathbf{A}_{p,|\lambda 0\rangle}\right\}. \qquad (7.55)$$

In particular, relation (7.55) yields a further physical interpretation of the "no privileged path condition" discussed in (7.54).

7.6.2 Other convolutions and random walks on group duals

Random walks on hypergroups, and specifically on group duals, have been actively studied in the seventies – see [89, Ch. 6]. Our aim in the sequel is to compare our definitions with those provided in this earlier literature, mainly by discussing the alternative physical meanings of the associated notion of convolution. We recall from Chapters 3 and 6 that, starting from the Wigner's D-matrices representation of $SO(3)$, we obtain the unitary equivalent reducible representations $\{D^{l_1}(g) \otimes D^{l_2}(g)\}$ and $\{\oplus_{l=|l_2-l_1|}^{l_2+l_1} D^l(g)\}$. Now note $\chi_l(g)$ the character of $D^l(g)$; for all $g \in SO(3)$, we have immediately

$$\chi_{l_1}(g)\chi_{l_2}(g) = \sum_{l=|l_2-l_1|}^{l_2+l_1} \chi_l(g) \, .$$

In [89, p. 222], an alternative class of Clebsch-Gordan coefficients $\{C_{l_1 l_2|G}^l : l_1, l_2, l \geq 0\}$ is defined by means of the identity

$$\frac{1}{2l_1+1}\chi_{l_1}(g)\frac{1}{2l_2+1}\chi_{l_2}(g) = \sum_l C_{l_1 l_2|G}^l \frac{1}{2l+1}\chi_l(g)$$

which leads to

$$C_{l_1 l_2|G}^l = \frac{2l+1}{(2l_1+1)(2l_2+1)} \{l_1 l_2 l\} \, ,$$

where we use the same notation as in [195] and in many other physical textbooks, i.e. we take $\{l_1 l_2 l\}$ to represent the indicator function of the event $|l_2 - l_1| \leq l \leq l_2 + l_1$. Of course

$$C_{l_1 l_2|G}^l = \sum_{l=|l_2-l_1|}^{l_2+l_1} \frac{2l+1}{(2l_1+1)(2l_2+1)} \equiv 1 \, . \tag{7.56}$$

As observed in [89], relation (7.56) can be used to endow $\widehat{SO(3)}$ with an hypergroup structure, via the formal addition $l_1 + l_2 \triangleq \sum_l l C_{l_1 l_2|G}^l$. Now let $\{\Gamma_l : l \geq 0\}$ be a collection of positive coefficients such that $\sum_l \Gamma_l = 1$. The convolutions and *-convolutions of the $\{\Gamma_l\}$ that are naturally associated with the above formal addition are given by

$$\widetilde{\Gamma}_{2,l} = \sum_{l_1,l_2} \Gamma_{l_1}\Gamma_{l_2}C_{l_1 l_2|G}^l \, , \quad \widetilde{\Gamma}_{3,l} = \sum_{L_1,l_3} \widetilde{\Gamma}_{2,L_1}\Gamma_{l_3}C_{L_1 l_3|G}^l, \, \cdots \tag{7.57}$$

$$\widetilde{\Gamma}_{q,l} = \sum_{L_1,l_q} \widetilde{\Gamma}_{q-1,L_{q-1}}\Gamma_{l_q}C_{L_{q-1}l_q|G}^l \, , \tag{7.58}$$

and, for $p \geq 2$,

$$\widetilde{\Gamma}_{p,l;l_1}^* = \sum_{l_2}\cdots\sum_{l_p} \Gamma_{l_2}\cdots\Gamma_{l_p} \sum_{L_1\ldots L_{p-2}} C_{l_1 l_2|G}^{L_1}C_{L_1 l_3|G}^{L_2}\cdots C_{L_{p-2}l_p|G}^l \, . \tag{7.59}$$

As shown in [89], the objects appearing in (7.57)-(7.59) can be used to define the law of a random walk $\widetilde{Z} = \{\widetilde{Z}_n : n \geq 1\}$ on $\mathbb{N}$ (regarded as an hypergroup isomorphic to $\widehat{SO(3)}$), exactly as we did in (7.36)-(7.37). In particular, since $\Gamma_* = \sum_l \Gamma_l = 1$, one has that $\widetilde{\Gamma}_{p,l;l_1}^* = \mathbb{P}\{\widetilde{Z}_p = l \mid \widetilde{Z}_1 = l_1\}$. Also, the convolutions (7.57)-(7.59) (and therefore the random walk $\widetilde{Z}$) enjoy a physical interpretation which is interesting to compare with our previous result. To see this, assume we have two mixed states Ξ_{l_1} and Ξ_{l_2}: in state Ξ_{l_1}, the particle has total angular momentum l_1 and its projection on the axis z takes values $m_1 = -l_1, ..., l_1$ with uniform (classical) probability $(2l_1 + 1)^{-1}$; analogous conditions are imposed for Ξ_{l_2}. Let us now compute the probability $\Pr\{l \mid \Xi_{l_1}, \Xi_{l_2}\}$ that the system will couple to form a particle with total angular momentum l and arbitrary projection on z. Start by observing that the probability that a particle in the state $|l_1 m_1\rangle$ will couple with another particle in the state $|l_2 m_2\rangle$ to yield the state $|lm\rangle$ is exactly given by $\{C_{l_1 m_1 l_2 m_2}^{lm}\}^2$. Hence, with straightforward notation,

$$
\begin{aligned}
\Pr\{l \mid \Xi_{l_1}, \Xi_{l_2}\} &= \sum_{m_1 m_2} \Pr\{l \mid |l_1 m_1\rangle, |l_2 m_2\rangle\} \Pr\{m_1, m_2\} \\
&= \sum_{m_1 m_2} \Pr\{l \mid |l_1 m_1\rangle, |l_2 m_2\rangle\} \frac{1}{2l_1 + 1} \frac{1}{2l_2 + 1} \\
&= \sum_{m} \sum_{m_1 m_2} \{C_{l_1 m_1 l_2 m_2}^{lm}\}^2 \frac{1}{2l_1 + 1} \frac{1}{2l_2 + 1} \\
&= \sum_{m} \frac{\{l_1 l_2 l\}}{2l_1 + 1} \frac{1}{2l_2 + 1} = \frac{2l + 1}{2l_1 + 1} \frac{\{l_1 l_2 l\}}{2l_2 + 1} = C_{l_1 l_2 | G}^{l} . \quad (7.60)
\end{aligned}
$$

It follows from (7.60) that the quantity $\widetilde{\Gamma}_{q,l}$ can be obtained as follows:

(i) consider a system of q particles $\alpha_1, ..., \alpha_q$ such that each α_j is in the mixed state Ξ according to which a particle is in the state $|ku\rangle$, $u = -k, ..., k$, with probability $(2k + 1)^{-1} \Gamma_k / \Gamma_* \ (k \geq 0)$;

(ii) obtain $\widetilde{\Gamma}_{q,l}$ as the probability that the elements of this system are coupled pairwise to form a particle in the state $|lm\rangle$, any $m = -l, ..., l$.

To sum up, both convolutions $\widehat{\Gamma}$ and $\widetilde{\Gamma}$ can be interpreted in terms of random interacting quantum particles: $\widehat{\Gamma}$-type convolutions are obtained from particles in mixed states where the mixing is performed over pure states of the form $|k0\rangle$; on the other hand, $\widetilde{\Gamma}$-type convolutions are associated with mixed state particles where mixing is over pure states of the type $\{|ku\rangle : u = -k, ..., k\}$, uniformly in u for every fixed k.

7.7 Application: algebraic/exponential dualities

In this section we discuss explicit conditions on the angular power spectrum $\{C_l : l \geq 0\}$ of the Gaussian field T introduced in Section 7.3, ensuring that the CLT (7.23) may hold. Our results show that, if the power spectrum decreases exponentially, then a high-frequency CLT holds, whereas the opposite implication holds if the spectrum decreases as a negative power. This duality was established in [135] and mirrors analogous conditions previously established in the Abelian case by [134]. For simplicity, we stick to the case $q = 2$.

7.7.1 The exponential case

As usual, given any two sequences $\{a_l, b_l\}$ we write $a_l \approx b_l$ if there exist costants $c_1, c_2 > 0$ such that $c_1 a_l \leq b_l \leq c_2 a_l$, for all $l = 1, 2, \dots$. Assume

$$C_l \approx (l+1)^\alpha \exp(-l), \quad \alpha \in \mathbb{R}. \tag{7.61}$$

To prove that, in this case, (7.23) is verified for $q = 2$, we will prove that (7.41) holds (recall the definition of Γ_l given in (7.28)). For the denominator of the previous expression we obtain the lower bound

$$\sum_{l_1, l_2 = 1}^{\infty} \Gamma_{l_1} \Gamma_{l_2} (C_{l_1 0 l_2 0}^{l0})^2 \geq \sum_{l_1 = [l/3]}^{[2l/3]} \Gamma_{l_1} \Gamma_{l - l_1} (C_{l_1 0 l - l_1 0}^{l0})^2$$

$$\approx \exp(-l) l^{2(\alpha+1)} \sum_{l_1 = [l/3]}^{[2l/3]} (C_{l_1 0 l - l_1 0}^{l0})^2 \tag{7.62}$$

and in view of [195], equation 8.5.2.33, and Stirling's formula

$$(7.62) \approx \exp(-l) l^{2(\alpha+1)} \sum_{l_1 = [l/3]}^{[2l/3]} \left(\frac{l!}{l_1! (l - l_1)!} \right)^2 \left(\frac{(2l_1)!(2l - 2l_1)!}{(2l)!} \right)$$

$$\approx \exp(-l) l^{2(\alpha+1)} \sum_{l_1 = [l/3]}^{[2l/3]} \frac{l^{2l+1}}{l_1^{2l_1+1}(l - l_1)^{2l - 2l_1 + 1}}$$

$$\times \left(\frac{(2l_1)^{2l_1 + 1/2}(2l - 2l_1)^{2l - 2l_1 + 1/2}}{(2l)^{2l + 1/2}} \right)$$

$$\approx \exp(-l) l^{2(\alpha+1)} \sum_{l_1 = [l/3]}^{[2l/3]} \frac{l^{1/2}}{l_1^{1/2}(l - l_1)^{1/2}} \approx \exp(-l) l^{2(\alpha+1)} l^{1/2}.$$

On the other hand, recall that by the triangle conditions $\{C_{l_1 0 l_2 0}^{l0}\}^2 \equiv 0$ unless $l_1 + l_2 \geq l$. Hence

$$\sup_{l_1} \sum_{l_2} \Gamma_{l_1} \Gamma_{l_2} \{C_{l_1 0 l_2 0}^{l0}\}^2$$

$$\leq K \sup_{l_1} \exp(-l) l_1^{\alpha+1} \left\{ |l - l_1|^{\alpha+1} + \sum_{u=1}^{\infty} \exp(-u) |l_1 + u|^{\alpha+1} \right\}$$

$$\approx \exp(-l) l^{2(\alpha+1)}.$$

It is then immediate to see that that (7.41) is satisfied.

7.7.2 Regularly varying functions

For $q = 2$, we show below that the CLT fails for all sequences C_l such that:
(a) C_l is quasi monotonic, i.e. $C_{l+1} \leq C_l(1 + K/l)$, some $K > 0$, and (b) C_l is
such that $\liminf_{l \to \infty} C_l/C_{l/2} > 0$. In particular, a necessary condition for the
CLT (7.23) to hold is that $C_l/C_{l/2} \to 0$. This is exactly the same necessary
condition as was derived by [134] in the Abelian case. For the general case
$q \geq 2$, we expect that the CLT fails for all regularly varying angular power
spectra, i.e. for all C_l such that $\liminf_{\ell \to \infty} C_l/C_{\alpha l} > 0$ for all $\alpha > 0$. Note that
we are thus covering all polynomial forms for C_l^{-1}.

Since (7.41) only provides a sufficient condition for the CLT, we need to
analyze directly the more primitive condition (7.22) for $m = 0$. We consider
first an upper bound for the square root of the denominator of (7.22), which is
given by $\widetilde{C}_l^{(2)}$.

We have

$$
\begin{aligned}
\widetilde{C}_l^{(2)} &= \sum_{j_1, j_2} C_{j_1} C_{j_2} \frac{(2j_1 + 1)(2j_2 + 1)}{4\pi(2l + 1)} \left(C_{j_1 0 j_2 0}^{l0} \right)^2 \\
&\leq 2 \sum_{j_1, j_2} C_{j_1} C_{j_2} \frac{(2j_1 + 1)(2j_2 + 1)}{4\pi(2l + 1)} \left(C_{j_1 0 j_2 0}^{l0} \right)^2 \\
&= \frac{1}{2\pi} \sum_{j_1} C_{j_1} (2j_1 + 1) \sum_{j_2 = j_1}^{\infty} C_{j_2} \left(C_{j_1 0 l 0}^{j_2 0} \right)^2 \\
&\leq \frac{1}{2\pi} \sum_{j_1} C_{j_1} (2j_1 + 1) \left\{ \sup_{j_2 \geq j_1, \, j_1 + j_2 > l} C_{j_2} \right\} \sum_{j_2 = 0}^{\infty} \left(C_{j_1 0 l 0}^{j_2 0} \right)^2 \leq K C_{l/2} .
\end{aligned}
$$

where we have used the relation $\frac{2j_2 + 1}{2l + 1} (C_{j_1 0 j_2 0}^{l0})^2 = (C_{j_1 0 l 0}^{j_2 0})^2$, as well as

$$
\sup_{j_2 \geq j_1, \, j_1 + j_2 > l} C_{j_2} \leq K C_{l/2} , \text{ and } \sum_{l = |l_2 - l_1|}^{l_2 + l_1} \left(C_{l_1 0 l_2 0}^{l0} \right)^2 \equiv 1 .
$$

The numerator of (7.22) is greater than (for some $c > 0$)

$$
\begin{aligned}
&\sum_{j_1, j_2} C_{j_1} C_{j_2} \frac{(2j_1 + 1)(2j_2 + 1)}{(4\pi(2l + 1))^2} \left| \sum_{l_1} C_{l_1} (2l_1 + 1) C_{l_1 0 j_1 0}^{l0} C_{l_1 0 j_1 0}^{l0} C_{l_1 0 j_2 0}^{l0} C_{l_1 0 j_2 0}^{l0} \right|^2 \\
&\geq \frac{C_l^2}{(4\pi)^2} \left| \sum_{l_1} C_{l_1} (2l_1 + 1) \left\{ C_{l_1 0 l 0}^{l0} \right\}^4 \right|^2 \geq \frac{C_l^2}{(4\pi)^2} \left| 5 C_2 \left\{ C_{2 0 l 0}^{l0} \right\}^4 \right|^2 \geq c C_l^2 ,
\end{aligned}
$$

some $c > 0$, because (see [195], equation 8.5.2.32)

$$
\lim_{l \to \infty} \left\{ C_{2 0 l 0}^{l0} \right\}^4 = \lim_{l \to \infty} \left[\frac{(2l + 1) [(l + 1)!]^2}{[(l - 1)!]^2} \left\{ \frac{2! 2! (2l - 2)!}{(2l + 3)!} \right\} \right]^2
$$

$$= \lim_{l \to \infty} \left[\frac{4(2l+1)\,[l(l+1)]^2}{(2l+3)(2l+2)(2l+1)(2l)(2l-1)} \right]^2 = \frac{1}{16}\,.$$

Provided $\liminf_{l \to \infty} C_l/C_{l/2} > 0$ holds, the left-hand side of condition (7.22) is then immediately seen to be bounded away from zero, so that the CLT (7.23) cannot hold.

8

Asymptotics for the Sample Power Spectrum

8.1 Introduction

The primary goal of this monograph is to investigate the mathematical foundations of the analysis of spherical random fields, and because of this we neglect throughout this work a detailed discussion on practical data analysis. In this chapter, however, we do find it necessary to discuss some background issues on CMB data collection, as these topics are indeed useful for understanding the motivations of the techniques presented below.

As discussed earlier in the monograph, CMB (temperature) maps can be viewed as the single realization of an isotropic, scalar-valued spherical random field. Observations are provided by means of electromagnetic detectors (so-called *radiometers* and/or *bolometers*) which measure fluxes of incoming radiations (i.e. photons) on a range of different frequencies. For instance, the celebrated *WMAP* experiment is endowed with 16 detectors, centred at frequencies 40.7, 60.8 and 93.5 GHz, which are labelled the Q, V and W band, respectively. The ESA (European Space Agency) mission *Planck* (launched on May 14, 2009) is based upon 70 channels ranging from 30 GHz to 857 GHz. As the satellites scan the sky, observations are collected as a vector time series, the number of observations being in the order of 10^9 for *WMAP* and 5×10^{10} for *Planck*. A first issue then relates to the construction of spherical maps starting from the *Time Ordered Data vector* (*TOD*) provided by the satellite observations; this is the so-called *map-making* challenge, see for instance [109] and [45]. For brevity's sake, we shall provide only the basic framework, and refer to the literature for more details. In short, we can assume that in each of the p channels we actually observe

$$O_i(x) = T(x) + F_i(x) + N_i(x) , i = 1, ..., p , x \in S^2 ;$$

here, $T(.)$ denotes the CMB signal, $F_i(x)$ denotes so-called *foreground emis-*

sions by galactic and extragalactic sources of non-cosmological nature (for instance galaxies, quasars, intergalactic dusts and others), and $N_i(x)$ instrumental noise. The crucial point to be understood is that the dependence across the different frequency channels of CMB emission is known, and it is different from the pattern followed by other sources: this capital property makes *component separation* possible and allows the construction of filtered maps (see for instance [153] and the references therein). More precisely, a clear prediction from theoretical physics, confirmed to amazing accuracy (for instance) from the *FIRAS* instrument on *COBE* (see [187]), is that the CMB radiation should follow the *Planckian curve* of blackbody radiation, i.e. radiation should be distributed across frequencies v_i, $i = 1, ..., p$, according to the function

$$R(v; x) = \frac{8\pi h v^3}{c^3} \frac{1}{e^{hv/k_B T(x)} - 1} , \qquad (8.1)$$

where $R(v; x)$ denotes the emission at frequency v for the corresponding temperature $T(x)$ (measured in Kelvin), c is the speed of light in the vacuum ($= 2.99798 \times 10^8$ m/s), h is Planck's constant ($= 6.6261 \times 10^{-27}$ erg/s), and k_B is Boltzmann's constant ($= 1.3807 \times 10^{-16}$ erg/K), see for instance [127, Chapter 2]. Even on noisy data, hence, the determination of $T(x)$ is made possible by the inversion of (8.1): the blackbody pattern can be estimated due to the presence of multiple detectors and the fact that astrophysical emissions of non-cosmological nature are characterized by a different pattern of dependence across frequencies. In some regions, however, foreground emissions are so strong that component separation is still a difficult statistical problem; moreover, in some areas of the sky (for instance the Galactic plane, i.e. the line of sight of the Milky Way) the problem is considered to be largely unsolvable, so that there are missing observations in CMB maps. Although these unobserved regions are becoming smaller and smaller with more refined experiments, their treatment is still an open problem and this is indeed one of the motivations for the introduction of spherical wavelets in Chapters 10-12.

In this chapter, we avoid all issues related to component separation or unobserved regions, and we focus on a simpler case, i.e. the circumstances where the collected data are simply provided by (CMB) signal plus noise. As already mentioned, in any given experiment the sky is scanned by detectors and at each time t the radiation from a certain direction is observed. We can then define "regressors" $\{x_{it}\}$, where $i = 1, ..., N_p$ labels the directions on the sphere (*pixels*) that are actually observed (in the order of $10^5/10^6$ for *WMAP-Planck*) and $t = 1, ..., N_T$ denotes the total number of time-ordered observations (in the order of $10^9/10^{10}$ for the above-mentioned experiments). More precisely, we take $x_{it} = 1$ if location i is pointed at time t, $x_{it} = 0$ otherwise. Given observations $\{y_t\}_{t=1,2,...}$, we are left with the regression model

$$y = XT + n,$$

where y is a $N_t \times 1$ vector of observations, X is a $N_T \times N_p$ "pointing" matrix, with a highly sparse structure (for each row one and only one elements is different from zero), and T, the $N_p \times 1$ vector of (random!) parameters, is just the value of the signal at different locations; finally, n is a $N_t \times 1$ vector representing instrumental noise. Under Gaussianity of noise n, and conditioning on a fixed value realization of the field, the vector T can be recovered by a *GLS* –type procedure

$$\widehat{T} = (X'\Omega^{-1}X)^{-1}X'\Omega^{-1}y,$$

where $\Omega = E[nn']$ denotes the variance-covariance matrix of noise. This procedure actually has to face two kinds of difficulties, namely, the matrix Ω is unknown and the estimates are computationally very demanding, as the matrices to be inverted have dimensions in the order of 10^6. We refer again to [109] and [45], and the references therein, for alternative methods to cope with these challenges.

There are several other statistically interesting issues involved with the reconstruction of the CMB maps $T(x)$ from the observations; actually the real experimental set-up is more complicated (and interesting) than this, because each location is observed unevenly, i.e. the *scanning strategy* is such that some regions are more accurately measured than others. Also, the contaminating noise can have a time-dependent structure (there is indeed strong evidence for long memory behaviour, see for instance [145]); the possible existence of noise correlation across different channels will be discussed below. These experimental features have sparked in the cosmological literature a very lively statistical debate on filtering and image reconstruction. We shall not have enough space here to investigate these important problems, and from the next section we will focus more directly on angular power spectrum estimation issues.

8.2 Angular power spectrum estimation

Let us consider first the "ideal" circumstances of maps without noise. We shall consider a strongly isotropic, finite-variance zero-mean spherical random field $T = \{T(x) : x \in S^2\}$, whose harmonic expansion is given by

$$T(x) = \sum_{lm} a_{lm} Y_{lm}(x), \quad x \in S^2,$$

(as usual, when no specification is added, the sum is taken over all possible values of l, m). Consistently with the notation introduced in (6.12), we write $\{C_l : l \geq 0\}$ to indicate the angular power spectrum of T, that is $C_l = E|a_{lm}|^2$ (with no dependence on m).

As noted in the previous chapters, having observed the random field $T(x)$, the coefficients $\{a_{lm}\}$ can be recovered by means of the inverse Fourier transform provided in Proposition 3.29. In practice, with real data the integral is replaced by finite sums, using (exact or approximate) cubature formulae, which are implemented in standard packages for CMB data analysis such as *HealPix* or *GLESP* (see [53, 85]). The angular power spectrum can then be estimated by using the statistic

$$\widehat{C}_l = \frac{1}{2l+1} \sum_{m=-l}^{l} |a_{lm}|^2. \tag{8.2}$$

It is indeed readily seen that we have

$$E\widehat{C}_l = C_l, \text{ for all } l = 1, 2, \dots$$

As already discussed (see (6.20)), square-integrability and isotropy entail that

$$E(T(x)^2) = \sum_l \frac{2l+1}{4\pi} C_l < \infty,$$

implying that $C_l \to 0$, as $l \to \infty$, and also

$$E\left|\widehat{C}_l - C_l\right| \leq 2C_l \to 0.$$

Moreover, for any $\varepsilon > 0$,

$$P\left\{\limsup_{l\to\infty} \left|\widehat{C}_l - C_l\right| > \varepsilon\right\} \leq \lim_{l\to\infty} \sum_{\ell \geq l} P\left\{\left|\widehat{C}_\ell - C_\ell\right| \geq \varepsilon\right\}$$

$$\leq \frac{1}{\varepsilon} \lim_{l\to\infty} \sum_{\ell \geq l} C_\ell = 0,$$

and we infer (by reasoning as in the proof of the Borel-Cantelli Lemma) that $\widehat{C}_l, \left|\widehat{C}_l - C_l\right| \to 0$ almost surely for any square-integrable strongly isotropic spherical random field. It is important to note that these conclusions do not provide any information about the magnitude of the ratio $\left|\widehat{C}_l - C_l\right|/C_l$, so that they have a very limited value from the standpoint of statistical estimation.

Starting from these considerations, we see that it is indeed necessary to focus on *normalized* quantities, such as the sequence

$$\widetilde{C}_l = \frac{1}{2l+1} \sum_{m=-l}^{l} \frac{|a_{lm}|^2}{C_l} = \frac{\widehat{C}_l}{C_l}, \quad l \geq 0. \tag{8.3}$$

Of course $E\widetilde{C}_l = 1$ for all l. The sequence $\left\{\widetilde{C}_l : l \geq 0\right\}$ can be used in order to meaningfully evaluate the asymptotic performance of any statistical procedure based on $\widehat{C}_l$. The following definition uses the coefficients $\widetilde{C}_l$ in order to define ergodicity.

Definition 8.1 (HFE)   Let T be a strongly isotropic, finite-variance spherical random field with angular power spectrum $\{C_l : l \geq 0\}$. We shall say that T is **High-Frequency Ergodic (HFE** – or ergodic in the high-frequency sense) if

$$\lim_{l\to\infty} E\left\{\widetilde{C}_l - 1\right\}^2 = \lim_{l\to\infty} E\left\{\frac{\widehat{C}_l}{C_l} - 1\right\}^2 = 0. \tag{8.4}$$

Condition (8.4) implies of course that $\widetilde{C}_l = \widehat{C}_l/C_l$ converges in probability towards the constant 1.

Remark 8.2   Consider the Fourier component

$$T_l(x) = \sum_{m=-l}^{l} a_{lm} Y_{lm}(x), \, l = 1, 2, ...,$$

which can be viewed (see Section 3.4.3) as a random eigenfunction of the spherical Laplacian, that is,

$$\Delta_{S^2} T_l(x) = -l(l+1)T_l(x), \, x \in S^2.$$

It should be noted that, owing to Parseval's identity,

$$ET_l^2 = \frac{(2l+1)}{4\pi}C_l, \, \int_{S^2} T_l^2(x)d\sigma(x) = \sum_{lm} |a_{lm}|^2 = (2l+1)\widehat{C}_l,$$

that is, investigating the convergence (8.4) is equivalent to studying the high-frequency behaviour of the ratio

$$\frac{\int_{S^2} T_l^2(x)d\sigma(x)}{4\pi ET_l^2}$$

between sample and population moments of spherical Gaussian eigenfunctions. More results on the geometric features of $\{T_l\}$ (for instance, the behaviour of their nodal lines or their excursion sets) have been recently proved in [138, 206, 207].

In the Gaussian case, as $l \to \infty$

$$E\left\{\frac{\widehat{C}_l}{C_l} - 1\right\}^2 = \frac{1}{(2l+1)^2}E\left[\frac{a_{l0}^2}{C_l} - 1 + 2\left\{\sum_{m=1}^{l} \frac{|a_{lm}|^2}{C_l} - 1\right\}\right]^2$$

$$= \frac{2}{2l+1} = o(1),$$

because $a_{l0}^2/C_l \overset{law}{\sim} \chi_1^2$ and for $m = 1, ..., l$, the variables $2|a_{lm}|^2/C_l$ are independent and identically distributed with law $2a_{lm}^2/C_l \overset{law}{\sim} \chi_2^2$, where χ_n^2 denotes a standard chi-square random variable with n degrees of freedom. In the Gaussian case with fully observed maps, the issue of angular power spectrum estimation can thus be considered trivial, and indeed the previous expressions not only ensure consistency but they also provide exact confidence intervals: it is immediate to see that

$$\sum_{m=-l}^{l} |a_{lm}|^2 = \left\{ |a_{l0}|^2 + \sum_{m=1}^{l} 2|a_{lm}|^2 \right\} \overset{law}{\sim} C_l \times \chi_{2l+1}^2 .$$

In view of the techniques discussed in Chapter 4, rather than simply giving a convergence result in the high-resolution sense (i.e. for $l \to \infty$), very sharp estimates can be given on the total variation distance distance between the law of the normalized random variable $\sqrt{\frac{2l+1}{2}} \left\{ \widehat{C}_l / C_l - 1 \right\}$ from a Gaussian law, for any finite value of l. Let us recall that the total variation distance between the laws of two random variables X_1, X_2, written $d_{TV}(X_1, X_2)$, is given by

$$d_{TV}(X_1, X_2) := \sup_A |P(X_1 \in A) - P(X_2 \in A)| ,$$

where the supremum runs over all Borel sets A. We have the following statement.

Lemma 8.3 *Let the isotropic field T considered in this section be Gaussian. Then, for all l, we have*

$$d_{TV}\left(\sqrt{\frac{2l+1}{2}} \left\{ \frac{\widehat{C}_l}{C_l} - 1 \right\}, N(0,1) \right) \le \sqrt{\frac{8}{2l+1}} .$$

Proof We have

$$\sqrt{\frac{2l+1}{2}} \left\{ \frac{\widehat{C}_l}{C_l} - 1 \right\} = \frac{1}{\sqrt{2(2l+1)}} \left\{ \frac{a_{l0}^2}{C_l} + \sum_{m=1}^{l} 2\frac{\{\text{Re}a_{lm}\}^2 + \{\text{Im}a_{lm}\}^2}{C_l} - (2l+1) \right\}$$

$$= \frac{1}{\sqrt{(2l+1)}} \left\{ \sum_{m=1}^{2l+1} \frac{(x_{lm}^2 - 1)}{\sqrt{2}} \right\} ,$$

where $\{x_{lm}\}$ are a triangular array of i.i.d. Gaussian standardized random variables. Also,

$$Cum_4 \left\{ \frac{\sqrt{2}}{\sqrt{(2l+1)}} \left[\sum_{m=1}^{2l+1} \frac{(x_{lm}^2 - 1)}{2} \right] \right\} = \frac{12}{2l+1} .$$

Now we recall that Theorem 4.18 states that for all zero-mean, unit variance random variables F_q that belong to the q-th Wiener chaos of a Gaussian random measure, the following inequality holds for the total variational distance with respect to a $N(0, 1)$ random variable:

$$d_{TV}(F_q, N(0, 1)) \leq \frac{2\sqrt{q-1}}{\sqrt{3q}} \sqrt{EF^4 - 3}.$$

The result now follows immediately. □

Remark 8.4 (The role of non-commutativity) The convergence result we have just illustrated does not have any analogue in Abelian circumstances, such as stationary random fields defined on the circle S^1. Indeed, assume that we deal with an isotropic random field $T(\vartheta)$, admitting the spectral expansion

$$T(\vartheta) = \sum_{k=-\infty}^{\infty} a_k \exp(ik\vartheta), \, 0 \leq \vartheta < 2\pi, \, E|a_k|^2 = C_k.$$

Assume we have observations on $T(\vartheta)$ which allow us to evaluate

$$a_k = \int_0^{2\pi} T(\vartheta) \exp(-ik\vartheta)d\vartheta, \, k = \pm 1, ..., \pm K.$$

We can define an estimator $\widehat{C}_k = |a_k|^2$ which could be viewed as an analogue to C_l; however even in such idealistic circumstances it is obvious that the ratio $\widehat{C}_k/C_k$ does not converge to unity in any meaningful sense and no asymptotic theory can be developed. We have thus seen that the non-Abelian nature of spectral analysis on the sphere allows for the possibility of an asymptotic theory that could not be pursued e.g. on the circle.

Clearly, the results presented in this section rely heavily on the Gaussian assumption. For instance, Theorem 6.12 implies that (under isotropy) the coefficients $\{a_{lm}\}$ can only be independent over m in the Gaussian case, despite the fact that they are always uncorrelated by construction (see Proposition 6.6). In other words, sampling independent, non-Gaussian random coefficients to generate maps using the spherical harmonic expansion will always yield an anisotropic random field. In that same Chapter 6, we have also shown how the dependence structure of the coefficients $\{a_{lm}\}$ is in general rather complicated, depending heavily up group representation properties for $SO(3)$. In view of this, to derive any asymptotic result for $\widehat{C}_l$ under non-Gaussianity is by no means trivial; indeed, even the possible consistency (as $l \to \infty$) of the estimator (8.2) in non-Gaussian circumstances is still an open issue for research.

Before turning to the analysis of a non-Gaussian framework, we shall discuss some important practical issues, i.e. how to deal with observational noise and missing observations.

8.3 Interlude: some practical issues

8.3.1 Dealing with instrumental noise

We shall now try to make our analysis more realistic by considering the effect of instrumental noise. More precisely, we shall consider the case where we observe $O(x) := T(x) + N(x)$, with $N(x)$ denoting noise; for simplicity, and following much of the cosmological literature, we shall assume $\{N(x)\}_{x \in S^2}$ to be also a zero-mean, square-integrable and strongly isotropic random field on the sphere, thus admitting the spectral expansion

$$N(x) = \sum_{lm} a^N_{lm} Y_{lm}(x), \ x \in S^2, \ E\left|a^N_{lm}\right|^2 =: C^N_l.$$

Whereas the assumptions of zero-mean and square-integrability are basically immaterial, isotropy of the noise may need to be relaxed if the sky is unevenly observed.

In this section, we shall label $\left\{a^T_{lm}\right\}$ the random spherical harmonic coefficients of the field $T(x)$, with $E\left|a^T_{lm}\right|^2 =: C^T_l$. We shall also assume that $T(.)$ and $N(.)$ are independent (as processes). Performing the spherical harmonic transform, we obtain, in an obvious notation

$$a_{lm} = \int_{S^2} \{T(x) + N(x)\} \overline{Y}_{lm}(x) d\sigma(x) =: a^T_{lm} + a^N_{lm},$$

which leads to

$$\widehat{C}_l = \frac{1}{2l+1} \left[\sum_{m=-l}^{l} |a^T_{lm}|^2 + \sum_{m=-l}^{l} |a^N_{lm}|^2 + 2\mathrm{Re}\left\{ \sum_{m=-l}^{l} a^T_{lm} \overline{a}^N_{lm} \right\} \right].$$

It is immediate to see that the estimator $\left\{\widehat{C}_l\right\}$ is biased for the parameters of interest $\left\{C^T_l\right\}$, indeed $E\widehat{C}_l = C^T_l + C^N_l$. Under Gaussianity, we have also

$$Var\left\{\widehat{C}_l\right\} = \frac{2\left\{C^T_l + C^N_l\right\}}{2l+1}. \tag{8.5}$$

In the cosmological literature, the standard procedure to address this bias is to assume that the noise correlation structure can be derived by Monte Carlo simulations or instrumental calibration; under this assumption, it is possible to subtract the bias from $\widehat{C}_l$ and obtain a correct estimator with variance (8.5). An obvious question is then to test whether the assumption that C^N_l is known does not introduce some spurious effect into the analysis (namely, some unaccounted bias). A proposal in this direction was put forward in [167]. To understand this idea, we must get back to the multi-channel setting, where we observe

$$O_i(x) := T(x) + N_i(x) , i = 1, ..., p ,$$

which in the harmonic domain leads to

$$a_{i;lm} := a_{lm}^T + a_{lm}^{N_i} .$$

Here, the subscript i labels the different frequencies at which CMB radiation is measured along the Planckian curve, compare our discussion in the Introduction to this chapter. Note that the signal component (i.e., CMB radiation) of the random spherical harmonics coefficients does not depend on the observing channel. We assume that the noise is independent over channels, which is believed to be consistent with the actual experimental set-ups of current datasets. Testing noise correlation across different channels is yet another open challenge for research. For a given noise structure, an obvious estimator for C_l is

$$\widetilde{C}_l^A := \frac{1}{p} \sum_{i=1}^p \left\{ \widehat{C}_{il} - C_l^{N_i} \right\} , \tag{8.6}$$

$$\widehat{C}_{il} := \frac{1}{2l+1} \sum_{m=-l}^l |a_{i;lm}|^2 .$$

The estimator $\widetilde{C}_l^A$ is known in the literature as the *auto-power spectrum*. Simple computations yield (see [167])

$$E\widetilde{C}_l^A = C_l , \quad Var\{\widetilde{C}_l^A\} = \frac{2}{2l+1} \left\{ C_l^2 + \frac{2C_l}{p^2} \sum_{i=1}^p C_l^{N_i} + \frac{1}{p^4} \sum_{i,j=1}^p C_l^{N_i} C_l^{N_j} \right\} .$$

Of course, the natural question that arises at this stage is the possible existence of misspecification, i.e., some errors in the bias-correction term $C_l^{N_i}$. A solution for this issue was proposed in [167]. The idea is to focus on the *cross-power spectrum* estimator

$$\widetilde{C}_l^{CP} = \frac{2}{p(p-1)} \sum_{i=1}^{p-1} \sum_{j=i+1}^p \left(\frac{1}{2l+1} \sum_{m=-l}^l a_{i;lm} \overline{a}_{j;lm} \right) .$$

The underlying rationale for $\widetilde{C}_l^{CP}$ is easy to gather: under the assumption that noise is independent across different channel, the estimator is unbiased, regardless of the value of the $C_l^{N_i}$. More precisely,

$$E\widetilde{C}_l^{CP} = \frac{2}{p(p-1)} \sum_{i=1}^{p-1} \sum_{j=i+1}^p \left(\frac{1}{2l+1} \sum_{m=-l}^l E\left(a_{lm}^T + a_{lm}^{N_i} \right) \left(\overline{a}_{lm}^T + \overline{a}_{lm}^{N_j} \right) \right)$$

$$= \frac{2}{p(p-1)} \sum_{i=1}^{p-1} \sum_{j=i+1}^{p} C_l^T = C_l^T .$$

Similar manipulations yield

$$Var\left\{\widetilde{C}_l^{CP}\right\} = \frac{2}{2l+1} \left\{ C_l^2 + \frac{2C_l}{p^2} \sum_{i=1}^{p} C_l^{N_i} + \frac{1}{p^2(p-1)^2} \sum_{i=1}^{p-1} \sum_{j=i+1}^{p} C_l^{N_i} C_l^{N_j} \right\} .$$

Merely for notational simplicity, we also assume that the noise variance is constant across detectors. It is then readily seen that

$$Var\left\{\widetilde{C}_l^{CP}\right\} - Var\left\{\widetilde{C}_l^{A}\right\} = \frac{2}{2l+1} \left\{ \frac{1}{p^2(p-1)} \left(C_l^N\right)^2 \right\} .$$

More explicitly, the auto-power spectrum estimator is more efficient that the cross-power spectrum; however the latter is robust to noise misspecification. This is the classical setting which makes the implementation of an Hausman-type test for misspecification feasible (see [95]). Indeed, it is possible to consider the statistic

$$H_l = \left[Var\left\{\widetilde{C}_l^{CP} - \widetilde{C}_l^{A}\right\} \right]^{-1/2} \left\{\widetilde{C}_l^{CP} - \widetilde{C}_l^{A}\right\} ,$$

$$Var\left\{\widetilde{C}_l^{CP} - \widetilde{C}_l^{A}\right\} = \frac{2}{2l+1} \left\{ \frac{1}{p^4} \sum_{i=1}^{p} \left\{C_l^{N_i}\right\}^2 + \frac{2}{(p-1)^2} \sum_{i=1}^{p-1} \sum_{j=i+1}^{p} C_l^{N_i} C_l^{N_j} \right\} .$$

Under the null hypothesis of exact bias correction, it is readily seen that $H_l \to_d$ $N(0, 1)$, as $l \to \infty$. On the other hand, in the presence of misspecification, i.e. when the actual noise variance is equal to $C_l^{N_i} + \delta$ for some i, $\delta > 0$, then we expect EH_l to diverge with rate $\sqrt{l}\delta$ as $l \to \infty$.

It is also possible to consider a functional form of the same test, focussing on

$$B_L(r) := \frac{1}{\sqrt{L}} \sum_{l=1}^{[Lr]} H_l , \ r \in [0, 1] .$$

Standard computations prove that $B_L(r)$ converges weakly to standard Brownian motion, as $L \to \infty$. A test for noise misspecification can then be constructed along the lines of standard Kolmogorov-Smirnov or Cramer-Von Mises statistics. We refer again to [167] for a much more detailed discussion and an extensive simulation study.

In the next subsection, we provide some preliminary remarks on another practical issue, i.e. the analysis of data from partially observed maps. This topic can be dealt with in much greater depth using the material introduced in Chapters 10-12, where spherical wavelets methods are discussed.

8.3.2 First remarks on missing observations

The presence of missing observations, i.e. regions of the sky where the CMB is deeply contaminated by astrophysical foregrounds, generates serious challenges for angular power spectrum estimation. The first consequence is that the sample spherical harmonics coefficients

$$a_{lm}^G := \int_{S^2 \backslash G} T(x) \overline{Y}_{lm}(x) d\sigma(x) ,$$

lose their uncorrelation properties (here, G denotes the unobserved region and for notational simplicity we came back to the case of a single detector with no instrumental noise). Indeed we have

$$
\begin{aligned}
E a_{l_1 m_1}^G & \overline{a}_{l_2 m_2}^G \\
&= E\left\{ \left(\int_{S^2 / G} T(x) \overline{Y}_{l_1 m_1}(x) d\sigma(x) \right) \left(\int_{S^2 / G} T(y) Y_{l_2 m_2}(y) d\sigma(y) \right) \right\} \qquad (8.7) \\
&= \sum_{l_1 m_1} \sum_{l_2 m_2} E a_{lm} \overline{a}_{l'm'} \left(\int_{S^2 \backslash G} Y_{lm}(x) \overline{Y}_{l_1 m_1}(x) d\sigma(x) \right) \left(\int_{S^2 \backslash G} \overline{Y}_{l'm'}(y) Y_{l_2 m_2}(y) d\sigma(y) \right) \\
&= \sum_{lm} C_l W_{lm l_1 m_1} W_{lm l_2 m_2} , \qquad (8.8)
\end{aligned}
$$

where $\{W_{lm l_1 m_1}\}$ denotes the so-called *coupling factors*

$$W_{lm l_1 m_1} := \int_{S^2 \backslash G} Y_{lm}(x) \overline{Y}_{l_1 m_1}(x) d\sigma(x) .$$

In case the spherical random field is fully observed, then $G = \emptyset$ (the empty set) and by standard orthonormality properties of the spherical harmonics Y_{lm} we obtain $W_{lm l_1 m_1} = \delta_{l_1}^l \delta_{m_1}^m$, $E a_{l_1 m_1} \overline{a}_{l_2 m_2} = C_l \delta_{l_1}^{l_2} \delta_{m_1}^{m_2}$, as expected. In the presence of missing observations, the (observed) random coefficients are no longer uncorrelated neither over l nor over m. In the physical literature, the values of $\{W_{lm l_1 m_1}\}$ are computed numerically exploiting the *a priori* knowledge on the geometry of the unobserved regions; the resulting *coupling matrices* can then be used to deconvolve the estimated values $\widehat{C}_l$, a procedure which has become extremely popular under the name of *MASTER* (see [100] for details). In practice, it is not possible to identify by this method the value of the angular power spectrum at every single multipole l; it is then customary to proceed with *binning techniques*, where the values of C_l at nearby frequencies is averaged and only these smoothed values are actually estimated. Plots for the estimates of the C_l derived along these lines can be found for instance on the web site of *WMAP*. Later in this monograph (Chapters 10 and 11) we will introduce an alternative strategy, that was first put forward in [13].

8.4 Asymptotics in the non-Gaussian case

We shall now investigate conditions for the (high-frequency) asymptotic Gaussianity of a given spherical random fields. This issue, which is studied in great detail [135], has already been discussed in Chapter 7. Roughly speaking, we are studying circumstances under which, although the field T is clearly non-Gaussian, its components T_l asymptotically are: in order to refer to this phenomenon in a compact way, we introduce the following definition.

Definition 8.5 (HFG)   Let $T(x)$ be a strongly isotropic, finite-variance spherical random field. We say that $T(x)$ is **high-frequency Gaussian (HFG)** whenever

$$\frac{T_l(x)}{\sqrt{Var\{T_l(x)\}}} \overset{law}{\to} N(0,1), \text{ as } l \to \infty, \tag{8.9}$$

for every fixed $x \in S^2$.

Remark 8.6   It is more delicate to define HFG involving convergence in the sense of finite-dimensional distributions. Indeed, we know from Chapter 7 that, even if relation (8.9) holds, the finite-dimensional distributions of order ≥ 2 of the field $x \mapsto T_l(x)/\sqrt{Var\{T_l(x)\}}$ may not converge to any limit (simply because the covariances may not converge).

Following Marinucci and Peccati [137], our purpose here is to investigate to what extent conditions (8.4) and (8.9) are equivalent, despite their apparent logical independence. Indeed, our results below suggest that equivalence may turn out to be the case in a wide array of examples, entailing that ergodicity (and hence the possibility to draw asymptotically justifiable statistical inferences on the angular power spectrum) and asymptotic Gaussianity are very tightly related in a high-resolution setting. This may lead, we believe, to important characterizations of Gaussian random fields, and to a better understanding on the conditions for the validity of statistical inferences on observations drawn from a unique realizations of compactly supported random fields, as in the spherical case.

We shall hence focus on the convergence (8.4), relaxing the assumption that T, and hence the sequence $\{a_{lm}\}$, is Gaussian. More precisely, we shall focus on Gaussian subordinated fields, as follows. Recall first that, for any isotropic random field,

$$T_l(\vartheta, \varphi) \overset{law}{=} T_l(\mathbf{e}_3) = \sum_{lm} a_{lm} Y_{lm}(\mathbf{e}_3) \overset{law}{=} a_{l0} \sqrt{\frac{2l+1}{4\pi}},$$

where we denote by $\mathbf{e}_3 = (0,0)$ (in spherical coordinates) the North Pole of the sphere. When T is Gaussian, we adopt the notation

$$H_q(T(x)) := T(x;q) = \sum_{l=0}^{\infty} T_l(x;q), \quad x \in S^2, q \geq 2, \tag{8.10}$$

with $H_q(.)$ denoting the qth Hermite polynomial, and where

$$T_l(x;q) = \sum_{m=-l}^{l} a_{lm;q} Y_{lm}(x) \tag{8.11}$$

is the lth frequency component of $T(.;q)$, with $a_{lm;q}$ the associated harmonic coefficients. Also, we write $\{C_{l;q} : l \geq 0\}$ and $\{\mathcal{T}_{ll}''(L;q)\}$, respectively, for the power spectrum and for the cumulant trispectrum of $T(.;q)$, i.e. (see Definition 6.37)

$$Cum\{a_{l_1 m_1}, a_{l_2 m_2}, a_{l_3 m_3}, a_{l_4 m_4}\} \tag{8.12}$$

$$= \sum_{LM} (-1)^M \begin{pmatrix} l_1 & l_2 & L \\ m_1 & m_2 & M \end{pmatrix} \begin{pmatrix} l_3 & l_4 & L \\ m_3 & m_4 & -M \end{pmatrix} (2L+1) \mathcal{T}_{l_1 l_2}^{l_3 l_4}(L;q),$$

i.e., with the notation of Definition 6.37, in this case $P_{l_1 \dots l_4}^C(L) = (-1)^L \mathcal{T}_{l_1 l_2}^{l_3 l_4}(L;q)$. As anticipated, we shall now prove some connections between HFE and HFG spherical fields (see Definitions 8.1 and 8.5), in the special case of fields of the type $T(.;q)$, as defined in (8.10). Note that the conditions appearing in the following statement involve the coefficients $C_{l;q}$, which are completely determined by the power spectrum of the underlying Gaussian field T. For the result to follow, we shall need the notation

$$w_{1l}(L) := \left(C_{l0l0}^{L0}\right)^2 \quad \text{and} \quad w_{2l}(L) := \frac{(2L+1)}{(2l+1)^2},$$

so that

$$\sum_{L=0}^{2l} w_{1l}(L) = \sum_{L=0}^{2l} w_{2l}(L) = 1$$

(the fact that the first sum is equal to one is a by-product of the computations below).

Theorem 8.7   *Let $q \geq 2$, and define $T(.;q)$ according to (8.10), where T is Gaussian and isotropic. Let $\mathcal{T}_{l_1 l_2}^{l_3 l_4}(L;q)$ be the reduced cumulant trispectrum of T. Then, the following holds.*

(1) The random field $T(.;q)$ is high-frequency Gaussian if and only if

$$\lim_{l \to \infty} \sum_{L=0}^{2l} w_{1l}(L) \frac{\mathcal{T}_{ll}''(L;q)}{C_{l;q}^2} = 0. \tag{8.13}$$

(2) On the other hand, $T(.;q)$ is high-frequency ergodic if and only if

$$\lim_{l\to\infty} \sum_{L=0}^{2l} w_{2l}(L) \frac{\mathcal{T}_{ll}^{ll}(L;q)}{C_{l;q}^2} = 0. \tag{8.14}$$

Before proving Theorem 8.7, we shall note that $\left\{C_{l0l0}^{L0}\right\}^2$ is different from zero only for even L, and $\mathcal{T}_{ll}^{ll}(L;q)$ is not in general positive-valued. Moreover, in view of the forthcoming Lemma 8.8, also in (8.14) the sum runs only over even values of L.

Lemma 8.8 $\mathcal{T}_{ll}^{ll}(L;q)$ *is zero when L is odd.*

Proof From [102, equation (17)], we infer that, for general trispectrum,

$$\mathcal{T}_{l_1 l_2}^{l_3 l_4}(L) = (-1)^{l_1+l_2+L} \mathcal{T}_{l_1 l_2}^{l_3 l_4}(L) .$$

Considering the case $l_1 = l_2 = l_3 = l_4 = l$, we obtain the desired result. □

Proof of Theorem 8.7 (1) Consider the random spherical field

$$(\vartheta, \varphi) \mapsto \hat{T}_{l;q}(\vartheta, \varphi) := \frac{T_l(\vartheta, \varphi; q)}{\sqrt{Var\left\{T_{l;q}(\mathbf{e}_3)\right\}}}, \quad (\vartheta, \varphi) \in [0, \pi] \times [0, 2\pi),$$

where $\mathbf{e}_3$ is the North Pole of S^2, and observe that, by isotropy and for every (ϑ, φ),

$$\hat{T}_{l;q}(\vartheta, \varphi) \stackrel{law}{=} \frac{a_{l0}}{\sqrt{4\pi C_{l;q}}} .$$

The field $\hat{T}_{l;q}$ is mean-zero and has unit variance: since it also belong to the qth Wiener chaos associated with T, we can use Theorem 4.18 to infer that it is asymptotically Gaussian if and only if

$$\lim_{l\to\infty} \frac{1}{C_{l;q}^2} Cum_4\left\{a_{l0;q}\right\} = 0 .$$

As discussed in Chapter 6 (see also [102] and [136]), isotropy entails that we can write the fourth-order cumulant as

$$Cum_4\left\{a_{l0;q}\right\} = \sum_{LM} (-1)^M \begin{pmatrix} l & l & L \\ 0 & 0 & M \end{pmatrix} \begin{pmatrix} l & l & L \\ 0 & 0 & -M \end{pmatrix} (2L+1) \mathcal{T}_{ll}^{ll}(L;q)$$

$$= \sum_{L} \begin{pmatrix} l & l & L \\ 0 & 0 & 0 \end{pmatrix}^2 (2L+1) \mathcal{T}_{ll}^{ll}(L;q) ,$$

where the second equality follows because the corresponding Clebsch-Gordan coefficients are identically zero unless $M = 0$. Hence the field is asymptotically Gaussian if and only if

$$\lim_{l \to \infty} \frac{1}{C_{l;q}^2} \sum_L \begin{pmatrix} l & l & L \\ 0 & 0 & 0 \end{pmatrix}^2 (2L + 1) \mathcal{T}_{ll}^{ll}(L; q) = 0. \tag{8.15}$$

We recall that

$$\begin{pmatrix} l & l & L \\ 0 & 0 & 0 \end{pmatrix}^2 (2L + 1) = \left\{ C_{l0l0}^{L0} \right\}^2,$$

entailing in turn that

$$\sum_L \begin{pmatrix} l & l & L \\ 0 & 0 & 0 \end{pmatrix}^2 (2L + 1) = \sum_{L=0}^{2l} \left\{ C_{l0l0}^{L0} \right\}^2 = \sum_{L=0}^{2l} \sum_{M=-L}^{L} \left\{ C_{l0l0}^{LM} \right\}^2 \equiv 1,$$

where the second equality follows from the fact that Clebsch-Gordan coefficients $C_{l_1 m_1 l_2 m_2}^{l_3 m_3}$ are different from zero only for $m_3 = m_1 + m_2$, and the third equality is a consequence from the orthonormality properties of these coefficients (see again Chapter 3). We therefore have

$$\frac{1}{C_{l;q}^2} Cum_4 \left\{ a_{l0;q} \right\} = \frac{1}{C_{l;q}^2} \sum_L \left\{ C_{l0l0}^{L0} \right\}^2 \mathcal{T}_{ll}^{ll}(L; q),$$

yielding the desired conclusion.
(2) On the other hand, we obtain also

$$E \left\{ \frac{\widehat{C}_{l;q}}{C_{l;q}} - 1 \right\}^2 = Var \left\{ \frac{\widehat{C}_{l;q}}{C_{l;q}} - 1 \right\} \tag{8.16}$$

$$= \frac{1}{(2l + 1)^2} \frac{1}{C_{l;q}^2} \sum_{m_1 m_2} Cum \left\{ a_{lm_1;q}, \bar{a}_{lm_1;q}, a_{lm_2;q}, \bar{a}_{lm_2;q} \right\} \tag{8.17}$$

$$+ \frac{2}{(2l + 1)^2} \frac{1}{C_{l;q}^2} \sum_m \left\{ E |a_{lm;q}|^2 \right\}^2 \tag{8.18}$$

$$= \frac{1}{(2l + 1)^2} \frac{1}{C_{l;q}^2} \sum_{m_1 m_2} (-1)^{m_1 + m_2} Cum \left\{ a_{lm_1;q}, a_{l,-m_1;q}, a_{lm_2;q}, a_{l,-m_2;q} \right\}$$

$$+ \frac{2}{(2l + 1)} \tag{8.19}$$

$$= \frac{2}{(2l + 1)^2} \frac{1}{C_{l;q}^2} \sum_{m_1 m_2} \sum_{LM} (-1)^{M + m_1 + m_2} \begin{pmatrix} l & l & L \\ m_1 & m_2 & M \end{pmatrix} \tag{8.20}$$

$$\times \begin{pmatrix} l & l & L \\ -m_1 & -m_2 & -M \end{pmatrix} (2L + 1) \mathcal{T}_{ll}^{ll}(L; q) + \frac{2}{(2l + 1)} \tag{8.21}$$

$$= \frac{2}{(2l+1)^2} \frac{1}{C_{l;q}^2} \sum_{L=0}^{2l} (2L+1) \mathcal{T}_{ll}^{ll}(L;q) + \frac{2}{(2l+1)}. \tag{8.22}$$

□

Remark 8.9   Note that

$$\left\{ C_{l0l0}^{L0} \right\}^2 = \frac{(2L+1)(\frac{2l+L}{2}!)^2}{(\frac{L}{2}!)^2} \frac{(L!)^2(2l-L)!}{(2l+L+1)!} \le \frac{1}{(2L+1)}$$

$$w_{2l}(L) = \frac{(2L+1)}{(2l+1)^2} \le \frac{1}{2l+1}.$$

Note also that in the Gaussian case (e.g., $q = 1$) we have $\mathcal{T}_{ll}^{ll}(L;1) \equiv 0$, whence

$$E \left\{ \frac{\widehat{C_{l;q}}}{C_{l;q}} - 1 \right\}^2 = \frac{2}{2l+1} \rightarrow 0,$$

as expected.

The previous result was first given in [137] and strongly suggests that the conditions for asymptotic Gaussianity (HFG) and for ergodicity (HFE) should be tightly related. Indeed we conjecture that HFE and HFG are equivalent in the case of Hermite type Gaussian subordinations (and most probably even in more general circumstances). However, proving this claim seems analytically too demanding at this stage, so that for the rest of this chapter we shall focus on a detailed analysis of quadratic Gaussian subordinations. In particular, we believe that the content of the forthcoming section may provide the seed for a complete understanding of the HFG-HFE connection.

Remark 8.10   It should be noted that, in general, the reduced trispectrum satisfies (see [102, Eq. (16)]),

$$\mathcal{T}_{ll}^{ll}(L') = \sum_{L} (2L+1) \left\{ \begin{array}{ccc} l & l & L \\ l & l & L' \end{array} \right\} \mathcal{T}_{ll}^{ll}(L).$$

In the previous remark, we used once again the Wigner's $6j$ coefficients, see Chapters 3, 6, or [23], [195] for further properties and more discussion.

8.5 The quadratic case

8.5.1 The class $\mathfrak{D}$ and main results

As anticipated, the purpose of this section is to provide a more detailed and explicit analysis of the quadratic case $q = 2$. For simplicity, in the sequel we consider a centered Gaussian isotropic spherical field T such that $Var(T(x)) = \sum_l (2l + 1)C_l/4\pi = 1$, where $\{C_l\}$ is as before the power spectrum of T. We start by recalling the notation

$$T(x; 2) = H_2(T(x)) = \sum_{l_1, l_2=0}^{\infty} \sum_{m_1 m_2} a_{l_1 m_1} a_{l_2 m_2} Y_{l_1 m_1}(x) Y_{l_2 m_2}(x) - 1 , \qquad (8.23)$$

where T is isotropic, centered and Gaussian. Our first result was given for instance in [135] and [137].

Lemma 8.11   *The angular power spectrum of the squared random field (8.23) is given by*

$$C_{l;2} = E|a_{lm;2}|^2 = 2 \sum_{l_1 l_2} C_{l_1} C_{l_2} \begin{pmatrix} l_1 & l_2 & l \\ 0 & 0 & 0 \end{pmatrix}^2 \frac{(2l_1 + 1)(2l_2 + 1)}{4\pi} .$$

Proof   From (8.23), and in view of (8.11)) we have

$$a_{lm;2} = \int_{S^2} \sum_{l_1 l_2} \sum_{m_1 m_2} a_{l_1 m_1} a_{l_2 m_2} Y_{l_1 m_1}(x) Y_{l_2 m_2}(x) \overline{Y}_{lm}(x) d\sigma(x)$$

$$= (-1)^m \sum_{l_1, l_2=0}^{\infty} \sum_{m_1 m_2} a_{l_1 m_1} a_{l_2 m_2} \begin{pmatrix} l_1 & l_2 & l \\ m_1 & m_2 & -m \end{pmatrix}$$

$$\times \begin{pmatrix} l_1 & l_2 & l \\ 0 & 0 & 0 \end{pmatrix} \sqrt{\frac{(2l_1 + 1)(2l_2 + 1)(2l + 1)}{4\pi}}, \qquad (8.24)$$

where we have used (3.64) together with the definition of the $3j$ symbols. Note that the constant term -1 has no effect for $l \geq 1$, because

$$\int_{S^2} Y_{lm}(x) d\sigma(x) = 0 \text{ for all } l \geq 1 .$$

Since T is centered, isotropic and has unit variance, we have $Ea_{lm;2} = 0$ for all l. Furthermore

$$E\left|a_{lm;2}\right|^2$$

$$= E\left\{ \sum_{l_1 l_2} \sum_{m_1 m_2} a_{l_1 m_1} a_{l_2 m_2} \begin{pmatrix} l_1 & l_2 & l \\ m_1 & m_2 & -m \end{pmatrix} \right.$$

$$\begin{pmatrix} l_1 & l_2 & l \\ 0 & 0 & 0 \end{pmatrix} \sqrt{\frac{(2l_1 + 1)(2l_2 + 1)(2l + 1)}{4\pi}}$$

$$\times \sum_{l_1' l_2'} \sum_{m_1' m_2'} \bar{a}_{l_1' m_1'} \bar{a}_{l_2' m_2'} \begin{pmatrix} l_1' & l_2' & l \\ m_1' & m_2' & -m \end{pmatrix}$$

$$\times \begin{pmatrix} l_1' & l_2' & l \\ 0 & 0 & 0 \end{pmatrix} \sqrt{\frac{(2l_1' + 1)(2l_2' + 1)(2l + 1)}{4\pi}} \Bigg\}$$

$$= 2 \sum_{l_1 l_2} C_{l_1} C_{l_2} \sum_{m_1 m_2} \begin{pmatrix} l_1 & l_2 & l \\ m_1 & m_2 & -m \end{pmatrix}^2 \begin{pmatrix} l_1 & l_2 & l \\ 0 & 0 & 0 \end{pmatrix}^2 \frac{(2l_1 + 1)(2l_2 + 1)(2l + 1)}{4\pi}$$

$$= 2 \sum_{l_1 l_2} C_{l_1} C_{l_2} \begin{pmatrix} l_1 & l_2 & l \\ 0 & 0 & 0 \end{pmatrix}^2 \frac{(2l_1 + 1)(2l_2 + 1)}{4\pi} ,$$

and the proof is complete. $\qquad\qquad\qquad\qquad\qquad\qquad\qquad\qquad\qquad$ □

Remark 8.12 Note that

$$Var\{T^2(x)\} = \sum_l \frac{2l + 1}{4\pi} C_{l;2}$$

$$= 2 \sum_{l_1 l_2} C_{l_1} C_{l_2} \frac{(2l_1 + 1)(2l_2 + 1)}{4\pi} \left\{ \sum_l \frac{2l + 1}{4\pi} \begin{pmatrix} l_1 & l_2 & l \\ 0 & 0 & 0 \end{pmatrix}^2 \right\}$$

$$= 2 \sum_{l_1 l_2} C_{l_1} C_{l_2} \frac{(2l_1 + 1)(2l_2 + 1)}{(4\pi)^2} = 2 \left[Var\{T(x)\} \right]^2 ,$$

as expected from standard properties of Gaussian variables. Here we have used again

$$\sum_l (2l + 1) \begin{pmatrix} l_1 & l_2 & l \\ 0 & 0 & 0 \end{pmatrix}^2 \equiv 1 .$$

Our strategy is now the following. We shall first define a very general class, noted $\mathfrak{D}$, of quadratic models in terms of the power spectrum of the underlying Gaussian field, and then we shall show that the two notions of HFG and HFE coincide within $\mathfrak{D}$.

Definition 8.13 The centered Gaussian isotropic field T is said to belong to the class $\mathfrak{D}$ if there exist real numbers α, β such that

(1) $\alpha \in \mathbb{R}$ and $\beta \geq 0$

(2) $\sum_{l=0}^{\infty} l^{-\alpha+1} e^{-\beta l} < \infty$

(3) There exists constants $c_1, c_2 > 0$ such that

$$0 < c_1 \leq \lim_{l \to \infty} \frac{C_l}{l^{-\alpha} e^{-\beta l}} \leq \lim_{l \to \infty} \frac{C_l}{l^{-\alpha} e^{-\beta l}} \leq c_2 < \infty . \qquad (8.25)$$

Remark 8.14   As a first approximation, the class $\mathfrak{D}$ contains virtually all models that are relevant for CMB modeling in the case of a quadratic Gaussian subordination. For instance, Sachs-Wolfe models with the so-called Bardeen's potential entail a polynomial decay of the C_l ($\beta = 0$), whereas the so-called *Silk damping* effect entails an exponential decay of the power spectrum of primary CMB anisotropies at higher l. We refer again to textbooks such as [51, 60] for more discussion on these points.

Remark 8.15   Note that Condition 2 in the definition of $\mathfrak{D}$ implies that the parameters α, β must be such that either $\beta = 0$ and $\alpha > 2$, or $\beta > 0$ and $\alpha \in \mathbb{R}$ (with no restrictions).

The next statement is the main achievement of this section. It shows in particular, that the HFG and HFE classes exhibit the same phase transition within the class $\mathfrak{D}$

Theorem 8.16   *Let $T(.;2) = H_2(T)$, where the centered Gaussian isotropic field $T(.)$ is an element of the class $\mathfrak{D}$. Then, the following three conditions are equivalent*

(i) $T(.;2)$ *is HFG*

(ii) $T(.;2)$ *is HFE*

(iii) $\beta > 0$ *and* $\alpha \in \mathbb{R}$.

8.5.2 Proof of Theorem 8.16

From [135, Section 6], we already know that Conditions (i) and (iii) in the statement of Theorem 8.16 are equivalent. The proof of the remaining implication (ii) $\Longleftrightarrow$ (iii) is divided in several steps.

We start by showing that, if (iii) is not verified, then the angular power spectrum of the transformed field, under broad conditions, exhibits the same behaviour as the angular power spectrum of the subordinating field.

Lemma 8.17 *Suppose $\beta = 0$ and $\alpha > 2$, then*

$$\frac{3 \times 2^{\alpha}}{4\pi} C_l \frac{c_1^2}{c_2} \leq C_{l;2} \leq \frac{c_2^2}{c_1 \pi} \{2\zeta(\alpha - 1) + \zeta(\alpha)\} C_{l/2} = O(C_l),$$

where $\zeta(.)$ denotes the Riemann zeta function.

Proof We have

$$\sum_{l_1 l_2} C_{l_1} C_{l_2} \begin{pmatrix} l_1 & l_2 & l \\ 0 & 0 & 0 \end{pmatrix}^2 \frac{(2l_1 + 1)(2l_2 + 1)}{4\pi} \tag{8.26}$$

$$\leq 2 \sum_{l_1 \leq l_2} C_{l_1} C_{l_2} \begin{pmatrix} l_1 & l_2 & l \\ 0 & 0 & 0 \end{pmatrix}^2 \frac{(2l_1 + 1)(2l_2 + 1)}{4\pi} \tag{8.27}$$

$$\leq 2 \frac{c_2}{c_1} C_{l/2} \sum_{l_1 \leq l_2} C_{l_1} \begin{pmatrix} l_1 & l_2 & l \\ 0 & 0 & 0 \end{pmatrix}^2 \frac{(2l_1 + 1)(2l_2 + 1)}{4\pi}, \tag{8.28}$$

because $(l_1 \vee l_2) > l/2$ by the triangle conditions and $\sup_{l_2 \geq l/2} C_{l_2}/C_{l/2} \leq c_2/c_1$.
Now

$$\sum_{l_1 \leq l_2} C_{l_1} \begin{pmatrix} l_1 & l_2 & l \\ 0 & 0 & 0 \end{pmatrix}^2 \frac{(2l_1 + 1)(2l_2 + 1)}{4\pi}$$

$$\leq \sum_{l_1} C_{l_1} \frac{(2l_1 + 1)}{4\pi} \sum_{l_2} (2l_2 + 1) \begin{pmatrix} l_1 & l_2 & l \\ 0 & 0 & 0 \end{pmatrix}^2 = \sum_{l_1} C_{l_1} \frac{(2l_1 + 1)}{4\pi} < \infty.$$

More precisely

$$\sum_{l_1} C_{l_1} \frac{(2l_1 + 1)}{4\pi} \leq \frac{c_2}{4\pi} \sum_l (2l + 1) l^{-\alpha} \leq \frac{c_2}{4\pi} \{2\zeta(\alpha - 1) + \zeta(\alpha)\}.$$

Hence

$$C_{l;2} \leq \frac{c_2^2}{2c_1 \pi} \{2\zeta(\alpha - 1) + \zeta(\alpha)\} C_{l/2}.$$

The upper bound is then established. For the lower bound, it is sufficient to
show that

$$\sum_{l_1 l_2} C_{l_1} C_{l_2} \begin{pmatrix} l_1 & l_2 & l \\ 0 & 0 & 0 \end{pmatrix}^2 \frac{(2l_1 + 1)(2l_2 + 1)}{4\pi} \tag{8.29}$$

$$\geq \sum_{l_2} C_l C_{l_2} \begin{pmatrix} l_1 & l_2 & l \\ 0 & 0 & 0 \end{pmatrix}^2 \frac{3(2l_2 + 1)}{4\pi} \tag{8.30}$$

$$\geq 3 \times 2^{\alpha} C_l \frac{c_1^2}{c_2} \sum_{l_2} \begin{pmatrix} l_1 & l_2 & l \\ 0 & 0 & 0 \end{pmatrix}^2 \frac{(2l_2 + 1)}{4\pi} \geq \frac{3 \times 2^{\alpha}}{4\pi} C_l \frac{c_1^2}{c_2}, \tag{8.31}$$

as claimed.   □

Loosely, the previous Lemma 8.17 states that, under algebraic decay, the rate of convergence to zero of the angular power spectrum is not affected by a quadratic transformation, i.e. $C_{l;2} \simeq C_l$. The following result holds for fixed l, and it is therefore not related to the high-frequency asymptotic behaviour of the power spectrum $\{C_l\}$. Note that we use the notation

$$\widehat{C}_{l;2} = \frac{1}{2l+1} \sum_{m=-l}^{l} |a_{lm;2}|^2, \ \widetilde{C}_{l;2} = \frac{\widehat{C}_{l;2}}{C_{l;2}}.$$

Lemma 8.18   *Let $T(.;2)$ be defined by (8.23). Then we have*

$$E\left\{\widetilde{C}_{l;2} - 1\right\}^2$$

$$= \frac{16}{C_{l;2}^2} \sum_{l_1 l_2 l_3} C_{l_1}^2 C_{l_2} C_{l_3} \begin{pmatrix} l_1 & l_2 & l \\ 0 & 0 & 0 \end{pmatrix}^2 \begin{pmatrix} l_1 & l_3 & l \\ 0 & 0 & 0 \end{pmatrix}^2$$

$$\times \frac{(2l_1 + 1)(2l_2 + 1)(2l_3 + 1)}{(4\pi)^2} + R(l),$$

where for all $l = 1, 2, \dots$

$$0 \le R(l) \le \frac{4}{2l+1}.$$

Proof   In the sequel, we shall use repeatedly the unitary properties of Clebsch-Gordan coefficients, i.e.

$$\sum_{m_1 m_2} \begin{pmatrix} l & l & L \\ m_1 & m_2 & M \end{pmatrix} \begin{pmatrix} l & l & L' \\ m_1 & m_2 & M' \end{pmatrix} = \frac{\delta_L^{L'} \delta_M^{M'}}{2L+1}. \qquad (8.32)$$

Recalling (8.18)–(8.19), we need to evaluate

$$\frac{1}{(2l+1)^2 C_{l;2}^2} \sum_{m_1 m_2} Cum\{a_{lm_1}, \overline{a}_{lm_1}, a_{lm_2}, \overline{a}_{lm_2}\}$$

$$= \frac{1}{(2l+1)^2 C_{l;2}^2} \sum_{m_1 m_2} (-1)^{m_1 + m_2} Cum\{a_{lm_1}, a_{l,-m_1}, a_{lm_2}, a_{l,-m_2}\}.$$

Now

$$Cum\{a_{lm_1}, a_{l,-m_1}, a_{lm_2}, a_{l,-m_2}\}$$

$$= Cum\left\{\sum_{l_1 l_2} \sum_{\mu_1 \mu_2} a_{l_1 \mu_1} a_{l_2 \mu_2} \begin{pmatrix} l_1 & l_2 & l \\ \mu_1 & \mu_2 & m_1 \end{pmatrix}\right.$$

$$\times \begin{pmatrix} l_1 & l_2 & l \\ 0 & 0 & 0 \end{pmatrix} \sqrt{\frac{(2l_1+1)(2l_2+1)(2l+1)}{4\pi}},$$

$$\sum_{l_3 l_4} \sum_{\mu_3 \mu_4} a_{l_3 \mu_3} a_{l_4 \mu_4} \begin{pmatrix} l_3 & l_4 & l \\ \mu_3 & \mu_4 & -m_1 \end{pmatrix}$$

$$\times \begin{pmatrix} l_3 & l_4 & l \\ 0 & 0 & 0 \end{pmatrix} \sqrt{\frac{(2l_3+1)(2l_4+1)(2l+1)}{4\pi}},$$

$$\sum_{l_5 l_6} \sum_{\mu_5 \mu_6} a_{l_5 \mu_5} a_{l_6 \mu_6} \begin{pmatrix} l_5 & l_6 & l \\ \mu_5 & \mu_6 & m_2 \end{pmatrix}$$

$$\times \begin{pmatrix} l_5 & l_6 & l \\ 0 & 0 & 0 \end{pmatrix} \sqrt{\frac{(2l_5+1)(2l_6+1)(2l+1)}{4\pi}},$$

$$\sum_{l_7 l_8} \sum_{\mu_7 \mu_8} a_{l_7 \mu_7} a_{l_8 \mu_8} \begin{pmatrix} l_7 & l_8 & l \\ \mu_7 & \mu_8 & -m_2 \end{pmatrix}$$

$$\times \begin{pmatrix} l_7 & l_8 & l \\ 0 & 0 & 0 \end{pmatrix} \sqrt{\frac{(2l_7+1)(2l_8+1)(2l+1)}{4\pi}} \Bigg\}$$

and counting equivalent permutations

$$= 8 \sum_{l_1 l_2 l_3 l_4} \sum_{\mu_1 \mu_2 \mu_3 \mu_4} (-1)^{\mu_1+\mu_2+\mu_3+\mu_4} C_{l_1} C_{l_2} C_{l_3} C_{l_4} \begin{pmatrix} l_1 & l_2 & l \\ \mu_1 & \mu_2 & m_1 \end{pmatrix}$$

$$\times \begin{pmatrix} l_1 & l_2 & l \\ 0 & 0 & 0 \end{pmatrix} \frac{(2l+1)^2 \prod_{i=1}^{4}(2l_i+1)}{(4\pi)^2}$$

$$\times \begin{pmatrix} l_1 & l_3 & l \\ -\mu_1 & -\mu_3 & -m_1 \end{pmatrix} \begin{pmatrix} l_1 & l_3 & l \\ 0 & 0 & 0 \end{pmatrix} \begin{pmatrix} l_4 & l_3 & l \\ \mu_4 & \mu_3 & m_2 \end{pmatrix}$$

$$\times \begin{pmatrix} l_4 & l_3 & l \\ 0 & 0 & 0 \end{pmatrix} \begin{pmatrix} l_4 & l_2 & l \\ -\mu_4 & -\mu_2 & m_2 \end{pmatrix} \begin{pmatrix} l_4 & l_2 & l \\ 0 & 0 & 0 \end{pmatrix}$$

$$+8 \sum_{l_1 l_2 l_3 l_4} \sum_{\mu_1 \mu_2 \mu_3 \mu_4} (-1)^{\mu_1+\mu_2+\mu_3+\mu_4} C_{l_1} C_{l_2} C_{l_3} C_{l_4} \begin{pmatrix} l_1 & l_2 & l \\ \mu_1 & \mu_2 & m_1 \end{pmatrix}$$

$$\times \begin{pmatrix} l_1 & l_2 & l \\ 0 & 0 & 0 \end{pmatrix} \frac{(2l+1)^2 \prod_{i=1}^{4}(2l_i+1)}{(4\pi)^2}$$

$$\times \begin{pmatrix} l_3 & l_4 & l \\ \mu_3 & \mu_4 & -m_1 \end{pmatrix} \begin{pmatrix} l_3 & l_4 & l \\ 0 & 0 & 0 \end{pmatrix} \begin{pmatrix} l_1 & l_3 & l \\ -\mu_1 & -\mu_3 & m_2 \end{pmatrix}$$

$$\times \begin{pmatrix} l_1 & l_3 & l \\ 0 & 0 & 0 \end{pmatrix} \begin{pmatrix} l_4 & l_2 & l \\ -\mu_4 & -\mu_2 & -m_2 \end{pmatrix} \begin{pmatrix} l_4 & l_2 & l \\ 0 & 0 & 0 \end{pmatrix}$$

$$+8 \sum_{l_1 l_2 l_3 l_4} \sum_{\mu_1 \mu_2 \mu_3 \mu_4} (-1)^{\mu_1+\mu_2+\mu_3+\mu_4} C_{l_1} C_{l_2} C_{l_3} C_{l_4} \begin{pmatrix} l_1 & l_2 & l \\ \mu_1 & \mu_2 & m_1 \end{pmatrix}$$

$$\times \begin{pmatrix} l_1 & l_2 & l \\ 0 & 0 & 0 \end{pmatrix} \begin{pmatrix} l_1 & l_3 & l \\ -\mu_1 & \mu_3 & -m_1 \end{pmatrix} \begin{pmatrix} l_1 & l_3 & l \\ 0 & 0 & 0 \end{pmatrix}$$

$$\times \begin{pmatrix} l_2 & l_4 & l \\ -\mu_2 & \mu_4 & m_2 \end{pmatrix} \begin{pmatrix} l_2 & l_4 & l \\ 0 & 0 & 0 \end{pmatrix} \begin{pmatrix} l_3 & l_4 & l \\ -\mu_3 & -\mu_4 & -m_2 \end{pmatrix}$$

$$\times \begin{pmatrix} l_4 & l_2 & l \\ 0 & 0 & 0 \end{pmatrix} \frac{(2l+1)^2 \prod_{i=1}^{4}(2l_i+1)}{(4\pi)^2}$$

$$=: 8\{A(m_1,-m_1,m_2,-m_2) + B(m_1,-m_1,m_2,-m_2) + C(m_1,-m_1,m_2,-m_2)\}.$$

For the first term, note first that $(-1)^{m_1+m_2+\mu_1+\mu_2+\mu_3+\mu_4} \equiv 1$, because the exponent is necessarily even by the properties of Wigner's coefficients. Moreover, applying iteratively (8.32)

$$\sum_{m_1 m_2} A(m_1,-m_1,m_2,-m_2)$$

$$= \sum_{l_1 l_2 l_3 l_4} \sum_{m_2 \mu_2 \, \mu_3 \mu_4} C_{l_1} C_{l_2} C_{l_3} C_{l_4} \begin{pmatrix} l_1 & l_2 & l \\ 0 & 0 & 0 \end{pmatrix} \begin{pmatrix} l_1 & l_3 & l \\ 0 & 0 & 0 \end{pmatrix} \frac{(2l+1)^2 \prod_{i=1}^{4}(2l_i+1)}{(4\pi)^2}$$

$$\times \begin{pmatrix} l_4 & l_3 & l \\ \mu_4 & \mu_3 & m_2 \end{pmatrix} \begin{pmatrix} l_4 & l_3 & l \\ 0 & 0 & 0 \end{pmatrix} \begin{pmatrix} l_4 & l_2 & l \\ -\mu_4 & -\mu_2 & m_2 \end{pmatrix} \begin{pmatrix} l_4 & l_2 & l \\ 0 & 0 & 0 \end{pmatrix} \frac{\delta_{\mu_3}^{\mu_2} \delta_{l_2}^{l_3}}{2l_3+1}$$

$$= \sum_{l_1 l_2 l_4} C_{l_1} C_{l_2}^{2} C_{l_4} \begin{pmatrix} l_1 & l_2 & l \\ 0 & 0 & 0 \end{pmatrix}^2 \begin{pmatrix} l_4 & l_2 & l \\ 0 & 0 & 0 \end{pmatrix}^2 \frac{(2l_1+1)(2l_2+1)(2l_4+1)(2l+1)^2}{(4\pi)^2}.$$

Likewise, for the second term we note that $(-1)^{\mu_1+\mu_2+\mu_3+\mu_4} \equiv 1$, and using the definition of Wigner's $6j$ coefficients

$$\sum_{m_1 m_2} (-1)^{m_1+m_2} B(m_1,-m_1,m_2,-m_2)$$

$$= \sum_{l_1 l_2 l_3 l_4} C_{l_1} C_{l_2} C_{l_3} C_{l_4} \begin{Bmatrix} l_1 & l_3 & l \\ l_4 & l_2 & l \end{Bmatrix} \begin{pmatrix} l_1 & l_2 & l \\ 0 & 0 & 0 \end{pmatrix} \begin{pmatrix} l_3 & l_4 & l \\ 0 & 0 & 0 \end{pmatrix}$$

$$\times \begin{pmatrix} l_1 & l_3 & l \\ 0 & 0 & 0 \end{pmatrix} \begin{pmatrix} l_4 & l_2 & l \\ 0 & 0 & 0 \end{pmatrix} \frac{(2l+1)^2 \prod\limits_{i=1}^{4}(2l_i+1)}{(4\pi)^2}.$$

Now by the Cauchy-Schwartz inequality and recalling that

$$\left| \left\{ \begin{array}{ccc} l_1 & l_3 & l \\ l_2 & l_4 & l \end{array} \right\} \right| \le \frac{1}{2l+1} \text{ for all } l_1, l_2, l_3, l_4 ,$$

the previous quantity can be bounded by

$$\frac{1}{2l+1} \sum_{l_1 l_2 l_3 l_4} C_{l_1} C_{l_2} C_{l_3} C_{l_4} \begin{pmatrix} l_1 & l_2 & l \\ 0 & 0 & 0 \end{pmatrix} \begin{pmatrix} l_3 & l_4 & l \\ 0 & 0 & 0 \end{pmatrix}$$

$$\times \begin{pmatrix} l_1 & l_3 & l \\ 0 & 0 & 0 \end{pmatrix} \begin{pmatrix} l_4 & l_2 & l \\ 0 & 0 & 0 \end{pmatrix} \frac{(2l+1)^2 \prod\limits_{i=1}^{4}(2l_i+1)}{(4\pi)^2}$$

$$\le \left[\sum_{l_1 l_2 l_3 l_4} C_{l_1} C_{l_2} C_{l_3} C_{l_4} \begin{pmatrix} l_1 & l_2 & l \\ 0 & 0 & 0 \end{pmatrix}^2 \begin{pmatrix} l_3 & l_4 & l \\ 0 & 0 & 0 \end{pmatrix}^2 \frac{(2l+1)^2 \prod\limits_{i=1}^{4}(2l_i+1)}{(4\pi)^2} \right]^{1/2}$$

$$\times \left[\sum_{l_1 l_2 l_3 l_4} C_{l_1} C_{l_2} C_{l_3} C_{l_4} \begin{pmatrix} l_1 & l_3 & l \\ 0 & 0 & 0 \end{pmatrix}^2 \begin{pmatrix} l_2 & l_4 & l \\ 0 & 0 & 0 \end{pmatrix}^2 \frac{(2l+1)^2 \prod\limits_{i=1}^{4}(2l_i+1)}{(4\pi)^2} \right]^{1/2}$$

$$= \frac{2l+1}{4} C_{l;2}^2 ,$$

whence

$$\left| \frac{8}{(2l+1)^2 C_{l;2}^2} \sum_{m_1,m_2} B(m_1, -m_1, m_2, -m_2) \right| \le \frac{2}{2l+1} .$$

It is easy to see that $\sum_{m_1 m_2} A(m_1, -m_1, m_2, -m_2) = \sum_{m_1 m_2} C(m_1, -m_1, m_2, -m_2)$. In view of 8.18, 8.19, the statement of the Lemma follows easily. □

The proof of Theorem 8.16 is now concluded by the following Lemma.

Lemma 8.19 *If $\beta = 0$ and $\alpha > 2$, then*

$$\liminf_{l\to\infty} E\left\{ \widetilde{C}_{l;2} - 1 \right\}^2 \ge C_2^2 \left\{ \frac{c_2^3}{c_1^2} \{2\zeta(\alpha-1) + \zeta(\alpha)\} 2^\alpha \right\}^{-2} > 0.$$

If $\beta > 0$ and α is real, then

$$\lim_{l\to\infty} E\left\{\widetilde{C}_{l;2} - 1\right\}^2 = 0 \, .$$

Proof For the first part, from Lemma 8.18 we can focus on

$$\frac{1}{C_{l;2}^2} \sum_{l_1 l_2 l_3} C_{l_1}^2 C_{l_2} C_{l_3} \begin{pmatrix} l_1 & l_2 & l \\ 0 & 0 & 0 \end{pmatrix}^2 \begin{pmatrix} l_1 & l_3 & l \\ 0 & 0 & 0 \end{pmatrix}^2 \frac{(2l_1+1)(2l_2+1)(2l_3+1)}{(4\pi)^2}$$

$$= \frac{1}{C_{l;2}^2} \sum_{l_1 l_2} (2l_1+1)(2l_2+1) C_{l_1} C_{l_2} \begin{pmatrix} l_1 & l_2 & l \\ 0 & 0 & 0 \end{pmatrix}^2 \sum_{l_3} C_{l_1} C_{l_3} \begin{pmatrix} l_1 & l_3 & l \\ 0 & 0 & 0 \end{pmatrix}^2 \frac{(2l_3+1)}{(4\pi)^2} \, ,$$

which is larger than

$$\frac{1}{C_{l;2}^2} \sum_{l_2} (2l_2+1) C_2 C_{l_2} \begin{pmatrix} 2 & l_2 & l \\ 0 & 0 & 0 \end{pmatrix}^2 \sum_{l_3} C_2 C_{l_3} \begin{pmatrix} 2 & l_3 & l \\ 0 & 0 & 0 \end{pmatrix}^2 \frac{(2l_3+1)}{(4\pi)^2}$$

$$\geq \frac{C_2^2 C_{l+2}^2}{C_{l;2}^2} \sum_{l_2} (2l_2+1) \begin{pmatrix} 2 & l_2 & l \\ 0 & 0 & 0 \end{pmatrix}^2 \sum_{l_3} \begin{pmatrix} 2 & l_3 & l \\ 0 & 0 & 0 \end{pmatrix}^2 \frac{(2l_3+1)}{(4\pi)^2} = \frac{C_2^2 C_{l+2}^2}{C_{l;2}^2 (4\pi)^2} \, .$$

Now we have proved earlier that in the polynomial case, $C_{l;2} \simeq C_l \simeq l^{-\alpha}$, so the previous ratio does not converge to zero and $\widehat{C}_{l;2}$ cannot be ergodic; the lower bound provided in the statement of the Lemma follows from previous computations and easy manipulations.

For the second part of the statement, it is sufficient to note that

$$\frac{1}{C_{l;2}^2} \sum_{l_1 l_2} (2l_1+1)(2l_2+1) C_{l_1} C_{l_2} \begin{pmatrix} l_1 & l_2 & l \\ 0 & 0 & 0 \end{pmatrix}^2 \sum_{l_3} C_{l_1} C_{l_3} \begin{pmatrix} l_1 & l_3 & l \\ 0 & 0 & 0 \end{pmatrix}^2 \frac{(2l_3+1)}{(4\pi)^2}$$

$$\leq \frac{\sup_{l_1} (2l_1+1)^{-1} \sum_{l_3} \Gamma_{l_1} \Gamma_{l_3} \left\{C_{l_1 0 l_3 0}^{l0}\right\}^2}{\sum_{l_1 l_3} \Gamma_{l_1} \Gamma_{l_3} \left\{C_{l_1 0 l_3 0}^{l0}\right\}^2} \leq \frac{\sup_{l_1} \sum_{l_3} \Gamma_{l_1} \Gamma_{l_3} \left\{C_{l_1 0 l_3 0}^{l0}\right\}^2}{\sum_{l_1 l_3} \Gamma_{l_1} \Gamma_{l_3} \left\{C_{l_1 0 l_3 0}^{l0}\right\}^2} \, ,$$

so the result follows from previous manipulations. □

Remark 8.20 By inspection of the previous proof, we note that we have shown how the sufficient condition for asymptotic Gaussianity (HFG) is also such for ergodicity (HFE). More precisely, we have proved that

$$\limsup_{l\to\infty} \frac{\Gamma_\lambda \sum_{l_2} \Gamma_{l_2} \left\{C_{l_1 0 l_2 0}^{l0}\right\}^2}{\sum_{l_1 l_2} \Gamma_{l_1} \Gamma_{l_2} \left\{C_{l_1 0 l_2 0}^{l0}\right\}^2} = \limsup_{l\to\infty} P(Z_1 = \lambda | Z_2 = l) = 0 \, ,$$

where $\{Z_l\}$ is the Markov chain defined in Remark (7.9), is a sufficient condition for the HFG (see Chapter 7 or [135, Proposition 9]) and also a sufficient condition to have $\lim_{l\to\infty} E\left\{\widetilde{C}_l - 1\right\}^2 = 0$.

Remark 8.21 In principle, the case $q = 3$ can be dealt along similar lines.

Remark 8.22 (*On Cosmic Variance*) Loosely speaking, the epistemological status of cosmological research has occasionally been the object of some debate, as in some sense one is dealing with a science based on a single observation (our observed Universe). In the CMB community, this issue has been somewhat rephrased in terms of so-called *Cosmic Variance* – i.e., it is taken as common knowledge that parameters relating only to lower multipoles (such as the value of C_l, for small values of l) are inevitably affected by an intrinsic uncertainty which cannot be eliminated (the variability due to the peculiar realization of the random field that we are able to observe), whereas this effect is regarded as vanishing at higher l (implicitly assuming that something like the HFE should always hold). As noted in [137], the previous results seem to point out the very profound role that the assumption of Gaussianity may play in this environment. In particular, for general non-Gaussian fields there is no guarantee that angular power spectra and related parameters can be consistently estimated, even at high multipoles – i.e., the Cosmic Variance does not decrease at high frequencies for general non-Gaussian models.

8.6 Discussion

This chapter leaves many directions open for further research. We believe the results of the previous two sections point out a very strong connection between conditions for High-Frequency Ergodicity (HFE) and High-Frequency Gaussianity (HFG) for isotropic spherical random fields. It is natural to suggest that equivalence may hold for Gaussian subordinated fields of any order q, or even more broadly for general Gaussian subordinated fields on homogeneous spaces of compact groups. Indeed, in this broader framework it is shown in [11] that independence of Fourier coefficients implies Gaussianity, which is the heuristic rationale behind our results here.

The connection between the HFE property can also be studied under a different environment than Gaussian subordination. Consider for instance the class of completely random spherical fields, which was recently introduced in [47, 48]. Following the definition therein, we shall say that a spherical random field is completely random if for each l we have that the vector $a_{l.} = (a_{l,-l}, ..., a_{ll})$ is invariant with respect to the action of all matrices belonging to $SU(2l + 1)$ and verifies $a_{lm} = (-1)^m \overline{a}_{lm}$. Because of this, the vector a_l is clearly uniformly distributed on the manifold of random radius $\left\{ \sum |a_{lm}|^2 \right\}^{1/2}$, or equivalently, introducing the $(2l + 1)$-dimensional vector U_l

$$U_l = \frac{1}{\sqrt{2l+1}} \left\{ \frac{\sqrt{2}\text{Re }a_{l1}}{\sqrt{\widehat{C_l}}}, \frac{\sqrt{2}\text{Re }a_{l2}}{\sqrt{\widehat{C_l}}}, ..., \frac{a_{l0}}{\sqrt{\widehat{C_l}}}, \frac{\sqrt{2}\text{Im }a_{l1}}{\sqrt{\widehat{C_l}}}, ..., \frac{\sqrt{2}\text{Im }a_{ll}}{\sqrt{\widehat{C_l}}} \right\}$$

(8.33)

it holds that, for l large, we have approximately $U_l \overset{law}{\sim} U(S^{2l})$, i.e. U_l it is asymptotically uniformly distributed on the unit sphere of $\mathbb{R}^{2l+1}$. Under these conditions, it is simple to show that HFE $\Rightarrow$ HFG, i.e.

$$\left\{ \lim_{l\to\infty} E \left\{ \widehat{C}_l - 1 \right\}^2 = 0 \right\} \Rightarrow \left\{ \frac{T_l(x)}{\sqrt{Var(T_l)}} \overset{law}{\to} N(0, 1) \text{, as } l \to \infty \right\}.$$

Indeed, it is sufficient to note that, as before

$$\frac{T_l(x)}{\sqrt{Var(T_l)}} = \frac{\sqrt{4\pi}T_l}{\sqrt{(2l+1)C_l}} \overset{law}{=} \frac{a_{l0}}{\sqrt{C_l}},$$

which we can write as

$$\frac{a_{l0}}{\sqrt{C_l}} = \frac{a_{l0}}{\sqrt{\widehat{C}_l}} \sqrt{\frac{\widehat{C}_l}{C_l}} = \frac{a_{l0}}{\sqrt{\widehat{C}_l}} \sqrt{\widetilde{C}_l}.$$

Now, as $l \to \infty$

$$\frac{a_{l0}}{\sqrt{\widehat{C}_l}} \overset{law}{\to} N(0, 1),$$

because the left-hand side can be viewed as the marginal distribution for a uniform law on a sphere of growing dimension; the latter is asymptotically Gaussian, as a consequence of Poincaré Lemma (see [50]). We do not investigate this issue more fully here, and we leave for future research the determination of general conditions such that (compare with (8.33))

the law of U_l and $U(S^{2l})$ are asymptotically close as $l \to \infty$.   (8.34)

Obviously, for all fields such that (8.34) holds (i.e. those that are asymptotically completely random, to mimic the terminology of [47, 48]), by the same argument as before we have that

$$\left\{ \sqrt{\frac{\widehat{C}_l}{C_l}} \overset{p}{\to} 1 \right\} \Rightarrow \left\{ \frac{\sqrt{4\pi}T_l}{\sqrt{(2l+1)C_l}} \overset{law}{\to} N(0, 1) \right\},$$

where $\xrightarrow{p}$ indicates convergence in probability. To conclude this work, we wish to provide two (somewhat pathological) examples where the HFE and HFG property are indeed not equivalent. As much of this chapter, this example is taken from [137]. Consider first the (anisotropic) field

$$h(x) = \sum_{lm} \xi_{lm} Y_{lm}(x) \,, \text{ where } \xi_{lm} = \left\{ \begin{array}{l} \xi_l \,, \text{ for } m = 0 \\ 0 \,, \text{ otherwise} \end{array} \right. ,$$

and the random variables ξ_l verifies the assumption

$$E\xi_l = 0, \quad \sum_l E\xi_l^2 < \infty \quad \text{and} \quad E\xi_l^4 < \infty \,.$$

The field can be made isotropic by taking a random rotation $T(x) = h(g^{-1}x)$, where g is a random, uniformly distributed element of $SO(3)$. We have as usual $T(x) = \sum_l \sum_{m=-l}^l a_{lm} Y_{lm}(x)$, where

$$a_{lm} \overset{law}{=} \sum_{m'=-l}^l D_{m'm}^l(g)\xi_{lm'} \overset{law}{=} \sqrt{\frac{4\pi}{2l+1}} Y_{lm}(ge_3)\xi_l \,,$$

and the first identity in law was discussed in Chapter 6. Note that

$$\sum_{m=-l}^l |a_{lm}|^2 \overset{law}{=} \frac{4\pi}{2l+1}\xi_l^2 \sum_{m=-l}^l |Y_{lm}(ge_3)|^2 = \xi_l^2 \,,$$

as expected, because the sample angular power spectrum is invariant to rotations. Of course in this case we do not have ergodicity in general, i.e. it may happen that

$$\frac{\sum_{m=-l}^l |a_{lm}|^2}{E\sum_{m=-l}^l |a_{lm}|^2} = \frac{\xi_l^2}{E\xi_l^2} \nrightarrow 1$$

and indeed for general sequences $\{\xi_l\}$

$$E\left\{ \frac{\xi_l^2}{E\xi_l^2} - 1 \right\}^2 = E\left\{ \frac{\xi_l^2}{E\xi_l^2} \right\}^2 - 1 \neq 0 \,.$$

However, in the special case where

$$\xi_l = \left\{ \begin{array}{l} e^{-l} \text{ with probability } \frac{1}{2} \\ -e^{-l} \text{ with probability } \frac{1}{2} \end{array} \right. ,$$

we obtain easily that $E\left\{ \frac{\xi_l^2}{E\xi_l^2} - 1 \right\}^2 \equiv 0$, while asymptotic Gaussianity fails. Hence, we have constructed an example where the HFE property holds but the HFG property does not. Note that the support of the vector $\{a_l\}$ is concentrated on a small subset of the sphere S^{2l}; heuristically, this is what prevents Poincarè-like arguments to go through.

Now let $T(x) = \sum_l T_l(x)$ a mean-square continuous, isotropic Gaussian field, and define

$$h(x) := \sum_l h_l(x) = \sum_l \eta_l T_l(x) ,$$

where $\{\eta_l\}$ is a sequence of independent random variables such that $\{\eta_l\} \perp \{T_l(x)\}$ and

$$\eta_l = \begin{cases} 1 & \text{w.p. } 1 - \frac{1}{l^2} \\ l & \text{w.p. } \frac{1}{2l^2} \\ -l & \text{w.p. } \frac{1}{2l^2} \end{cases} , \text{ whence } E\eta_l = 0 , \, E\eta_l^2 = 1 , \, E\eta_l^4 = l^2 .$$

It is trivial to verify that the field $\{h(.)\}$ is isotropic, mean-square continuous, and the HFG property holds, i.e., for any $x \in S^2$, as $l \to \infty$,

$$\frac{h_l(x)}{\sqrt{Var(h_l)}} \overset{law}{\to} N(0, 1) .$$

On the other hand,

$$Var\left\{\frac{\widehat{C_l}}{C_l}\right\} = E\eta_l^4 \times E\left\{\frac{\widehat{C_l}}{C_l}\right\}^2 - 1 = l^2\left\{1 + \frac{2}{2l+1}\right\} - 1 ,$$

whence the HFE clearly fails. Note, though, that here

$$\frac{\widehat{C_l}}{C_l} \overset{a.s.}{\to} 1 , \text{ as } l \to \infty ,$$

that is, convergence in the almost sure sense holds, while convergence in the mean square sense (which we used to define the HFE property) fails.

9

Asymptotics for Sample Bispectra

9.1 Introduction

As discussed in the previous chapters (see, in particular, Section 6.3), if an isotropic field is Gaussian, its dependence structure is completely identified by the angular correlation function and its harmonic transform, that is, the angular power spectrum. For non-Gaussian fields, the dependence structure becomes much richer, and higher order correlation functions are of interest. In turn, this leads to the analysis of so-called higher order angular power spectra, which we investigated in Chapter 6. Cumulant angular power spectra are identically zero for Gaussian fields, and hence they also provide natural tools to test for non-Gaussianity: this is a topic of the greatest importance in modern cosmological data analysis (see [51, 60]). Indeed, the validation of the Gaussian assumption is urged by the necessity to provide firm grounds to statistical inference on cosmological parameters, which is dominated by likelihood approaches. More importantly, tests for Gaussianity are needed to discriminate among competing scenarios for the physics of the primordial epochs: here, the currently favoured inflationary models predict (very close to) Gaussian CMB fluctuations, whereas other models yield different observational consequences (see [8, 18, 19, 20, 67, 184]). Tests for non-Gaussianity are also powerful tools to detect systematic effects in the outcome of the experiments. For these reasons, very many papers have focussed on testing for non-Gaussianity on CMB, some of them by means of topological properties of Gaussian fields, some others through spherical wavelets, or by harmonic space methods, see for instance [18, 35, 36, 37, 117, 119, 130, 164, 165, 175, 176, 198], and the references therein.

In this chapter, we investigate the asymptotic properties for the observed bispectrum of spherical Gaussian fields, and we analyze its use as a probe of non-Gaussian features. The (sample) bispectrum (defined in Section 2) is probably the single most popular statistic to search for non-Gaussianity in CMB data; on one hand, in fact, working on harmonic space is extremely convenient, and the bispectrum is the simplest harmonic space statistic which is sensitive to non-Gaussian features. On the other hand, it is possible to derive analytically the behaviour of the bispectrum for non-Gaussian fields of physical interests.

As described in the next section, the sample bispectrum is a function of the spherical harmonic coefficients a_{lm}'s; in the presence of an ideal experiment, we know that the latter are easily derived from a map of CMB fluctuations by an harmonic transform performed on the observed data. However, as mentioned in Chapter 8 it is important to stress that in realistic situations the a_{lm}'s are affected by instrumental noise, missing observations and many other sources. In the CMB literature, these perturbing effects are usually corrected by means of *ad hoc* techniques and huge numerical simulations (see for instance [119] and [198]). In this chapter we assume that the a_{lm}'s are observed without error; evidence on the performance of the procedures we advocate in the presence of noise and missing observations is provided for instance by [36]. Later in this book we shall discuss approaches such as the *needlets bispectrum* (see [123, 164, 165, 175, 176]), where the bispectrum has been combined with a wavelets-based approach, in order to yield a procedure which is more robust to the presence of unobserved regions and observational noise.

9.2 Sample bispectra

9.2.1 Preliminary considerations

As usual in this monograph, we try to motivate our ideas by first considering the simpler case of the sample bispectrum for a stationary process $\{X_t : t = 1, ..., n\}$ defined on the cyclic group $\{1, ..., n\}$ (endowed with the group operation $t \circ s = ts = (t + s) \bmod(n)$). Under these circumstances, the bispectrum is given by (with $\lambda_j = 2\pi j/n$ as before)

$$
\begin{aligned}
I_{j_1 j_2 j_3} &= a_{j_1} a_{j_2} a_{j_3} \\
&= \frac{1}{\sqrt{(2\pi n)^3}} \sum_{t_1,t_2,t_3=1}^{n} X_{t_1} X_{t_2} X_{t_3} e^{(-i\{\lambda_{j_1} t_1 + \lambda_{j_2} t_2 + \lambda_{j_3} t_3\})} \delta^0_{\lambda_{j_1} + \lambda_{j_2} + \lambda_{j_3}}.
\end{aligned}
$$

This random variable is immediately seen to be invariant with respect to the choice of coordinates. Indeed, if we define $t' = \tau t = (t + \tau) \bmod(n)$, and if we denote by $I'_{j_1 j_2 j_3}$ the sample bispectrum associated with the translated process $t \mapsto X_{\tau t}$, we obtain

$$I'_{j_1 j_2 j_3} = a'_{j_1} a'_{j_2} a'_{j_3}$$

$$= \frac{1}{\sqrt{(2\pi n)^3}} \sum_{t'_1, t'_2, t'_3 = 1}^{n} X_{t'_1} X_{t'_2} X_{t'_3} e^{(-i\{\lambda_{j_1} t'_1 + \lambda_{j_2} t'_2 + \lambda_{j_3} t'_3\})} \delta^0_{\lambda_{j_1} + \lambda_{j_2} + \lambda_{j_3}}$$

$$= \frac{1}{\sqrt{(2\pi n)^3}} \sum_{t_1, t_2, t_3 = 1}^{n} X_{t_1} X_{t_2} X_{t_3} e^{(-i\{\lambda_{j_1} t_1 + \lambda_{j_2} t_2 + \lambda_{j_3} t_3\})}$$

$$\times e^{(-i\tau\{\lambda_{j_1} + \lambda_{j_2} + \lambda_{j_3}\})} \delta^0_{\lambda_{j_1} + \lambda_{j_2} + \lambda_{j_3}}$$

$$= \frac{1}{\sqrt{(2\pi n)^3}} \sum_{t_1, t_2, t_3 = 1}^{n} X_{t_1} X_{t_2} X_{t_3} e^{(-i\{\lambda_{j_1} t_1 + \lambda_{j_2} t_2 + \lambda_{j_3} t_3\})} \delta^0_{\lambda_{j_1} + \lambda_{j_2} + \lambda_{j_3}}$$

$$= I_{j_1 j_2 j_3} .$$

In conclusion, we have proved that the bispectrum is in this case invariant with respect to the action of the cyclic group on the process X.

The next statement shows that, in a very general framework, the combination of isotropy and invariance produces statistics with smaller fluctuations (the result is certainly known, but we failed to locate any reference and we report it for completeness).

Lemma 9.1 *Let* $T := \{T_1, ..., T_n\}$ *be a collection of random variables on which the topological compact group G acts in such a way that*

$$g \cdot T \overset{law}{=} T , \text{ for all } g \in G .$$

Assume also that $S = S(T) = S(T_1, ..., T_n)$ *is square-integrable. Then,*

$$E\widetilde{S} = ES , \ Var(\widetilde{S}) \le Var(S) ,$$

where we used the notation

$$\widetilde{S} = \int_G S(g \cdot T) dg ,$$

(which can be viewed as the average of the statistic S over the compact group G, endowed with the Haar measure dg).

Proof We have that

$$E\widetilde{S} = E \int_G S(g \cdot T) dg = \int_G ES(g \cdot T) dg$$

$$= \int_G ES(T) dg = \int_G dg \{ES(T)\} = ES ,$$

where the interchange of the two integrals can be justified by Fubini's Theorem. On the other hand, by Jensen inequality,

$$
\begin{aligned}
Var\{\widetilde{S}\} &= E\left\{\int_G [S(g \cdot T) - ES]\,dg\right\}^2 \\
&\le E\left\{\int_G [S(g \cdot T) - ES]^2\,dg\right\} \\
&= \int_G E[S(g \cdot T) - ES]^2\,dg = \int_G E[S(T) - ES]^2\,dg \\
&= E[S(T) - ES]^2 = Var[S(T_1, ..., T_n)] \ .
\end{aligned}
$$

$\square$

Remark 9.2   A special case of Lemma 9.1 is the well-known result that, for exchangeable random variables, a statistic can be efficient only if it is a symmetric function of its arguments. Indeed exchangeability of a finite vector can be viewed as (strong) isotropy with respect to the action of the permutation group.

9.2.2 Definitions

In this section, we consider a random field $T = \{T(x) : x \in S^2\}$, which is centered, isotropic, and with finite moments up to the order three. We recall from Chapter 6 that the finiteness on the second moments and the isotropy assumption imply that the paths of T are almost surely in $L^2(S^2, d\sigma\}$, in such a way that T admits the harmonic expansion

$$
T(x) = \sum_{l=0}^{\infty} \sum_{m=-l}^{l} a_{lm} Y_{lm}(x) = \sum_{l=0}^{\infty} T_l(x) , \quad x \in S^2, \tag{9.1}
$$

$$
a_{lm} = \int_{S^2} T(x)\overline{Y_{lm}}(x)d\sigma(x). \tag{9.2}
$$

Combining e.g. Proposition 6.6 and (6.21) yields the identity

$$
E[T_l(x)^2] = C_l \frac{2l+1}{4\pi}, \tag{9.3}
$$

where $\{C_l : l \ge 0\}$ is the angular power spectrum of T. Also, according to Example (6.33), the (reduced) bispectrum $\{b_{l_1 l_2 l_3}\}$ of T is defined by the relation

$$
Ea_{l_1 m_1} a_{l_2 m_2} a_{l_3 m_3} = \begin{pmatrix} l_1 & l_2 & l_3 \\ m_1 & m_2 & m_3 \end{pmatrix} \begin{pmatrix} l_1 & l_2 & l_3 \\ 0 & 0 & 0 \end{pmatrix} b_{l_1 l_2 l_3}
$$

$$= \begin{pmatrix} l_1 & l_2 & l_3 \\ m_1 & m_2 & m_3 \end{pmatrix} B_{l_1 l_2 l_3} ,$$

where

$$B_{l_1 l_2 l_3} := \sum_{m_1 m_2 m_3} \begin{pmatrix} l_1 & l_2 & l_3 \\ m_1 & m_2 & m_3 \end{pmatrix} E a_{l_1 m_1} a_{l_2 m_2} a_{l_3 m_3} ,$$

and we used Wigner's, rather than Clebsch-Gordan coefficients for convenience. In the literature, $B_{l_1 l_2 l_3}$ is sometimes labeled angle averaged bispectrum, however for brevity's sake we shall simply call it bispectrum.

Inspired by the rationale implicit in Lemma 9.1, our aim is now to define an estimator of $B_{l_1 l_2 l_3}$ that is unbiased and invariant with respect to the action of $SO(3)$. We start with the proof of a useful identity, involving the class Wigner D matrices $\{D^l : l \geq 0\}$.

Lemma 9.3 *For all $g \in SO(3)$, $l_1 + l_2 + l_3$ even, we have*

$$\sum_{m_1 m_2 m_3} \begin{pmatrix} l_1 & l_2 & l_3 \\ m_1 & m_2 & m_3 \end{pmatrix} D^{l_1}_{m'_1 m_1}(g) D^{l_2}_{m'_2 m_2}(g) D^{l_3}_{m'_3 m_3}(g) \equiv \begin{pmatrix} l_1 & l_2 & l_3 \\ m'_1 & m'_2 & m'_3 \end{pmatrix}.$$

Proof By using several times Proposition 3.43, as well as the definition of $3j$ coefficients in terms of Clebsch-Gordan coefficients, we have that

$$\sum_{m_1 m_2 m_3} \begin{pmatrix} l_1 & l_2 & l_3 \\ m_1 & m_2 & m_3 \end{pmatrix} D^{l_1}_{m'_1 m_1}(g) D^{l_2}_{m'_2 m_2}(g) D^{l_3}_{m'_3 m_3}(g)$$

$$= \left\{ \begin{pmatrix} l_1 & l_2 & l_3 \\ 0 & 0 & 0 \end{pmatrix} \sqrt{\frac{(2l_1 + 1)(2l_2 + 1)(2l_3 + 1)}{4\pi}} \right\}^{-1}$$

$$\times \sum_{m_1 m_2 m_3} D^{l_1}_{m'_1 m_1}(g) D^{l_2}_{m'_2 m_2}(g) D^{l_3}_{m'_3 m_3}(g) \int_{S^2} Y_{l_1 m_1} Y_{l_2 m_2} Y_{l_3 m_3} d\sigma$$

$$= \left\{ \begin{pmatrix} l_1 & l_2 & l_3 \\ 0 & 0 & 0 \end{pmatrix} \sqrt{\frac{(2l_1 + 1)(2l_2 + 1)(2l_3 + 1)}{4\pi}} \right\}^{-1}$$

$$\times \int_{S^2} \left[\sum_{m_1 m_2 m_3} D^{l_1}_{m'_1 m_1}(g) D^{l_2}_{m'_2 m_2}(g) D^{l_3}_{m'_3 m_3}(g) Y_{l_1 m_1} Y_{l_2 m_2} Y_{l_3 m_3} \right] d\sigma$$

$$= \left\{ \begin{pmatrix} l_1 & l_2 & l_3 \\ 0 & 0 & 0 \end{pmatrix} \sqrt{\frac{(2l_1 + 1)(2l_2 + 1)(2l_3 + 1)}{4\pi}} \right\}^{-1}$$

$$\times \int_{S^2} \left[Y_{l_1 m'_1} Y_{l_2 m'_2} Y_{l_3 m'_3} \right] d\sigma = \begin{pmatrix} l_1 & l_2 & l_3 \\ m'_1 & m'_2 & m'_3 \end{pmatrix}.$$

$\square$

Similarly to the discussion on the cyclic group at the beginning of this section, for observations on stationary stochastic processes defined on $\mathbb{Z}^k$ or $\mathbb{R}^k$, $k \geq 1$, the bispectrum is a well-known and extensively studied statistic, see for instance [29]. Under those circumstances, standard (Euclidean) Fourier transforms are implemented; the bispectrum is then defined to be invariant under the action of the corresponding Abelian group of translations and the asymptotic theory is developed in the usual large sample framework, i.e. limit results are derived under the assumption that a larger and larger span of observations becomes available.

As already discussed, our situation is quite different, since our limit theorems are in the *high-frequency sense*, that is, they are obtained by letting the frequency indices l_i diverge to infinity in an appropriate sense. As an estimator for $B_{l_1 l_2 l_3}$, we shall therefore consider the *sample bispectrum* , which we define to be

$$\widehat{B}_{l_1 l_2 l_3} := \sum_{m_1=-l_1}^{l_1} \sum_{m_2=-l_2}^{l_2} \sum_{m_3=-l_3}^{l_3} \begin{pmatrix} l_1 & l_2 & l_3 \\ m_1 & m_2 & m_3 \end{pmatrix} (a_{l_1 m_1} a_{l_2 m_2} a_{l_3 m_3}) ,$$

where $l_1 + l_2 + l_3$ is constrained to be even. See for instance [102, 131] for a detailed discussion of the statistical properties of this estimator. Here, we shall only record the following facts:

- The statistic $\widehat{B}_{l_1 l_2 l_3}$ is real-valued and it is invariant with respect to permutation of its arguments l_1, l_2, l_3.
- $E\widehat{B}_{l_1 l_2 l_3} = B_{l_1 l_2 l_3}$, that is, $\widehat{B}_{l_1 l_2 l_3}$ is an unbiased estimator of the true reduced bispectrum (to see this, just use the definition of $B_{l_1 l_2 l_3}$ given above, and apply the "Rule n. 2" of subsection 4.5.6).
- The statistic $\widehat{B}_{l_1 l_2 l_3}$ is invariant under rotations and it is therefore of "minimal variance", in the sense of Lemma 9.1.

The last point of the previous list is better detailed in the next statement.

Lemma 9.4 $\widehat{B}_{l_1 l_2 l_3}$ *is invariant under the action of the group SO(3).*

Proof Recall from Chapter 6 that under a rotation $g \in SO(3)$ the spherical harmonic coefficients transform as

$$a'_{lm} = \sum_{m'} D^l_{m'm}(g) a_{lm'} .$$

Hence

$$\widehat{B}'_{l_1 l_2 l_3} := \sum_{m_1 m_2 m_3} \begin{pmatrix} l_1 & l_2 & l_3 \\ m_1 & m_2 & m_3 \end{pmatrix} (a'_{l_1 m_1} a'_{l_2 m_2} a'_{l_3 m_3})$$

$$= \sum_{m_1 m_2 m_3} \begin{pmatrix} l_1 & l_2 & l_3 \\ m_1 & m_2 & m_3 \end{pmatrix} \sum_{m'_1 m'_2 m'_3} D^l_{m'_1 m_1}(g) D^l_{m'_2 m_2}(g) D^l_{m'_3 m_3}(g) (a_{l_1 m'_1} a_{l_2 m'_2} a_{l_3 m'_3})$$

$$= \sum_{m'_1 m'_2 m'_3} \left\{ \sum_{m_1 m_2 m_3} \begin{pmatrix} l_1 & l_2 & l_3 \\ m_1 & m_2 & m_3 \end{pmatrix} D^l_{m'_1 m_1}(g) D^l_{m'_2 m_2}(g) D^l_{m'_3 m_3}(g) \right\} (a_{l_1 m'_1} a_{l_2 m'_2} a_{l_3 m'_3})$$

$$= \sum_{m'_1 m'_2 m'_3} \begin{pmatrix} l_1 & l_2 & l_3 \\ m'_1 & m'_2 & m'_3 \end{pmatrix} (a_{l_1 m'_1} a_{l_2 m'_2} a_{l_3 m'_3}) = \widehat{B}_{l_1 l_2 l_3} ,$$

in view of Lemma 9.3. Hence the sample bispectrum is invariant under the action of the group $SO(3)$. □

In the Sections to follow, we shall also focus on a normalized version of the angular bispectrum, which is defined by

$$I_{l_1 l_2 l_3} := \frac{\widehat{B}_{l_1 l_2 l_3}}{\sqrt{C_{l_1} C_{l_2} C_{l_3}}} . \tag{9.4}$$

This is the statistic considered for instance by [131, 132], on which this chapter is partially based. In practice, $I_{l_1 l_2 l_3}$ is unfeasible because C_l is unknown. As discussed in the previous chapter, and neglecting measurement errors, a feasible estimator for C_l is

$$\widehat{C}_l := \frac{1}{2l+1} \sum_{m=-l}^{l} |a_{lm}|^2 , l = 1, 2, \dots ;$$

thus $I_{l_1 l_2 l_3}$ can be replaced by the feasible statistic

$$\widehat{I}_{l_1 l_2 l_3} := \frac{\widehat{B}_{l_1 l_2 l_3}}{\sqrt{\widehat{C}_{l_1} \widehat{C}_{l_2} \widehat{C}_{l_3}}} . \tag{9.5}$$

Remark 9.5 Recall that we defined

$$T_l(x) = \sum_m a_{lm} Y_{lm}(x) ,$$

and note that

$$\int_{S^2} T_{l_1}(x) T_{l_2}(x) T_{l_3}(x) d\sigma(x)$$

$$= \int_{S^2} \sum_{m_1 m_2 m_3} a_{l_1 m_1} a_{l_2 m_2} a_{l_3 m_3} Y_{l_1 m_1}(x) Y_{l_2 m_2}(x) Y_{l_3 m_3}(x) d\sigma(x)$$

$$= \sum_{m_1 m_2 m_3} a_{l_1 m_1} a_{l_2 m_2} a_{l_3 m_3} \int_{S^2} Y_{l_1 m_1}(x) Y_{l_2 m_2}(x) Y_{l_3 m_3}(x) d\sigma(x)$$

$$= \sum_{m_1 m_2 m_3} a_{l_1 m_1} a_{l_2 m_2} a_{l_3 m_3} \begin{pmatrix} l_1 & l_2 & l_3 \\ m_1 & m_2 & m_3 \end{pmatrix}$$

$$\times \begin{pmatrix} l_1 & l_2 & l_3 \\ 0 & 0 & 0 \end{pmatrix} \sqrt{\frac{(2l_1 + 1)(2l_2 + 1)(2l_3 + 1)}{4\pi}} \, .$$

Hence

$$\frac{1}{4\pi} \frac{\int_{S^2} T_{l_1}(x) T_{l_2}(x) T_{l_3}(x) d\sigma(x)}{\sqrt{Var(T_{l_1}) Var(T_{l_2}) Var(T_{l_3})}}$$

$$= \frac{\sqrt{4\pi} \int_{S^2} T_{l_1}(x) T_{l_2}(x) T_{l_3}(x) d\sigma(x)}{\sqrt{(2l_1 + 1)(2l_2 + 1)(2l_3 + 1)} C_{l_1} C_{l_2} C_{l_3}} \tag{9.6}$$

$$= \sum_{m_1 m_2 m_3} \frac{a_{l_1 m_1} a_{l_2 m_2} a_{l_3 m_3}}{\sqrt{C_{l_1} C_{l_2} C_{l_3}}} \begin{pmatrix} l_1 & l_2 & l_3 \\ m_1 & m_2 & m_3 \end{pmatrix} \begin{pmatrix} l_1 & l_2 & l_3 \\ 0 & 0 & 0 \end{pmatrix}$$

$$= I_{l_1 l_2 l_3} \times \begin{pmatrix} l_1 & l_2 & l_3 \\ 0 & 0 & 0 \end{pmatrix} . \tag{9.7}$$

In other words, the sample bispectrum is proportional to the sample third moment of the Fourier components of the random field, averaged over the sphere.

9.3 A central limit theorem

In this section, we shall investigate the behaviour of the higher order moments for the normalized bispectrum (9.4), under the following assumption: *the spherical field T is centered, Gaussian and isotropic.*

We assume (without loss of generality) that $l_1 \leq_1 l_2 \leq l_3$, and define

$$\Delta_{l_1 l_2 l_3} := 1 + \delta_{l_1}^{l_2} + \delta_{l_2}^{l_3} + 3\delta_{l_1}^{l_3} = \begin{cases} 1, & \text{for } l_1 < l_2 < l_3 \\ 2, & \text{for } l_1 = l_2 < l_3 \text{ or } l_1 < l_2 = l_3 \\ 6, & \text{for } l_1 = l_2 = l_3 \end{cases} ;$$

here and in the sequel, δ_a^b denotes Kronecker's delta, that is $\delta_a^b = 1$ for $a = b$, zero otherwise.

For the computations to follow, the following Lemma will prove to be extremely useful.

Lemma 9.6 *For all l_1, l_2, l_3, the sample bispectrum $I_{l_1 l_2 l_3}$ is equal in law to a random variable living in the third Wiener chaos associated with a standard Brownian motion. In particular, $EI_{l_1 l_2 l_3} = 0$*

Proof For simplicity, we assume that the three indices l_i are ≥ 1; the case where at least one of the l_i's is zero is dealt with analogously, and left to the reader as an exercise. As before, we write H_i in order to indicate the ith Hermite polynomial. For $l_1 < l_2 < l_3$, by independence, we have clearly that $T_{l_1}(x)T_{l_2}(x)T_{l_3}(x)$ belongs to the third Wiener chaos associated with the underlying Gaussian field (that we can always assume to be generated by a standard Brownian motion). For $l_1 + l_2 + l_3$ even, in view of (9.7), we have that

$$
I_{l_1 l_2 l_3} = \begin{pmatrix} l_1 & l_2 & l_3 \\ 0 & 0 & 0 \end{pmatrix}^{-1} \frac{\int_{S^2} \tilde{T}_{l_1}(x)\tilde{T}_{l_2}(x)\tilde{T}_{l_3}(x)d\sigma(x)}{4\pi},
$$

where $\tilde{T}_l = T_l / \sqrt{Var(T_l)}$. Hence, by linearity, $I_{l_1 l_2 l_3}$, belongs to the third Wiener chaos as well (by virtue e.g. of the multiplication formula (4.12)). When two indexes are equal, we write

$$
\int_{S^2} \tilde{T}_{l_1}^2(x)\tilde{T}_{l_2}(x)d\sigma(x) = \int_{S^2} \left\{ T_{l_1}^2(x) - 1 \right\} T_{l_2}(x)d\sigma(x)
$$

$$
= \int_{S^2} H_2(\tilde{T}_{l_1}(x))H_1(\tilde{T}_{l_2}(x))d\sigma(x) ,
$$

the last identity following from

$$
\int_{S^2} H_1(\tilde{T}_{l_2}(x))d\sigma(x) = \int_{S^2} \tilde{T}_{l_2}(x)d\sigma(x) = 0,
$$

and the conclusion is deduced from (4.18) and the multiplication formula (4.12). Finally, when the three indexes are equal

$$
\int_{S^2} \tilde{T}_l^3(x)d\sigma(x) = \int_{S^2} \left\{ \tilde{T}_l^3(x) - 3\tilde{T}_l(x) \right\} d\sigma(x) = \int_{S^2} H_3(\tilde{T}_l(x))d\sigma(x),
$$

and the conclusion follows from (4.18). $\square$

Under Gaussianity, it is obvious that the expectation of all odd powers of $I_{l_1 l_2 l_3}$ is zero (every random variable living in an odd Wiener chaos has indeed vanishing odd moments). To analyze the behaviour of even powers, we shall use extensively the approximation results for Gaussian subordinated processes introduced in Chapter 4. We have the following result.

Theorem 9.7 *For all $l_1 \leq l_2 \leq l_3$ we have*

$$
EI_{l_1 l_2 l_3}^2 = \Delta_{l_1 l_2 l_3}; \tag{9.8}
$$

moreover, for any $p \geq 2$,

$$
EI_{l_1 l_2 l_3}^{2p-1} = 0 , \tag{9.9}
$$

$$EI^{2p}_{l_1 l_2 l_3} = (2p-1)!!\Delta^p_{l_1 l_2 l_3} + O(l_1^{-1/2}),\qquad(9.10)$$

as $l_1 \to \infty$.

Proof For notational simplicity, we only consider the case $l_1 < l_2 < l_3$; the remaining cases can be established with analogous arguments. For (9.8), it suffices to notice that

$$EI^2_{l_1 l_2 l_3} = \sum_{m_1 m_2 m_3}\sum_{m'_1 m'_2 m'_3}\begin{pmatrix} l_1 & l_2 & l_3 \\ m_1 & m_2 & m_3 \end{pmatrix}\begin{pmatrix} l_1 & l_2 & l_3 \\ m'_1 & m'_2 & m'_3 \end{pmatrix}$$
$$\times\frac{E(a_{l_1 m_1} a_{l_2 m_2} a_{l_3 m_3})(a_{l_1 m'_1} a_{l_2 m'_2} a_{l_3 m'_3})}{C_{l_1} C_{l_2} C_{l_3}}$$

$$= \sum_{m_1 m_2 m_3}\sum_{m'_1 m'_2 m'_3}\begin{pmatrix} l_1 & l_2 & l_3 \\ m_1 & m_2 & m_3 \end{pmatrix}\begin{pmatrix} l_1 & l_2 & l_3 \\ m'_1 & m'_2 & m'_3 \end{pmatrix}$$
$$\times(-1)^{m'_1+m'_2+m'_3}\delta^{-m'_1}_{m_1}\delta^{-m'_2}_{m_2}\delta^{-m'_3}_{m_3}$$

$$= \sum_{m_1 m_2 m_3}\begin{pmatrix} l_1 & l_2 & l_3 \\ m_1 & m_2 & m_3 \end{pmatrix}\begin{pmatrix} l_1 & l_2 & l_3 \\ -m_1 & -m_2 & -m_3 \end{pmatrix}$$

$$= (-1)^{l_1+l_2+l_3}\sum_{m_1 m_2 m_3}\begin{pmatrix} l_1 & l_2 & l_3 \\ m_1 & m_2 & m_3 \end{pmatrix}^2 = 1,$$

where we used the orthonormality of Wigner's coefficients and the constraint $l_1 + l_2 + l_3$ is even. Relation (9.9) is trivial, as these expressions entail the expectation of an odd number of coefficients $\{a_{lm}\}$. To prove (9.10), use Proposition 4.23 in order to deduce that

$$\left|EI^{2p}_{l_1 l_2 l_3} - (2p-1)!!\right| \le c_p\sqrt{EI^4_{l_1 l_2 l_3} - 3},\qquad(9.11)$$

where

$$c_p = (2p-1)\frac{2^{p-1}}{3}\left(\sqrt{\frac{(4p-4)!}{(2p-2)!}} + (4p-5)^{3p-3}\right).$$

Now, applying the diagram formula in Proposition 4.15 (that we apply separately to the independent real and imaginary parts of the harmonic coefficients), we infer that

$$EI^4_{l_1 l_2 l_3} - 3 = Cum_4(I_{l_1 l_2 l_3})$$

$$= 6\sum_{m_1 m_2 m_3}\sum_{m_4 m_5 m_6}\begin{pmatrix} l_1 & l_2 & l_3 \\ m_1 & m_2 & m_3 \end{pmatrix}\begin{pmatrix} l_1 & l_2 & l_3 \\ m_1 & m_4 & m_5 \end{pmatrix}$$

$$\times \begin{pmatrix} l_1 & l_2 & l_3 \\ m_6 & m_2 & m_5 \end{pmatrix} \begin{pmatrix} l_1 & l_2 & l_3 \\ m_6 & m_4 & m_3 \end{pmatrix}$$

$$+ 6 \sum_{m_1 m_2 m_3} \sum_{m_4 m_5 m_6} \begin{pmatrix} l_1 & l_2 & l_3 \\ m_1 & m_2 & m_3 \end{pmatrix} \begin{pmatrix} l_1 & l_2 & l_3 \\ m_1 & m_4 & m_3 \end{pmatrix}$$

$$\times \begin{pmatrix} l_1 & l_2 & l_3 \\ m_5 & m_4 & m_6 \end{pmatrix} \begin{pmatrix} l_1 & l_2 & l_3 \\ m_5 & m_2 & m_6 \end{pmatrix}$$

$$+ 6 \sum_{m_1 m_2 m_3} \sum_{m_4 m_5 m_6} \begin{pmatrix} l_1 & l_2 & l_3 \\ m_1 & m_2 & m_3 \end{pmatrix} \begin{pmatrix} l_1 & l_2 & l_3 \\ m_1 & m_2 & m_4 \end{pmatrix}$$

$$\times \begin{pmatrix} l_1 & l_2 & l_3 \\ m_5 & m_6 & m_4 \end{pmatrix} \begin{pmatrix} l_1 & l_2 & l_3 \\ m_5 & m_6 & m_3 \end{pmatrix}$$

$$+ 6 \sum_{m_1 m_2 m_3} \sum_{m_4 m_5 m_6} \begin{pmatrix} l_1 & l_2 & l_3 \\ m_1 & m_2 & m_3 \end{pmatrix} \begin{pmatrix} l_1 & l_2 & l_3 \\ m_4 & m_2 & m_3 \end{pmatrix}$$

$$\times \begin{pmatrix} l_1 & l_2 & l_3 \\ m_4 & m_5 & m_6 \end{pmatrix} \begin{pmatrix} l_1 & l_2 & l_3 \\ m_1 & m_5 & m_6 \end{pmatrix} .$$

The key argument for obtaining the previous equality is the following: once a non-flat connected diagram, associated with a table with four lines of three elements, is reduced to a graph (according to the procedure explained in Section 4.5.1), then either it is isomorphic to (a) in Fig. 9.1 (that is, to a clique) or it is isomorphic to (b) in the same Fig. 9.1. Note also that all factors of the type $(-1)^{\eta}$ cancel out, due to the implicit constraints on l_1, l_2, l_3 and m_1, m_2, m_3.

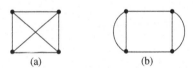

(a)   (b)

Figure 9.1 Two isomorphism classes for connected graphs

Finally, combining the graphical rules n.3 (see subsection 4.5.7) and n.4 (see subsection 4.5.8) with the estimate (4.42), gives

$$Cum_4(I_{l_1 l_2 l_3}) = 6 \left\{ \begin{matrix} l_1 & l_2 & l_3 \\ l_1 & l_2 & l_3 \end{matrix} \right\} + \frac{6}{2l_1 + 1} + \frac{6}{2l_2 + 1} + \frac{6}{2l_3 + 1}$$

$$\leq \frac{12}{l_1},$$

and therefore the desired conclusion.   □

Remark 9.8   Although sufficient for our purposes, the estimate (9.10) is not optimal. Indeed, in [132] the following relation is proved by means of the graphical methods of Section 4.5:

$$EI^{2p}_{l_1 l_2 l_3} = (2p-1)!!\Delta^p_{l_1 l_2 l_3} + O(l_1^{-1}) \, . \tag{9.12}$$

Using Theorem 9.7 and the Theorem 4.18 for $q = 3$ we obtain the following Central Limit Theorem for the bispectrum:

Theorem 9.9   *For any $l_1 \leq l_2 \leq l_3$ we have*

$$d_{TV}\left(\frac{I_{l_1 l_2 l_3}}{\sqrt{\Delta_{l_1 l_2 l_3}}}, Z\right) \leq \sqrt{\frac{32}{3l_1}}, \quad Z \overset{law}{=} N(0,1) \, ,$$

and hence, as $l_1 \rightarrow \infty$,

$$\frac{I_{l_1 l_2 l_3}}{\sqrt{\Delta_{l_1 l_2 l_3}}} \overset{law}{\rightarrow} N(0,1), \tag{9.13}$$

in the sense of the total variation distance.

Remark 9.10   A multivariate Central Limit Theorem for the angular bispectrum was first given in [132]. The proof was based on an evaluation of tight bounds for cumulants of arbitrary order by graphical techniques, and it was hence much more complicated than the argument here; no result on total variation distance was provided. Using Theorem 4.21, as well as combinatorial arguments as the ones above, we could obtain a multivariate limit theorem with uniform bounds on classes of twice differentiable functions. We leave this task to the reader.

9.4 Limit theorems under random normalizations

In this section we focus on the more realistic case where the angular power spectrum is unknown and estimated from the data; so we consider $\widehat{I}_{l_1 l_2 l_3}$ rather than $I_{l_1 l_2 l_3}$. As before, under Gaussianity of the underlying field $T(x)$

$$E\widehat{I}^{2p-1}_{l_1 l_2 l_3} = 0 \, , \, p = 1, 2, ...,$$

for instance by a simple symmetry argument. Now note that

$$\left(\frac{|a_{l0}|^2}{\widehat{C}_l}, \frac{2|a_{l1}|^2}{\widehat{C}_l}, \cdots, \frac{2|a_{ll}|^2}{\widehat{C}_l}\right)$$

$$= (2l+1)\left(\frac{|a_{l0}|^2}{|a_{l0}|^2 + \sum_{m=1}^{l} 2|a_{lm}|^2}, \frac{2|a_{l1}|^2}{|a_{l0}|^2 + \sum_{m=1}^{l} 2|a_{lm}|^2}, \cdots, \frac{2|a_{ll}|^2}{|a_{l0}|^2 + \sum_{m=1}^{l} 2|a_{lm}|^2}\right)$$

$$:= (2l+1)(\xi_{l0}, \cdots, \xi_{ll}) \overset{law}{=} (2l+1)Dir\left(\frac{1}{2}, 1, \cdots, 1\right);$$

here, $\overset{law}{=}$ denotes as usual equality in distribution and $Dir(\theta_0, ..., \theta_p)$ a Dirichlet distribution with parameters $(\theta_0, ..., \theta_p)$ (see for instance [106]) . Define

$$u_{lm} = \frac{a_{lm}}{\sqrt{C_l}}, \widehat{u}_{lm} = \frac{a_{lm}}{\sqrt{\widehat{C}_l}}, m = 0, 1, ..., l; \qquad (9.14)$$

we have the following result (see [131]).

Proposition 9.11   *Let l and p be positive integers, and define*

$$g(l; p) = \prod_{k=1}^{p}\left\{\frac{2l+1}{2l+2k-1}\right\}.$$

Now for u and $\widehat{u}$ defined by (9.14), we have

$$E\left\{\overbrace{\widehat{u}_{l0}\cdots\widehat{u}_{l0}}^{q_0}\overbrace{\widehat{u}_{l1}\cdots\widehat{u}_{l1}}^{q_1}\overbrace{\widehat{u}_{l1}^*\cdots\widehat{u}_{l1}^*}^{q_1'}\cdots\overbrace{\widehat{u}_{lk}\cdots\widehat{u}_{lk}}^{q_k}\overbrace{\widehat{u}_{lk}^*\cdots\widehat{u}_{lk}^*}^{q_k'}\right\}$$

$$= E\left\{\overbrace{u_{l0}\cdots u_{l0}}^{q_0}\overbrace{u_{l1}\cdots u_{l1}}^{q_1}\overbrace{u_{l1}^*\cdots u_{l1}^*}^{q_1'}\cdots\overbrace{u_{lk}\cdots u_{lk}}^{q_k}\overbrace{u_{lk}^*\cdots u_{lk}^*}^{q_k'}\right\}g(l; q_0+q_1+\cdots+q_k').$$

Proof   By symmetry, it is easy to see that both sides are zero unless $q_0 = 2p_0$ (say) is even and $q_i = q_i' = p_i$ (say), for $i = 1, ..., k$. The u_{lm} are independent over different m's, and thus we have

$$E\left\{\overbrace{u_{l0}\cdots u_{l0}}^{2p_0}\overbrace{u_{l1}\cdots u_{l1}}^{q_1}\overbrace{u_{l1}^*\cdots u_{l1}^*}^{q_1'}\cdots\overbrace{u_{lk}\cdots u_{lk}}^{q_k}\overbrace{u_{lk}^*\cdots u_{lk}^*}^{q_k'}\right\}$$

$$= Eu_{l0}^{2p_0}Eu_{l1}^{p_1}(u_{l1}^*)^{p_1}\cdots Eu_{lk}^{p_k}(u_{lk}^*)^{p_k} = \begin{cases} (2p_0-1)!! \prod_{i=1}^{k} p_i! \text{ for } p_0 > 0 \\ \prod_{i=1}^{k} p_i!g(l; p) \text{ for } p_0 = 0 \end{cases},$$

because

$$u_{l0} \overset{law}{=} N(0, 1) \text{ and } u_{lm}u_{lm}^* = |u_{lm}|^2 \overset{law}{=} \exp(1),$$

where $\exp(1)$ denotes an exponential random variable with parameter 1. Now write $p = p_0 + \cdots + p_k$, and note that ([106], p.233)

$$E\left\{\widehat{u}_{l0}^{2p_0}\widehat{u}_{l1}^{p_1}(\widehat{u}_{l1}^*)^{p_1}\cdots\widehat{u}_{lk}^{p_k}(\widehat{u}_{lk}^*)^{p_k}\right\} = \frac{(2l+1)^p}{2^{p_1+\cdots+p_k}}E\xi_{l0}^{p_0}\cdots\xi_{lk}^{p_k}$$

$$= \frac{(2l+1)^p}{2^{p_1+\cdots+p_k}}\frac{\Gamma(l+1/2)}{\Gamma(l+p+1/2)}\frac{\Gamma(p_0+1/2)\Gamma(p_2+1)\cdots\Gamma(p_k+1)}{\Gamma(1/2)}$$

$$= \frac{(2p_0-1)!!p_1!\times\cdots\times p_k!}{(2l+1)\times\cdots\times(2l+2p-1)}(2l+1)^p$$

$$= \begin{cases} (2p_0-1)!!\prod_{i=1}^{k}p_i!g(l;p) \text{ for } p_0 > 0 \\ \prod_{i=1}^{k}p_i!g(l;p) \text{ for } p_0 = 0 \end{cases},$$

as claimed. $\qquad\qquad\qquad\qquad\qquad\qquad\qquad\qquad\qquad\qquad\qquad\qquad\qquad\square$

Some special cases of the previous statement are the following:

$$E\frac{|a_{l0}|^2}{\widehat{C}_l} = E\frac{|a_{lm}|^2}{\widehat{C}_l} = 1 , m = \pm 1, ..., \pm l ,$$

$$E\left\{\frac{|a_{l0}|^2}{\widehat{C}_l}\right\}^p = \frac{(2p-1)!!}{(2l+1)\times...\times(2l+2p-1)}(2l+1)^p$$

$$E\left\{\frac{|a_{lm}|^2}{\widehat{C}_l}\right\}^p = \frac{(2l+1)^p}{2^p}\frac{\Gamma(l+1/2)}{\Gamma(l+p+1/2)}\frac{\Gamma(1/2)\Gamma(p+1)}{\Gamma(1/2)}$$

$$= \frac{p!}{(2l+1)\times\cdots\times(2l+2p-1)}(2l+1)^p,$$

and for $p = p_1 + p_2, p_1, p_2 > 0$

$$E\left\{\left(\frac{|a_{l0}|^2}{\widehat{C}_l}\right)^{p_1}\left(\frac{|a_{l1}|^2}{\widehat{C}_l}\right)^{p_2}\right\} = \frac{(2l+1)^p}{2^p}\frac{\Gamma(l+1/2)}{\Gamma(l+p+1/2)}\frac{\Gamma(p_1+1/2)\Gamma(p_2+1)}{\Gamma(1/2)}$$

$$= \frac{(2p_1-1)!!p_2!}{(2l+1)\times\cdots\times(2l+2p-1)}(2l+1)^p;$$

$$E\left\{\left(\frac{|a_{l1}|^2}{\widehat{C}_l}\right)^{p_1}\left(\frac{|a_{l2}|^2}{\widehat{C}_l}\right)^{p_2}\right\}$$

$$= \frac{(2l+1)^p}{2^p}\frac{\Gamma(l+1/2)}{\Gamma(l+p+1/2)}\frac{\Gamma(1/2)\Gamma(p_1+1)\Gamma(p_2+1)}{\Gamma(1/2)}$$

$$= \frac{p_1!p_2!}{(2l+1)\times\cdots\times(2l+2p-1)}(2l+1)^p.$$

By Proposition 9.11, it is possible to establish a simple relationship between the normalized bispectrum with known or unknown angular power spectrum. More precisely, it is immediate to see that for $l_1 < l_2 < l_3$

$$E\widehat{I}^{2p}_{l_1 l_2 l_3} = E I^{2p}_{l_1 l_2 l_3} \prod_{i=1}^{3} g(l_i; p)$$

that is, for instance

$$E\widehat{I}^{2}_{l_1 l_2 l_3} = E I^{2}_{l_1 l_2 l_3} = 1 \qquad (9.15)$$

and

$$E\widehat{I}^{4}_{l_1 l_2 l_3} = E I^{4}_{l_1 l_2 l_3} \left(1 - \frac{2}{2l_1 + 3}\right)\left(1 - \frac{2}{2l_2 + 3}\right)\left(1 - \frac{2}{2l_3 + 3}\right). \qquad (9.16)$$

Also, for $l_1 = l_2 < l_3$ and $l_1 < l_2 = l_3$

$$E\widehat{I}^{2p}_{l_1 l_1 l_3} = E I^{2p}_{l_1 l_1 l_3} g(l_1; 2p) g(l_3; p) \,, \ E\widehat{I}^{2p}_{l_1 l_3 l_2} = E I^{2p}_{l_1 l_3 l_3} g(l_1; p) g(l_3; 2p) \,;$$

finally, for $l_1 = l_2 = l_3 = l$

$$E\widehat{I}^{2p}_{lll} = E I^{2p}_{lll} g(l; 3p)$$

so that, for instance

$$E\widehat{I}^{2}_{lll} = 6\left(1 - \frac{2}{2l + 3}\right)\left(1 - \frac{4}{2l + 5}\right). \qquad (9.17)$$

It is interesting to note that

$$E\widehat{I}^{2p}_{l_1 l_2 l_3} \le E I^{2p}_{l_1 l_2 l_3} \ \text{and} \ \frac{E\widehat{I}^{2p}_{l_1 l_2 l_3}}{E I^{2p}_{l_1 l_2 l_3}} = 1 + O(l_1^{-1}) \,, \ \text{as } l_1 \to \infty \,, \qquad (9.18)$$

for all choices of (l_1, l_2, l_3) for which the bispectrum is well-defined. Also, we have shown in the previous Section that (see 9.11)

$$\left| E\left(\frac{I_{l_1 l_2 l_3}}{\sqrt{\Delta_{l_1 l_2 l_3}}}\right)^{2p} - (2p - 1)!! \right| \le \sqrt{\frac{c_p}{l_1}} \,, \ \text{some } c_p > 0 \,. \qquad (9.19)$$

An obvious consequence of (9.18) is hence the following result:

Theorem 9.12 *There exist an absolute constant* $c'_p > 0$ *such that, for all* $l_1 \le l_2 \le l_3$

$$\left| E\left(\frac{\widehat{I}_{l_1 l_2 l_3}}{\sqrt{\Delta_{l_1 l_2 l_3}}}\right)^{2p} - (2p - 1)!! \right| \le \sqrt{\frac{c'_p}{l_1}}, \qquad (9.20)$$

whence, as $l_1 \to \infty$,

$$\frac{\widehat{I}_{l_1 l_2 l_3}}{\sqrt{\Delta_{l_1 l_2 l_3}}} \overset{law}{\to} N(0, 1). \qquad (9.21)$$

Remark 9.13 In the previous discussion, the fact that convergence of moments is sufficient to have convergence in law to a Gaussian comes from the fact that the Gaussian distribution is determined by the moments.

Remark 9.14 Results (9.19)-(9.20) provide the rate of convergence of the moments of the bispectrum to the Gaussian values under random and non-random normalizations. This settles some questions raised in [118], where the moments of $I_{l_1 l_2 l_3}$ and $\widehat{I}_{l_1 l_2 l_3}$ where compared by means of Monte Carlo simulations. It can also be shown that the angular bispectrum ordinates at different multipoles are asymptotically independent, see [132].

9.5 Testing for non-Gaussianity

In this Section, we shall show how the previous results and their extensions can be used to implement non-Gaussianity tests for spherical random fields. To this purpose, we shall neglect many practical relevant questions, in particular the presence of unobserved regions in the field, which will make the exact evaluaton of spherical harmonic coefficients unfeasible. These difficulties are faced in the following sections, which are concerned with wavelet techniques.

More precisely, assume that the resolution of the experiment we are dealing with is such that it yields a maximum observable multipole equal to L. It is, in practice, unfeasible to take into account all available bispectrum ordinates for the implementation of a statistical procedure: indeed, for current satellite experiments such as *WMAP* or *Planck* these ordinates are in the order of $L^3 \sim 10^9$, and the evaluation of all these statistics is beyond the power of the fastest supercomputers for the near future. There are some numerical solutions to this problem which have been considered in the physical literature, for instance the widely popular *KSW* method by [198]. Here, we find it convenient to consider only a subset of bispectrum ordinates for the test, in some sort of the narrow-band approaches which have turned out to be so useful in the analysis of stationary and nonstationary stochastic processes (see [170, 171]).

There are, of course, several possible choices of configurations for the multipoles on which to focus our attention; we refer to [184] for discussion on the importance of the identification of alternative models for inflation dynamics. As in ([131]), we shall restrict our attention to two kinds of patterns; more precisely, for finite integers $l_0 \geq 2$, $K \geq 0$ we shall consider the processes

$$J_{1L;l_0,K}(r) = \frac{1}{\sqrt{L/2}} \sum_{l \text{ even}, l=l_0+K}^{[Lr]} \left\{ \frac{1}{\sqrt{K+1}} \sum_{u=0}^{K} (-1)^{3l/2} \frac{\widehat{I}_{l-u,l,l+u}}{\sqrt{\Delta_{l-u,l,l+u}}} \right\}, \quad (9.22)$$

and

$$J_{2L;l_0,K}(r) = \frac{1}{\sqrt{L}} \sum_{l=l_0+K+1}^{[Lr]-l_0-K} \left\{ \frac{1}{\sqrt{K+1}} \sum_{u=0}^{K} (-1)^{l_0+u+3l/2} \widehat{T_{l_0+u,l,l+l_0+u}} \right\}, \qquad (9.23)$$

where [.] denotes the integer part of a real number; $0 \le r \le 1$ and l_0 is an (arbitrary but fixed) value; the rationale for the phase factors $(-1)^{(l_1+l_2+l_3)/2}$ is understood from the computations below, see for instance (9.36) where these terms offset the alternating signs in Wigner's $3j$ coefficients. For cosmological applications l_0 can be taken for instance equal to two, because the so-called dipole $l = 1$ is usually discarded from CMB data, being associated with kinematic effects mainly due to the motion of the Milky Way and the local group of galaxies. As usual, the sums are taken to be equal to zero when the index set is empty. K is a fixed pooling parameter: for $K = 0$ we obtain the special cases

$$J_{1L;l_0}(r) = \frac{1}{\sqrt{L/2}} \sum_{l \text{ even}, l \ge l_0}^{[Lr]} (-1)^{3l/2} \frac{\widehat{T_{lll}}}{\sqrt{6}}, \qquad (9.24)$$

and

$$J_{2L;l_0}(r) = \frac{1}{\sqrt{L}} \sum_{l=l_0+1}^{[Lr]-l_0} (-1)^{l_0+u+3l/2} \widehat{T_{l_0,l,l+l_0}}. \qquad (9.25)$$

The normalizing factors are chosen to ensure an asymptotic unit variance for all summands. The processes $J_{1L;l_0,K}(r)$, $J_{2L;l_0,K}(r)$ can be viewed as a sort of boundary cases for the possible configurations of multipoles. More precisely, it may seem natural to restrict the attention on multipoles close or on the "main diagonal" $l_1 = l_2 = l_3 = l$, under the (very misleading) conjecture that the greatest part of the non-Gaussian signal should concentrate in that area. This rationale motivates $J_{1L;l_0,K}(r)$: we shall show below, though, how such a choice can be very far from optimal in relevant cases. On the other hand, $J_{2L;l_0,K}(r)$ is based on a sort of opposite strategy, that is, for a fixed l_0 we aim at maximizing the distance among multipoles, albeit preserving the triangle conditions $l_i \le l_j + l_k$. There are several alternative procedures one may wish to consider, but those we mentioned lend themselves to a simple analysis, while highlighting some quite unexpected features of asymptotics for fixed-radius random fields.

Theorem 9.15 *As $L \to \infty$, for any fixed integers $l_0 > 0$, $K \ge 0$*

$$J_{1L;l_0,K}(r), J_{2L;l_0,K}(r) \Rightarrow W(r), 0 \le r \le 1, \qquad (9.26)$$

where $\Rightarrow$ denotes weak convergence in the Skorohod space $D[0, 1]$ and $W(r)$ denotes standard Brownian motion.

Proof The proof for (9.24) is trivial, as we are considering partial sums of zero-mean, unit variance, finite fourth moment independent random variables; we can hence apply standard results on weak convergence, as presented for instance in the classical textbook [24]. For $J_{2L;l_0,K}(r)$, note that the set of random coefficients $\{a_{lm} : l = l_0, ..., l_0 + K, m = -l, ..., l\}$ belongs to each summand in (9.25), which makes the structure of dependence much more complicated. Denote by $\mathfrak{I}_l$ the filtration generated by the triangular array $\{a_{l,-l}, ..., a_{l,l}\}$, $l = 1, 2, ...,$ and define

$$X_{l,L} = \frac{1}{\sqrt{K+1}} \sum_{u=0}^{K} (-1)^{l_0+u+3l/2} \widehat{I}_{l_0+u,l,l+l_0+u} \, , \, l = l_0 + K + 1, l_0 + K + 2, ... \, ,$$

that is

$$J_{2L;l_0,K}(r) = \sum_{l=l_0+K+1}^{[Lr]-l_0-K} X_{l,L} \, .$$

Now we note first that

$$E\{X_{l,L}|\mathfrak{I}_{l-k}\} = \frac{1}{\sqrt{K+1}} \sum_{u=0}^{K} (-1)^{l_0+u+3l/2} E\widehat{I}_{l_0+u,l,l+l_0+u} = 0 \, , \, k \geq 1 \, . \quad (9.27)$$

Equation (9.27) does not imply that the triangular array $\{X_{l,L}\}_{l=2,3,...}$ obeys a Martingale difference property, because the sequence $X_{l,L}$ is not adapted to the filtration $\mathfrak{I}_l$. However, (9.27) proves that the pair sequences $\{X_{l,L}, \mathfrak{I}_l\}_{l=2,3,...}$ do satisfy a mixingale property (see [43]), that is

$$\left[E \left(E\{X_{l,L}|\mathfrak{I}_{l-k}\} \right)^2 \right]^{1/2} \leq c_1 \frac{k^{-\phi}}{\sqrt{L}}, \text{ for } k \geq 1 \, , \quad (9.28)$$

$$\left[E \left(X_{l,L} - E\{X_{l,L}|\mathfrak{I}_{l+k}\} \right)^2 \right]^{1/2} \leq c_2 \frac{k^{-\phi}}{\sqrt{L}}, \text{ for } k \geq 1 \, , \quad (9.29)$$

$$\text{for some } c_1, c_2, \phi \qquad > 0 \, ; \quad (9.30)$$

actually the left-hand sides of (9.28)-(9.29) are identically zero for k larger than $l_0 + K$, so that, for suitable choices of the constants c_1, c_2, the bounds on the right-hand sides hold for an arbitrary large ϕ. Note that

$$\left\{ X_{l,L}^2 \, , \, l = 1, 2, ..., L \, , \, L = 1, 2, ..., \, \right\}$$

is a uniformly integrable array, because $\widehat{I}_{l_1 l_2 l_3}$ has finite fourth-order moments which are uniformly bounded. Also, it is readily seen that

$$\sup_{0 \leq r_1 < r_2 \leq 1} \limsup_{L \to \infty} \frac{\sum_{l=[Lr_1]}^{[Lr_2]} EX_{l,L}^2}{r_2 - r_1} < \infty \, , \, \lim_{L \to \infty} \max_{l=1,...,L} EX_{l,L}^2 = 0 \, .$$

Using a classical Functional Central Limit Theorem for mixingales (see for instance ([43]), to complete the proof, we need only to show that

$$\lim_{L\to\infty} E\left|E\left\{\left(\sum_{l=[Ls]}^{[Lt]} X_{l,L}\right)^2 |\mathfrak{I}_{[Lr]}\right\} - (t-s)\right| = 0 \text{ for any } r < s < t,$$

a task which is easily accomplished by simple algebraic manipulations, see [131] for more details and extensions. □

Theorem 9.15 can be immediately applied to derive the asymptotic distribution under the null of several non-Gaussianity tests. For instance, we might focus on $\sup_{0\le r\le 1} J_{aL;l_0,K}(r)$; by the Continuous Mapping Theorem we obtain (for $x \ge 0$, $a = 1, 2$,)

$$\lim_{L\to\infty} P\left\{\sup_{0\le r\le 1} J_{aL;l_0,K}(r) \le x\right\} = P\left\{\sup_{0\le t\le 1} W_t \le x\right\} = 2\Phi(x) - 1, \quad (9.31)$$

where $\Phi(.)$ denotes the cumulative distribution function of a standard Gaussian variable (for the last equality, see for instance [24]).

We shall now discuss the behaviour of the previously proposed procedures under some examples of non-Gaussian spherical fields. In particular, we focus on the (extremely popular) Sachs-Wolfe bispectrum, i.e.

$$B_{l_1 l_2 l_3} = G f_{NL} h_{l_1 l_2 l_3} \begin{pmatrix} l_1 & l_2 & l_3 \\ 0 & 0 & 0 \end{pmatrix} \{C_{l_1} C_{l_2} + C_{l_2} C_{l_3} + C_{l_1} C_{l_3}\}, \quad (9.32)$$

where G is a positive constant and

$$h_{l_1 l_2 l_3} = \left(\frac{(2l_1 + 1)(2l_2 + 1)(2l_3 + 1)}{4\pi}\right)^{1/2}.$$

As discussed in Chapter 6, neglecting lower order terms (9.32) represents the bispectrum of quadratic field under Gaussian subordination,

$$T = T_G + f_{NL}\left(T_G^2 - ET_G^2\right). \quad (9.33)$$

We consider for simplicity a power-like behaviour

$$C_l \approx l^{-\alpha}, \alpha > 2. \quad (9.34)$$

The assumption described by (9.34) is not unreasonable: indeed, for the standard cosmological models and best fit parameters from *WMAP* data one has $0.2c \times l^{-2} \le C_l \le 5c \times l^{-2}$, some $c > 0$, for all l up to 2×10^{-3}, approximately. For simplicity, let us assume that the normalizing angular power spectrum is non-random, that is, known *a priori*; without loss of generality, we take $K = 0$. Recall that (see Chapter 3 or [195, Eqs. 8.1.2.12 and 8.5.2.32])

$$\begin{pmatrix} l_1 & l_2 & l_3 \\ 0 & 0 & 0 \end{pmatrix} = \frac{(-1)^{(l_1+l_2+l_3)/2} [(l_1 + l_2 + l_3)/2]!}{[(l_1 + l_2 - l_3)/2]! [(l_1 - l_2 + l_3)/2]! [(-l_1 + l_2 + l_3)/2]!}$$

$$\times \left\{ \frac{(l_1 + l_2 - l_3)!(l_1 - l_2 + l_3)!(-l_1 + l_2 + l_3)!}{(l_1 + l_2 + l_3 + 1)!} \right\}^{1/2}.$$

Thus, for fixed $l_0 \geq 2$,

$$\begin{pmatrix} l_0 & l & l+l_0 \\ 0 & 0 & 0 \end{pmatrix} = \frac{(-1)^{l_0+l}(l + l_0) \times \cdots \times (l+1)}{l_0!} \frac{\sqrt{(2l_0)!}}{\sqrt{(2l+1) \times \cdots \times (2l + 2l_0 + 1)}}$$

$$= C \frac{(-1)^{l_0+l}}{\sqrt{l}} + O\left(\frac{1}{l^{3/2}}\right),$$

for some $C > 0$ which depends on l_0 but not on l. Then we have easily that

$$EJ_{2L}(r) \approx \frac{f_{NL}}{\sqrt{L}} \sum_{l=l_0+1}^{[Lr]} \sqrt{\frac{(2l_0 + 1)(2l + 1)(2l + 2l_0 + 1)}{4\pi}} \frac{1}{\sqrt{l}} \tag{9.35}$$

$$\times \left[\sqrt{\frac{C_{l_0}C_l}{C_{l+l_0}}} + \sqrt{\frac{C_{l_0}C_{l+l_0}}{C_l}} + \sqrt{\frac{C_l C_{l+l_0}}{C_{l_0}}} \right]$$

$$\approx \frac{f_{NL}}{\sqrt{L}} \sum_{l=l_0+1}^{[Lr]} \sqrt{l} \sqrt{\frac{C_{l_0}C_l}{C_{l+l_0}}} \approx \frac{f_{NL}}{\sqrt{L}} \sum_{l=l_0+1}^{[Lr]} \sqrt{l} \approx f_{NL}L. \tag{9.36}$$

Of course, (9.36) diverges as the number of observed multipoles increases ($L \to \infty$), that is, as the resolution of the experiment improves. The constants of proportionality are typically small; for the model we adopted in the simulations below, they are in the order of 10^{-4}. On the other hand, we recall that for $l_1 = l_2 = l_3 = l$

$$\begin{pmatrix} l & l & l \\ 0 & 0 & 0 \end{pmatrix} = (-1)^{3l/2} \frac{[3l/2]!}{[(l/2)!]^3} \left\{ \frac{[l!]^3}{(3l+1)!} \right\}^{1/2}$$

$$\approx (-1)^{3l/2} \frac{[3l]^{(3l+1)/2}}{l^{3/2(l+1)}} \left\{ \frac{l^{3l+3/2}}{(3l)^{3(l+1/2)}} \right\}^{1/2} \approx \frac{(-1)^{3l/2}}{l},$$

where we have used Stirling's formula

$$n!/(\sqrt{2\pi}n^{n+1/2}e^{-n}) = 1 + O(12n^{-1}).$$

Hence we have

$$EJ_{1L;l_0,K}(r) = O\left(\frac{1}{\sqrt{L}} \sum_{l=l_0+K}^{[Lr]} \sqrt{\frac{(2l+1)^3}{4\pi}} \begin{pmatrix} l & l & l \\ 0 & 0 & 0 \end{pmatrix} \frac{C_l^2}{\sqrt{C_l^3}}\right)$$

$$= O\left(\frac{1}{\sqrt{L}} \sum_{l=l_0+K}^{[Lr]} \sqrt{lC_l}\right) = o(1), \text{ as } L \to \infty.$$

Numerical evidence on the performance of this procedure is given for instance in [131] and in [36]. In short, the bispectrum can be shown to perform extremely satisfactorily in the idealistic circumstances of fully observed spherical maps with no instrumental noise. Indeed, in such cases some calculations suggest that at the resolution of *Planck* the bispectrum may have power against alternatives where the non-Gaussian components $f_{nl}(T^2 - 1)$ is in the order of 10^{-4} with respect to the Gaussian part T. More precisely, the standard parametrization used in the astrophysical literature entails a variance for T in the order of $Var(T) = ET^2 \simeq 10^{-8}$. We can then present a rough relationship between the value of the nonlinearity parameter f_{NL} and the relative amount of the non-Gaussian signal, namely

$$\frac{\sqrt{Var\{f_{NL}T^2\}}}{\sqrt{Var\{T\}}} = \sqrt{2}f_{NL}\sqrt{Var\{T\}} \simeq f_{NL} \times 10^{-4}. \tag{9.37}$$

In the absence of gaps and noise, the averaged bispectrum has power larger than 50% for f_{nl} as small as 20 at the WMAP resolution $L = \max l \simeq 700$. This performance deteriorates drastically under more realistic circumstances, and a crucial issue relates to the possibility to entertain corrections to take this feature into account. Recently, the physical literature has focussed on the possibility to calibrate by means of Monte Carlo simulations the effect of instrumental noise and missing data, much the same way as we discussed for *MASTER* in Chapter 8. This has led to the so-called *KSW* estimator (see [198]), currently very popular and leading to estimates $\widehat{f_{NL}} \simeq 30$ on *WMAP* 5 years data; the resulting standard deviation is estimated by simulations to be in the order of 30, so that the Gaussian case $f_{NL} = 0$ cannot be ruled out at 95% confidence level. In the chapters to follow, we shall pursue an alternative strategy to correct for the effect of missing observations, namely we shall focus on a combination of ideas from the bispectrum literature and a newly established wavelet system on the spheres, called needlets.

Remark 9.16 A remarkable consequence of the results in this Section is the following: Equation (9.36) suggests that a testing procedure in harmonic space can yield consistent tests of Gaussianity even for a single realization of an isotropic random field. This is to some extent an unexpected result, which yields some important insights into the validity and relevance of the high-frequency asymptotics approach.

Remark 9.17 It is important to stress the huge impact of the choice of combined angular scales on the expected power under non-Gaussian alternatives. It is noteworthy that the choice of a (close to) "main diagonal" configuration can yield negligible power, the expected value of the non-Gaussian signal decreasing to zero as the resolution of the experiment improves. The fact that for "local" models of non-Gaussianity the signal is mainly concentrated on the *squeezed* or *collapsed* configurations, i.e. those where one of the three multipoles l_1 is very small, was noted and discussed in a huge number of papers, see for instance [8, 19, 20, 36, 67, 68, 131, 182]. Other forms of non-Gaussianity may tend to favour different configurations, for instance *equilateral* shapes where the three multipoles are roughly of the same order, see again [67, 68, 184], and the references therein. More generally, the determination of the triples of angular scales (l_1, l_2, l_3) where the largest part of the non-Gaussian signal is to be expected, for a given class of models, represents an issue of great importance for future cosmological data analysis.

10

Spherical Needlets and their Asymptotic Properties

10.1 Introduction

In the last few years, a rather extensive literature has been developed on the construction of wavelets systems on the sphere; a very incomplete list of references includes for instance [6, 42, 70, 101, 172, 202]. Some of these attempts have been explicitly motivated applications in astronomy and/or cosmology (see [141]), as those we have described in the previous chapters.

In this area, the interest for spherical wavelets is easily understood. Indeed, as mentioned earlier, theoretical predictions involving CMB are almost exclusively elaborated in harmonic spaces, focussing on quantities such as the angular power spectra $\{C_l\}$ or the bispectra $\{B_{l_1 l_2 l_3}\}$ (which we discussed in the previous chapters). We have already considered statistical procedures for the estimation of angular power spectra and bispectra and developed their asymptotic theory in the high-frequency sense. These procedures were all based upon spherical harmonic transforms and manipulations of the random coefficients $\{a_{lm}\}$. While these techniques have been shown to enjoy a number of attractive properties, their practical applications is hindered by some important difficulties. In particular, the evaluation of the exact inverse transforms $a_{lm} = \int_{S^2} T(x)\overline{Y}_{lm}(x)d\sigma(x)$ (where T is a given spherical field) requires that the spherical random field is fully observed. Unfortunately, this is not the case in practice. For instance, in the framework of CMB data analysis the presence of foreground emissions from the Milky Way and other astrophysical sources prevents the exact evaluation of spherical harmonic transforms on approximately 20% of the entire sphere. As noted in the previous chapter, the coefficients $\{a_{lm}\}$ derived from such incomplete maps lose many of their standard properties. For instance, we recall that sample coefficients from incomplete maps are no longer uncorrelated. In the cosmological literature, this problem is well-known, and indeed enormous efforts have been entertained in order to address these difficulties. In particular, huge amounts of simulations are usually realized to reconcile the theory (for fully observed maps) with the practice of data analysis techniques. This has led, for instance, to the well-established methods we mentioned earlier, such as (to mention just a few) *MASTER* for the estimation of angular power spectra (see [100]) and *KSW* for the estimation of bispectra (see [119]). In these approaches, the biases and correlations which are introduced by missing observations are basically corrected by means of Monte Carlo simulations. It is easy to understand that the problem is mainly generated by the lack of any form of real space localization in the spherical harmonic transforms: this indeed implies that the effect of missing data is spread throughout the $\{a_{lm}\}$ array, introducing distortions and correlations which are very difficult to handle. In view of this, it is natural to expect that wavelets systems, characterized (in a broad sense) by double localization properties in real and harmonic space, may provide a very natural alternative for statistical applications.

It is also important to stress how localization in real space may represent a very valuable statistical feature, even in the presence of fully observed maps. For instance, most of this monograph has been developed under the assumptions of isotropy: this is indeed a cornerstone of modern astrophysics, as a consequence of Einstein's *Cosmological Principle* on the symmetry of physical laws (see [154, 161]). However, in the last few years one of the main themes of CMB data analysis has been the search for violations of isotropy in the observed data. A celebrated example is the possible existence of localized features such as the well-known *Cold Spot*, which have raised enormous debate – see for instance [40, 41, 62, 93, 196] for further details and discussions. It is therefore of the greatest importance to develop statistical tools which enjoy at the same time localization properties in the real and frequency space. This requirement naturally leads to the analysis of spherical wavelets.

In this monograph, among spherical wavelets we shall focus the so-called *needlets*, which were introduced into the functional analysis literature by Narcowich, Petrushev and Ward [143, 144]. Their statistical properties were first considered by Baldi, Kerkyacharian, Marinucci and Picard [13, 14], see also [123, 124, 140]. Needlets fit especially well in the perspective of this book, due to their peculiar asymptotic properties, in the high-frequency sense. In particular, it has been shown in [13] that random needlet coefficients enjoy a capital uncorrelation property: namely, for any fixed angular distance, random needlets coefficients are asymptotically uncorrelated as the frequency parameter grows larger and larger. Note that the uncorrelation property of wavelets coefficients does not follow at all by their localization properties in real domain. Indeed, given the fixed-domain asymptotics we are considering, perfect localization in real space does not ensure any form of uncorrelation (all random values at different locations on the sphere have in general a non-zero correlation).

The meaning of this uncorrelation property must be carefully understood, given the specific setting of statistical inference in cosmology. Indeed, as we already argued in several occasions, the CMB radiation can be viewed as a single realization of an isotropic random field on a sphere of a finite radius ([51, 60]). The asymptotic theory we entertain is thus (as always in the previous chapters) in the high-frequency sense, i.e. it is considered that observations at higher and higher frequencies (smaller and smaller scales) become available with the growing sophistication of CMB satellite experiments. Of course, uncorrelation entails independence in the Gaussian case: as a consequence, from the above-mentioned uncorrelation property it follows that (at least in the Gaussian case) an increasing array of asymptotically i.i.d. coefficients can be derived from a single realization of a spherical random field. This of course makes possible the introduction of a variety of statistical procedures for testing non-Gaussianity, estimating the angular power spectrum, testing for asymmetries, implementing bootstrap techniques, testing for cross-correlation among CMB and Large Scale Structure data, and many others, see for instance [13, 14, 46, 65, 83, 87, 123, 124, 139, 140, 162, 163, 164, 165, 175, 176]. We shall review many of these statistical applications in the following Chapter 11.

In this chapter, we shall first review the main features of the needlet construction, and discuss their main properties, focussing in particular on the asymptotic uncorrelation discussed above. Given the capital importance of this uncorrelation property, it is natural to investigate to what extent this property should be considered unique for the construction in [143, 144], or else whether it is actually true in other frameworks. Later in the chapter, we address this question by investigating the stochastic properties of the so-called Mexican needlets , which have been introduced by Geller and Mayeli in [76, 77, 78], see also [70] for a related setting. We shall present both a positive and a negative result: namely, we will provide necessary and sufficient conditions for the Mexican needlets coefficients to be uncorrelated, depending on the behavior of the angular power spectrum of the underlying random fields. In particular, unlike for needlets (see [143, 144]), we shall show that there may be correlation of the random coefficients when the angular power spectrum is decaying faster than a certain limit. However, higher order versions (already considered in [76]) of the Mexican needlets can indeed provide uncorrelated coefficients, depending on a parameter which is related to the decay of the angular power spectrum.

In some sense, a heuristic rationale under these results can be explained as follows: the correlation among coefficients is due to the presence in each of these terms of random elements which are fixed (with respect to growing frequencies) in a given realization of the random field, because they depend only on very large scale behavior (as mentioned, this is known in the physical literature as the *Cosmic Variance effect*). Because of the needlets' compact support in the frequency domain (see again [143, 144]), these low-frequency components are always dropped and uncorrelation is ensured. On the other hand, the same components could be relevant for Mexican needlets, and in this case it is necessary to introduce suitably modified versions which are better localized in the frequency domain (i.e., they allow less weight on very low frequency components). It should be noted that in settings of interest for CMB data analysis Mexican needlets have been recently shown to perform very efficiently both in terms of localization and uncorrelation (see [179]).

10.2 The construction of spherical needlets

In this section we shall briefly recall the main features of the needlets construction. We follow mainly the discussion in [13, 14].

10.2.1 Definition of spherical needlets

We have proved earlier in this book that the space of square-integrable functions on the sphere, written $L^2(S^2, d\sigma) = L^2(S^2)$, can be decomposed into the direct sum of orthogonal spaces which are spanned by the spherical harmonics $\{Y_{lm}\}_{m=-l,...,l}$, $l = 0, 1, 2, ...$ This result can be written formally as

$$L^2(S^2) = \bigoplus_{l=0}^{\infty} \mathcal{H}_l \, ,$$

where $\mathcal{H}_l = \mathcal{H}_l(S^2) = span\{Y_{lm}, \ m = -l, ..., l\}$. The (least square) projector on $\mathcal{H}_l$ is given by a kernel operator $P_{\mathcal{H}_l}$ such that, for all $f \in L^2(S^2)$

$$P_{\mathcal{H}_l} f(x) = \int_{S^2} f(x) Z_l(\langle x, y \rangle) dy \, ,$$

where (compare with formula (3.42))

$$Z_l(\langle x, y \rangle) := \sum_m Y_{lm}(x) \overline{Y_{lm}}(y) = \frac{2l+1}{4\pi} P_l(\langle x, y \rangle) \, .$$

It is readily seen that the *reproducing kernel property* is satisfied, that is

$$\int_{S^2} Z_{l_1}(\langle x, y\rangle) Z_{l_2}(\langle y, z\rangle) dy$$

$$= \int_{S^2} \sum_{m_1} Y_{l_1 m_1}(x)\overline{Y_{l_1 m_1}}(y) \sum_{m_2} Y_{l_2 m_2}(y)\overline{Y_{l_2 m_2}}(z) dy$$

$$= \sum_{m_1} \sum_{m_2} Y_{l_1 m_1}(x)\overline{Y_{l_2 m_2}}(z) \int_{S^2} \overline{Y_{l_1 m_1}}(y) Y_{l_2 m_2}(y) dy$$

$$= \sum_{m_1} \sum_{m_2} Y_{l_1 m_1}(x)\overline{Y_{l_2 m_2}}(z)\delta_{l_1}^{l_2}\delta_{m_1}^{m_2} = \begin{cases} Z_{l_1}(\langle x, z\rangle), & \text{for } l_1 = l_2 \\ 0, & \text{otherwise} \end{cases}.$$

For the construction of needlets, we should first start to define $\mathcal{K}_l = \bigoplus_{k=0}^{l} \mathcal{H}_k$ as the space of the restrictions to the sphere S^2 of polynomials of degree less than l. The next ingredient are the set of *cubature points* and *cubature weights*. It is now a standard result (see again [143, 144]) that for all $j \in \mathbb{N}$, there exists a finite subset X_j of S^2 and positive real numbers $\lambda_{jk} > 0$, indexed by the elements of X_j, such that

$$\forall f \in \mathcal{K}_l, \quad \int_{S^2} f(x)dx = \sum_{\xi_{jk} \in X_j} \lambda_{jk} f(\xi_{jk}). \tag{10.1}$$

We denoted by ξ_{jk} the cubature points corresponding to a level j. It is known that the points in $\{X_j\}_{j=0}^{\infty}$ are almost uniformly ε_j–distributed with $\varepsilon_j := \kappa B^{-j}$, and the coefficients $\{\lambda_{jk}\}$ are such that $\lambda_{jk} \approx cB^{-2j}$, card $\{X_j\} \approx B^{2j}$. In other words, the cubature weights $\left\{\lambda_{jk}\right\}_{jk}$ and the cubature points $\left\{\xi_{jk}\right\}_{jk}$ are such that, for all polynomials $Q_l(x)$ of degree smaller than B^{j+1},

$$\sum_k Q_l(\xi_{jk})\lambda_{jk} = \int_{S^2} Q_l(x)dx.$$

Definition 10.1 (See [143, 144].) A family of **spherical needlets** $\{\psi_{jk}\}$ is defined by setting

$$\psi_{jk}(x) := \sqrt{\lambda_{jk}} \sum_l b\left(\frac{l}{B^j}\right) \sum_{m=-l}^{l} Y_{lm}(\xi_{jk})\overline{Y_{lm}}(x) \tag{10.2}$$

$$= \sqrt{\lambda_{jk}} \sum_l b\left(\frac{l}{B^j}\right) \frac{2l+1}{4\pi} P_l(\langle\xi_{jk}, x\rangle), \tag{10.3}$$

where $x \in S^2$, $\left\{\lambda_{jk}, \xi_{jk}\right\}$ are a set of cubature points and weights, $B > 1$ is a constant and $b(.)$ is a weight function satisfying three conditions, namely

(a) (compact support) $b(.) > 0$ in (B^{-1}, B), and it is equal to zero otherwise

(b) (partition of unity) for all $\xi \geq 1$

$$\sum_{j=0}^{\infty} b^2 \left(\frac{\xi}{B^j} \right) = 1 ,$$

(c) (smoothness) $b(.) \in C^M$, i.e. it is M times continuously differentiable, for some $M = 1, 2, ...$ or $M = \infty$.

10.2.2 Numerical recipes

We provide below two alternative recipes for the construction of a function $b(.)$ satisfying the previous assumptions.

A B-Spline approach

A general recipe to construct $b(.) \in C^M$ can be given in terms of a standard B-Spline approach, as follows. First recall that Bernstein polynomials are defined as

$$B_i^{(n)}(t) = \binom{n}{i} t^i (1 - t)^{n-i} ,$$

where $t \in [0, 1]$, $i = 0, ..., n$ and $n = 1, 2, ...$ For instance, we have

$$B_0^{(1)}(t) = (1 - t) , B_1^{(1)}(t) = t , \text{ for } n = 1 ,$$

$$B_0^{(2)}(t) = (1 - t)^2 , B_1^{(2)}(t) = 2t(1 - t) , B_2^{(2)}(t) = t^2 , \text{ for } n = 2$$

$$B_0^{(3)}(t) = (1 - t)^3 , B_1^{(3)}(t) = 3t(1 - t)^2 , B_2^{(3)}(t) = 3t^2(1 - t) ,$$

$$B_3^{(3)}(t) = t^3 , \text{ for } n = 3 .$$

We can hence define polynomials

$$p_{2k+1}(t) = \sum_{i=0}^{k} B_i^{(2k+1)}(t) .$$

More explicitly, for $k = 1, 2$

$$p_3(t) = \sum_{i=0}^{1} B_i^{(3)}(t) = (1 - t)^3 + 3t(1 - t)^2 ,$$

$$p_5(t) = \sum_{i=0}^{2} B_i^{(5)}(t) = (1 - t)^5 + 5t(1 - t)^4 + 10t^2(1 - t)^3 .$$

Note that

$$p_3(0) = p_5(0) = 1 , p_3(1) = p_5(1) = 0 ,$$

and

$$p_3'(t) = -6t(1 - t) , p_3'(1) = p_3'(0) = 0 ,$$

$$p_5'(t) = -30t^2(1 - t)^2 , p_5'(1) = p_5'(0) = 0 .$$

Likewise, the second order derivative of p_5 has a zero at $t = 0$ and $t = 1$. In general, it is readily checked that

$$p_{2k+1}^{(r)}(1) = p_{2k+1}^{(r)}(0) = 0 \text{ for } r = 1, ..., k .$$

Hence, for $B > 1$ let

$$t = \frac{x - 1/B}{1 - 1/B} ,$$

and define the function

$$\phi(x) := \begin{cases} 1 \text{ if } x \in [0, \frac{1}{B}] \\ p_{2k+1}(t) = p_{2k+1}\left(\frac{x-1/B}{1-1/B}\right) \text{ if } x \in [\frac{1}{B}, 1] \\ 0 \text{ if } x > 1 \end{cases} .$$

We can take

$$b(x) = \begin{cases} \sqrt{\phi(\frac{x}{B}) - \phi(x)} , \frac{1}{B} \le x \le B \\ 0 , \text{ otherwise} \end{cases} .$$

This function satisfies the three conditions (a)-(b)-(c) listed in the definition of spherical needlets, for $M < k + 1/2$.

A construction for $b(.) \in C^\infty$

A simple numerical recipe for $b(.) \in C^\infty$ is provided by [13]. The idea can be summarized in the following steps:

- STEP 1: Construct the function

$$\phi_1(t) = \begin{cases} \exp(-\frac{1}{1-t^2}), & -1 \le t \le 1 \\ 0, & \text{otherwise} \end{cases} .$$

It is immediate to check that the function $f(\cdot)$ is C^∞ and compactly supported in the interval $(-1, 1)$;

- STEP 2: Construct the function

$$\phi_2(u) = \frac{\int_{-1}^u f(t)dt}{\int_{-1}^1 f(t)dt} .$$

The function $\phi_2(\cdot)$ is again C^∞; it is moreover non-decreasing and normalized so that $\phi_2(-1) = 0, \phi_2(1) = 1;$

- STEP 3: Construct the function

$$
\phi_3(t) = \left\{ \begin{array}{ll} 1 & \text{if } 0 \le t \le \frac{1}{B} \\ \phi_2(1 - \frac{2B}{B-1}(t - \frac{1}{B})) & \text{if } \frac{1}{B} < t \le 1 \\ 0 & \text{if } t > 1 \end{array} \right. .
$$

Here we are simply implementing a change of variable so that the resulting function $\phi(\cdot)$ is constant on $(0, B^{-1})$ and monotonically decreasing to zero in the interval $(B^{-1}, 1)$. Indeed it can be checked that

$$
1 - \frac{2B}{B-1}(t - \frac{1}{B}) = \left\{ \begin{array}{ll} 1 & \text{for } t = \frac{1}{B} \\ -1 & \text{for } t = 1 \end{array} \right. .
$$

- STEP 4: As before, construct

$$
b^2(x) = \phi_3\left(\frac{x}{B}\right) - \phi_3(x), \quad -\infty < x < \infty,
$$

and for $b(x)$ take the positive root. Clearly $b(.) \in C^\infty$, and satisfies the three conditions (a)-(b)-(c) listed in the definition of spherical needlets, for all $M = 1, 2, 3... $.

From the computational point of view, we should stress that needlets are not only feasible, but indeed extremely convenient. In fact, standard packages for the analysis of spherical random fields (such as *HEALPIX* or *GLESP*, see [85] and [53]) provide not only the spherical harmonics, but also (exact or approximate) cubature points and cubature weights. The implementation of needlets can then be performed with a minimal effort, see for instance [139], where plots and numerical evidence on localization and uncorrelation are also provided; we refer also to [179] for further results, and to [166] for a publicly available software.

10.3 Properties of spherical needlets

10.3.1 Tight frames and reconstruction formulae

We start by recalling here a few basic properties of the needlets system. We note first that needlets are computationally very convenient, and rely naturally on the manifold structure of the sphere. Indeed, the implementation of needlets only requires the evaluation of weighted averages of spherical harmonics transforms, which are now available in several software packages; (approximate) cubature points and weights are also available in the toolbox of applied scientists, for instance in popular packages such as [53, 85]. It is also noteworthy how needlets do not require any form of tangent plane approximation, as it was the case for earlier attempts to build wavelets on the sphere by adapting popular techniques for $\mathbb{R}^2$.

Spherical needlet coefficients are provided by the analytical formula

$$\beta_{jk} = \int_{S^2} T(x)\psi_{jk}(x)dx = \sqrt{\lambda_{jk}} \sum_l b\left(\frac{l}{B^j}\right) \sum_{m=-l}^{l} a_{lm}Y_{lm}(\xi_{jk}) . \tag{10.4}$$

We have immediately

$$\sum_k \beta_{jk}\sqrt{\lambda_{jk}} = 0 , \tag{10.5}$$

i.e., the (weighted) sample mean of the needlets coefficients is identically zero at all levels j. The proof of this relation is trivial, because

$$\sum_k \beta_{jk}\sqrt{\lambda_{jk}} = \sum_{l=B^{j-1}}^{B^{j+1}} \sum_{m=-l}^{l} b\left(\frac{l}{B^j}\right) a_{lm} \left[\sum_k \lambda_{jk}Y_{lm}(\xi_{jk})\right]$$

$$= \sum_{l=B^{j-1}}^{B^{j+1}} \sum_{m=-l}^{l} b\left(\frac{l}{B^j}\right) a_{lm} \left[\int_{S^2} Y_{lm}(x)dx\right] = 0 .$$

A very important feature of needlets is that they make up a *tight frame*. Frames were introduced in functional analysis in the fifties, starting from the paper [58]. In some sense, they can be viewed as the closest system to a (redundant) basis. More precisely, given a Hilbert space of functions $\mathcal{H}$ with inner product $\langle .,. \rangle$, a countable family of functions $\{e_k : k = 1, 2, 3, ...\}$ is called a frame if there exist universal constants c, C (the "frame bounds") such that for all $f \in \mathcal{H}$

$$c \|f\|^2 \le \sum_k |\langle f, e_k \rangle|^2 \le C \|f\|^2 . \tag{10.6}$$

The frame is called tight if $c = C$. For our purposes, we take as the Hilbert space $L^2(S^2)$ with the usual inner product

$$\langle f, g \rangle := \int_{S^2} fg \, d\sigma ,$$

and we have easily the following result, proved in [143, 144]:

Proposition 10.2 *The needlet system provides a tight frame on $L^2(S^2)$ with frame bounds $c = C = 1$.*

Proof   The proof is straightforward, indeed it suffices to notice that

$$
\sum_{jk} \beta_{jk}^2 = \sum_{jk} \lambda_{jk} \left\{ \sum_l b\left(\frac{l}{B^j}\right) \sum_{m=-l}^{l} a_{lm} Y_{lm}(\xi_{jk}) \right\}^2
$$

$$
= \sum_j \sum_{l_1 l_2} b\left(\frac{l_1}{B^j}\right) b\left(\frac{l_2}{B^j}\right) \sum_{m_1=-l_1}^{l_1} a_{l_1 m_1} \overline{a_{l_2 m_2}} \sum_k \lambda_{jk} Y_{l_1 m_1}(\xi_{jk}) \overline{Y_{l_2 m_2}}(\xi_{jk})
$$

$$
= \sum_j \sum_{l_1 l_2} b\left(\frac{l_1}{B^j}\right) b\left(\frac{l_2}{B^j}\right) \sum_{m_1=-l_1}^{l_1} a_{l_1 m_1} \overline{a_{l_2 m_2}} \delta_{l_1}^{l_2} \delta_{m_1}^{m_2}
$$

$$
= \sum_j \sum_l b^2\left(\frac{l}{B^j}\right) \sum_{m_1=-l_1}^{l_1} |a_{lm}|^2 = \sum_l (2l+1)\widehat{C}_l = \|f\|^2 \,,
$$

because of course

$$
\|f\|^2 = \int_{S^2} f^2(x) d\sigma(x) = \int_{S^2} \left\{ \sum_l \sum_{m=-l}^{l} a_{lm} Y_{lm}(x) \right\}^2 d\sigma(x)
$$

$$
= \sum_{l_1 l_2} \sum_{m_1 m_2} a_{l_1 m_1} \overline{a_{l_2 m_2}} \int_{S^2} Y_{l_1 m_1}(x) \overline{Y_{l_2 m_2}}(x) dx
$$

$$
= \sum_{l_1 l_2} \sum_{m_1 m_2} a_{l_1 m_1} \overline{a_{l_2 m_2}} \delta_{l_1}^{l_2} \delta_{m_1}^{m_2}
$$

$$
= \sum_l (2l+1)\widehat{C}_l = \sum_{jk} \beta_{jk}^2 \,.
$$

$\square$

Heuristically, we have that the "energy" of the function is conserved when moving from its real space expression to its projection on the needlet frame – this property will play a crucial role when deriving needlet-based angular power spectrum estimators in the next chapter. From the mathematical point of view, the tight frame property is stating that the needlet frame defines an isometry $\Psi : L^2(S^2) \to \ell^2$ from the space of square-integrable functions $L^2(S^2)$ to the space of square-summable sequences ℓ^2. The correspondence between the two spaces is not one-to-one (see [97]), but the tight frame property is indeed strictly related to the following reconstruction formula, again a remarkable property of needlets. For all functions $f \in L^2(S^2)$ with corresponding needlet coefficients $\{\beta_{jk} = \beta_{jk}(f)\}$, we have

$$
\sum_{j,k} \beta_{jk} \psi_{jk}(x)
$$

$$
= \sum_j \sum_{l_1=B^{j-1}}^{B^{j+1}} \sum_{m_1=-l_1}^{l_1} b\left(\frac{l_1}{B^j}\right) b\left(\frac{l_2}{B^j}\right) a_{l_1 m_1} Y_{l_1 m_1}(x)
$$

$$\times \sum_{l_2=B^{j-1}}^{B^{j+1}} \sum_{m_2=-l_2}^{l_2} \sum_{k} Y_{l_1 m_1}(\xi_{jk}) \overline{Y}_{l_2 m_2}(\xi_{jk}) \lambda_{jk}$$

$$= \sum_{j} \sum_{l_1=B^{j-1}}^{B^{j+1}} \sum_{m_1=-l_1}^{l_1} b\left(\frac{l_1}{B^j}\right) b\left(\frac{l_2}{B^j}\right) a_{l_1 m_1} Y_{l_1 m_1}(x) \sum_{l_2=B^{j-1}}^{B^{j+1}} \sum_{m_2=-l_2}^{l_2} \delta_{l_1}^{l_2} \delta_{m_1}^{m_2}$$

$$= \sum_{l=1}^{\infty} \sum_{m=-l}^{l} a_{lm} Y_{lm}(x) ,$$

so that

$$f(x) = \sum_{j,k} \beta_{jk} \psi_{jk}(x) , \tag{10.7}$$

in the L^2 sense. The pair (10.4)-(10.7) can hence be viewed as a wavelet analogue of the direct and inverse Fourier transform (6.1)-(6.2).

Remark 10.3   A simple corollary of the tight frame property is the following:
for all j, k, we have $\left\| \psi_{jk}(.) \right\|^2 \le 1$. The proof is immediate: it suffices to take
$f(x) = \psi_{j^* k^*}(x)$ and $C = 1$ in the right-hand side of 10.6, to obtain

$$\sum_{jk} \left| \langle \psi_{j^* k^*}(x), \psi_{jk}(x) \rangle \right|^2 = \left\| \psi_{j^* k^*}(x) \right\|^4 + \sum_{jk \ne j^* k^*} \left| \langle \psi_{j^* k^*}(x), \psi_{jk}(x) \rangle \right|^2$$

$$\le \left\| \psi_{j^* k^*}(x) \right\|^2 .$$

This result provides further support to the interpretations of tight frames as a
redundant basis, indeed $\left\| \psi_{jk}(.) \right\|^2 = 1$ can only hold in the standard orthogonal
case (see [97]).

Remark 10.4   More generally, it was shown in [143, 144] that the L^p norm
of needlets is of order $\left\| \psi_{jk}(.) \right\|_{L^p} \approx B^{2j(\frac{1}{2}-\frac{1}{p})}$, where $\approx$ as before means that the
ratio of the left and right-hand sides is uniformly bounded above and below, see
also [15] for discussion and statistical applications. Thus, for $p = \infty$ we obtain
a diverging behavior with rate B^j, for $p = 1$ convergence to zero at rate B^{-j},
in both cases mimicking standard results for wavelets on $\mathbb{R}$. Careful evaluation
of the L^p norms of needlets are the basis for the development of a full theory
of *Besov spaces* over the sphere. These characterizations of functional spaces
are the basis for the construction of an adaptive theory of density estimation
and spherical deconvolution. These issues are outside the scope of the present
monograph: see again [15, 111, 112] for further details.

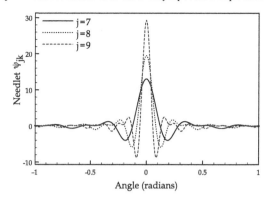

Figure 10.1 Needlets

10.3.2 Localization properties

In this subsection, we shall show that needlets enjoy excellent localization properties in the real domain, each $\psi_{jk}(x)$ being quasi-exponentially localized around its centre ξ_{jk}. We will exploit this property below, for instance to show that needlets coefficients are asymptotically unaffected by the presence of missing observations.

More precisely, the main localization property of $\{\psi_{jk}(x)\}$ is established in [143], where the following proposition is proved.

Proposition 10.5 (See [143])   *Consider the function $\psi_{jk}(x)$ defined in (10.1); then there exists a constant $c_M > 0$ such that, for every $x \in S^2$:*

$$\left|\psi_{jk}(x)\right| \leq \frac{c_M B^j}{(1 + B^j \arccos\langle \xi_{jk}, x \rangle)^M} \text{ uniformly in } (j, k).$$

Clearly for $b(.) \in C^\infty$ Proposition 10.5 holds for all $M \in \mathbb{N}$, i.e. needlets are almost exponentially localized around any cubature point, which motivates their name. As mentioned above, the original proof of this statement was given in [143], in the general framework of $L^2(S^n)$ spaces for arbitrary $n = 1, 2, 3, \ldots$. The argument in [143] is rather sophisticated, indeed it covers general spheres S^{n-1} in $\mathbb{R}^n$ and requires careful manipulations of Legendre polynomials by means of the so-called Mehler-Dirichlet representation formula. We defer the proof of Proposition 10.5 to the Appendix. The concentration phenomenon is illustrated in Fig. 10.1.

10.4 Stochastic properties of needlet coefficients

In the previous Section, we discussed the analytic properties of the needlet systems, in particular their localization properties in real and harmonic spaces. Here we shall discuss the properties of random needlet coefficients from an asymptotic point of view. These properties turn out to be to some extent surprising, in particular (as anticipated above) we shall show how needlet coefficients are asymptotically uncorrelated for any fixed angular distance, as the frequency goes to infinity. In what follows we considered a weakly isotropic centered random field $T = \{T(x) : x \in S^2\}$, with harmonic decomposition $T = \sum_{l=0}^{\infty} \sum_{m=-l}^{l} a_{lm} Y_{lm}$, and associated power spectrum $\{C_l : l \geq 0\}$.

First recall that the needlets system is compactly supported in the harmonic domain; as such, it is immediate to see for (10.4) that the random needlets coefficients are uncorrelated whenever $j - j' \geq 2$, i.e.

$$E\beta_{jk}\beta_{j'k'} = \sqrt{\lambda_{jk}\lambda_{j'k'}} \sum_{ll'} b\left(\frac{l}{B^j}\right) b\left(\frac{l'}{B^{j'}}\right) \sum_{mm'} Ea_{lm}a_{l'm'} Y_{lm}(\xi_{jk}) Y_{l'm'}(\xi_{j'k'})$$

$$= 0 .$$

For $j = j'$, the variance of the needlets coefficients is given by

$$E\beta_{jk}^2 = \lambda_{jk} \sum_{l=B^{j-1}}^{B^{j+1}} b^2\left(\frac{l}{B^j}\right) C_l \sum_{m=-l}^{l} Y_{lm}(\xi_{jk})\overline{Y_{lm}(\xi_{jk})}$$

$$= \lambda_{jk} \sum_{l=B^{j-1}}^{B^{j+1}} b^2\left(\frac{l}{B^j}\right) C_l \frac{2l+1}{4\pi} P_l(\cos 0)$$

$$= \lambda_{jk} \sum_{l=B^{j-1}}^{B^{j+1}} \frac{2l+1}{4\pi} b^2\left(\frac{l}{B^j}\right) C_l =: \sigma_{jk}^2 > 0 .$$

Note that we have $\sigma_{jk}^2 \approx: \sigma_j^2$ uniformly over k, where

$$\sigma_j^2 := \frac{1}{N_j} \sum_{l=B^{j-1}}^{B^{j+1}} \frac{2l+1}{4\pi} b^2\left(\frac{l}{B^j}\right) C_l , N_j := \text{card}\{X_j\} .$$

To investigate the correlation, we introduce now the same, mild regularity conditions on the angular power spectrum C_l of the random field $T(x)$ as in [13, 14], see also [123, 124, 140].

Condition 10.6 There exist $M \in \mathbb{N}, \alpha > 2$ and a sequence of functions $\{g_j(.)\}$ such that

$$C_l = l^{-\alpha} g_j\left(\frac{l}{B^j}\right) > 0 , \text{ for } B^{j-1} < l < B^{j+1} , \tag{10.8}$$

where

$$c_0^{-1} \le g_j \le c_0 \text{ for all } j \in \mathbb{N},$$

and for $r = 1, ..., M$

$$\sup_j \sup_{B^{-1} \le u \le B} \left| \frac{d^r}{du^r} g_j(u) \right| \le c_r \text{ some } c_1, ...c_M > 0.$$

Remark 10.7 Condition 10.6 entails a weak smoothness requirement on the behaviour of the angular power spectrum, which is trivially satisfied by cosmologically relevant models (where the angular power spectrum usually behaves as an inverse polynomial, see again [51]). The condition is fulfilled for instance by all models of the form

$$C_l = \frac{F_1(l)}{l^\beta F_2(l)},$$

where $F_1(l), F_2(l) > 0$ are polynomials of order q_1, q_2 and $\alpha = \beta + q_2 - q_1$.

The correlation coefficient can then be derived as

$$Corr\left(\beta_{jk}, \beta_{jk'}\right) = \frac{E\beta_{jk}\beta_{jk'}}{\sqrt{E\beta_{jk}^2 E\beta_{jk'}^2}}$$

$$= \frac{\sqrt{\lambda_{jk}\lambda_{jk'}}\sum_{l \ge 1} b^2(\frac{l}{B^j})\frac{2l+1}{4\pi}C_l P_l(\langle \xi_{jk}, \xi_{jk'} \rangle)}{\sqrt{\lambda_{j,k}\lambda_{j,k'}}\sum_{l \ge 1} b^2(\frac{l}{B^j})\frac{2l+1}{4\pi}C_l}.$$

The following result was first proved in [13].

Lemma 10.8 *Under Condition 10.6 the following inequality holds*

$$\left| Corr\left(\beta_{jk}, \beta_{jk'}\right) \right| \le \frac{C_M}{\left(1 + B^j d\left(\xi_{jk}, \xi_{jk'}\right)\right)^M}, \text{ some } C_M > 0, \tag{10.9}$$

where $d(\xi_{jk}, \xi_{jk'}) = \arccos(\langle \xi_{jk}, \xi_{jk'} \rangle)$ is the standard distance on the sphere.

Proof Observe first that, as we assumed $c_1 l^{-\alpha} \le C_l \le c_2 l^{-\alpha}$, we have

$$c_1 B^{(2-\alpha)j} \le \sum_l b^2\left(\frac{l}{B^j}\right) C_l \frac{2l+1}{4\pi} \le c_2 B^{(2-\alpha)j}, \tag{10.10}$$

whence

$$\sigma_j^2 = \frac{1}{N_j} \sum_{l=B^{j-1}}^{B^{j+1}} \frac{2l+1}{4\pi} b^2\left(\frac{l}{B^j}\right) C_l \approx B^{-\alpha j}. \tag{10.11}$$

Introduce now the sequence of functions $\{\phi_j\}$, $\phi_j := b^2(x)x^{-\alpha}g_j(x)$; notice that they uniformly satisfy the assumptions of Proposition (10.5), so that

$$\left| \sum_l \phi_j \left(\frac{l}{B^j} \right) (2l+1) P_l(\langle x,y \rangle) \right| \leq \frac{c_M B^{2j}}{(1 + B^j d(x,y))^M} ,$$

where c_M depends only on $\sup_{j \geq 1, r \leq M} \left\| \phi_j^{(r)} \right\|$, $\phi_j^{(r)}$ denoting as usual the rth derivative of ϕ_j, see again [143]. Using this result and 10.10, we obtain

$$\left| \frac{\sqrt{\lambda_{jk} \lambda_{jk'}} \sum_{l \geq 1} b^2 (\frac{l}{B^j})(2l+1) C_l P_l(\langle \xi_{jk}, \xi_{jk'} \rangle)}{\sqrt{\lambda_{j,k} \lambda_{j,k'}} \sum_{l \geq 1} b^2 (\frac{l}{B^j})(2l+1) C_l} \right|$$

$$= \left| \frac{\sum_{l \geq 1} b^2 (\frac{l}{B^j})(2l+1) l^{-\alpha} g_j(l/B^j) P_l(\langle \xi_{jk}, \xi_{jk'} \rangle)}{\sum_{l \geq 1} b^2 (\frac{l}{B^j})(2l+1) C_l} \right|$$

$$= \left| \frac{\sum_{l \geq 1} b^2 (\frac{l}{B^j})(l/B^j)^{-\alpha} g_j(l/B^j)(2l+1) P_l(\langle \xi_{jk}, \xi_{jk'} \rangle)}{B^{j\alpha} \sum_{l \geq 1} b^2 (\frac{l}{B^j})(2l+1) C_l} \right| \qquad (10.12)$$

$$\leq \frac{C_M}{\left(1 + B^j d \left(\xi_{jk}, \xi_{jk'} \right) \right)^M} . \qquad (10.13)$$

$\square$

In words, (10.9) is stating that as the frequency increases, spherical needlets coefficients are asymptotically uncorrelated (and hence, in the Gaussian case, independent), for any given angular distance. This property is of course of the greatest importance when investigating the asymptotic behaviour of statistical procedures: loosely speaking, it states that it is possible to derive an infinitely growing array of asymptotically independent "observations" (the needlets coefficients) out of a single realization of a continuous random field on a compact domain. It should be stressed that this property is not by any means a consequence of the localization properties of the needlets frame. As a counterexample, it is easy to construct spherical frames having bounded support in real space, whereas the corresponding random coefficients are not at all uncorrelated (recall the angular correlation function can be taken to be bounded from below at any distance on the sphere). More discussion on this in the sections to follow, where we consider the class of Mexican needlets.

10.5 Missing observations

We pointed out above, and we shall further discuss in the next sections, how the capital uncorrelation property of needlet coefficients cannot be viewed as a consequence of localization by itself. Localization, however, does of course make up a fundamental property of the needlet frame, and indeed it ensures that random coefficients are asymptotically unaffected by the presence of missing observations/ masked regions over the sphere where data cannot be collected (i.e., in a CMB framework, the regions where the background radiation is obscured by the Milky Way and other foreground sources). In this Section, we shall investigate more rigorously these robustness properties.

In particular, following [13], let us assume we observe $\widetilde{T}(x) = T(x)_{S^2 \backslash G} \mathbb{I}(x)$, i.e. there exists a region G where the observations are corrupted and hence they are just set equal to zero. We write $G_\varepsilon := \{x \in S^2 : d(x, G) \le \varepsilon\}$ for the neighbourhood of radius ε around the set $G \subset S^2$. Also, we write $\{\beta^*_{jk}\}$ for the random needlet coefficients evaluated on the partially observed sphere, so that

$$\beta^*_{jk} = \int_{S^2} \widetilde{T}(x)\psi_{jk}(x)dx = \int_{S^2 \backslash G} T(x)\psi_{jk}(x)dx$$

$$= \beta_{jk} - \int_G T(x)\psi_{jk}(x)dx \,.$$

Proposition 10.9   *For $\xi_{jk} \in S^2 \backslash G_\varepsilon$, we have*

$$\sqrt{E\left\{\beta^*_{jk} - \beta_{jk}\right\}^2} \le \frac{c_M B^j \sigma(G)}{(1 + B^j \varepsilon)^M} \sqrt{ET^2(x)} \,.$$

where $\sigma(G)$ denotes the Lebesgue measure of $G \subset S^2$.

Proof   By the localization property and Cauchy-Schwartz inequality, we have easily

$$\left\{\beta^*_{jk} - \beta_{jk}\right\}^2 \le \left\{\int_G T(x)\psi_{jk}(x)dx\right\}^2$$

$$\le \left\{\frac{c_M B^j}{(1 + B^j \varepsilon)^M}\right\}^2 \sigma(G) \left\{\int_G T^2(x)dx\right\}$$

whence

$$E\left\{\beta^*_{jk} - \beta_{jk}\right\}^2 \le \left\{\frac{c_M B^j}{(1 + B^j \varepsilon)^M}\right\}^2 \sigma(G) \int_G ET^2(x)dx$$

$$= \left\{\frac{c_M B^j \sigma(G)}{(1 + B^j \varepsilon)^M}\right\}^2 ET^2(x) \,.$$

$\square$

10.6 Mexican needlets

A key feature of needlets is the compact support in harmonic space which is ensured by the properties of the weight function $b(.)$. This compact support on one hand allows for the construction of exact cubature formulae, as the resulting needlets turn out to be linear combinations of finite-order trigonometric (Legendre) polynomials. Moreover, as we shall see below compact support in harmonic space plays a crucial role to grant asymptotic uncorrelation properties for a general form of the angular power spectrum.

It is of obvious interest to investigate what properties one might obtain by considering the case where the support of $b(.)$ is no longer compact. This is the route undertaken by Geller and Mayeli in [76, 77, 78], see also [70]. As explained, in these circumstances we can no longer expect an exact reconstruction formula or the tight frame property, because an exact cubature formula is unfeasible in the presence of polynomials of arbitrarily large order, for any fixed j. Likewise, exact localization in the frequency domain is no longer achievable. It should be added, however, that the approach initiated by [76, 77, 78] enjoys some undeniable strong points: firstly, it yields Gaussian localization properties in the real domain, i.e. much better concentration in real space than standard needlets. Moreover it can be formulated in terms of an explicit recipe in real space, a feature which is certainly valuable for practical data analysis. In particular, as we report below, in the high-frequency limit the Mexican needlets are asymptotically close to the Spherical Mexican Hat Wavelets, which have been exploited in several cosmological papers but still lack a complete stochastic investigation.

More precisely, the authors of [76] propose to use in (10.3) a kernel of the form

$$b\left(\frac{l}{B^j}\right) = \left(\frac{l(l+1)}{B^{2j}}\right)^p \varphi\left(\frac{l(l+1)}{B^{2j}}\right) ,$$

where $\varphi(.)$ belongs to the Schwartz space $\mathcal{S}(\mathbb{R}^+)$ (see [190]) and is not necessarily of bounded support, while the sequence $\{-l(l+1)\}_{l=1,2,...}$ represents the eigenvalues of the Laplacian operator Δ_{S^2}. In particular, Mexican needlets can be obtained by taking $\varphi(s) = \exp(-s)$, so to obtain

$$\psi_{jk}(x;p) := \sqrt{\lambda_{jk}} \sum_{l \geq 1} \left(\frac{l(l+1)}{B^{2j}}\right)^p e^{-l(l+1)/B^{2j}} \frac{2l+1}{4\pi} P_l(\langle x, \xi_{jk}\rangle) .$$

The resulting functions make up a frame which is not tight, but very close to, in a sense which is made rigorous in [76]. Exact cubature formulae cannot hold (in particular, $\{\lambda_{jk}\}$ are not exact cubature weights in this case), because polynomials of infinitely large order are involved in the construction, but again this entails very minor approximations in practical terms (see the discussion in [76] and the numerical evidence provided by [179]). The random Mexican needlet coefficients are immediately seen to be given by

$$
\beta_{jk}(p) = \int_{S^2} T(x)\,\psi_{jk}(x; p)\,dx = \int_{S^2} \sum_{l,m} a_{lm} Y_{lm}(x)\,\psi_{jk}(x; p)\,dx
$$

$$
= \sqrt{\lambda_{jk}} \sum_{l \geq 1} \left(\frac{l(l+1)}{B^{2j}} \right)^{p} e^{-l(l+1)/B^{2j}} \sum_{m} a_{lm} Y_{lm}\left(\xi_{jk} \right),
$$

whence their covariance is

$$
E\beta_{jk}(p)\beta_{jk'}(p)
$$

$$
= \sqrt{\lambda_{jk}\lambda_{jk'}} \sum_{l \geq 1} \left(\frac{l(l+1)}{B^{2j}} \right)^{2p} e^{-2l(l+1)/B^{2j}} \frac{2l+1}{4\pi} C_l P_l(\langle \xi_{jk}, \xi_{jk'} \rangle).
$$

The results in [76, 77, 78] are indeed much broader than reported here, for instance they cover general oriented manifold rather than S^2; moreover, as we wrote, they consider general smooth weighting functions that need not be of the form $s\exp(-s)$. However this choice lends itself to very neat results and seems the most natural for practical applications. In particular, it makes possible a clear interpretation of the final results, i.e. the effect of varying p on the structure of dependence is immediately understood. We believe that this feature is also a valuable asset for practitioners. In the sequel, we shall drop the subscript p, whenever this is possible without risk of confusion.

Remark 10.10　It is suggested from results in [76] that Mexican needlets for $p = 1$ provide as j diverges a very good approximation to so-called Spherical Mexican Hat Wavelets (SMHW), which have been used in many physical papers; when centred on the North Pole $(0, 0)$, SMHW can be written in discretized form as

$$
\Psi_{jk}(\vartheta; B) = \frac{1}{(2\pi)^{\frac{1}{2}} \sqrt{2}B^{-j}(1 + B^{-2j} + B^{-4j})^{\frac{1}{2}}} [1 + (\tfrac{y}{2})^2]^2 [2 - \frac{y^2}{2t^2}] e^{-y^2/4B^{-2j}},
$$

where $y = 2\tan\frac{\theta}{2}$ follows from stereographic projection on the tangent plane.

As mentioned earlier (and as detailed in Chapter 11), the correlation inequality (10.9) opens several avenues for developments in the statistical analysis of spherical random fields. It is therefore a very important question to establish under what circumstances these results can be extended to other constructions, such as Mexican needlets. To analyze this issue, we start by writing the expression for the correlation coefficients, which is given by

$$Corr\left(\beta_{jk_1},\beta_{jk_2}\right) = \frac{\sum_{l\geq 1}\left(\frac{l(l+1)}{B^{2j}}\right)^{2p}e^{-2l(l+1)/B^{-2j}}(2l+1)C_l P_l(\langle\xi_{jk_1},\xi_{jk_2}\rangle)}{\sum_{l\geq 1}\left(\frac{l(l+1)}{B^{2j}}\right)^{2p}e^{-2l(l+1)/B^{2j}}(2l+1)C_l}.$$

We now provide upper bounds on the correlation of random coefficients, this result has been proved by [124] and [140] under slightly different conditions.

Proposition 10.11 *Assume Condition 10.6 holds with $\alpha < 4p + 2$ and $M \geq 4p + 2 - \alpha$; then there exist some constant $C_M > 0$ such that*

$$\left|Corr\left(\beta_{jk_1;p},\beta_{jk_2;p}\right)\right| \leq \frac{C_M}{\left(1 + j^{-1}B^j d\left(\xi_{jk_1},\xi_{jk_2}\right)\right)^{(4p+2-\alpha)}}. \tag{10.14}$$

The previous result shows that Mexican needlets can enjoy the same uncorrelation properties as standard needlets, in the circumstances where the angular power spectrum is decaying "slowly enough", that is $\alpha < 4p + 2$. It should be noted that for cosmological applications $\alpha \simeq 2$ provides an excellent fit to the data, and therefore in these circumstances asymptotic uncorrelation is granted even for $p = 1$. Indeed, numerical results in [179] show that for CMB-like power spectra the uncorrelation properties of Mexican needlet coefficients can even be superior to those for standard needlets. The extra j^{-1} term in (10.14) is a consequence of some standard technical difficulty when dealing with boundary cases such as $M = 4p + 2 - \alpha$.

It is also possible to complete the previous analysis, establishing indeed that the random Mexican needlets coefficients are necessarily correlated at some angular distance in the presence of faster memory decay. This is different from needlets, which are always uncorrelated. As mentioned earlier, the heuristic rationale behind this duality can be explained as follows: it should be stressed that we are focussing on high-resolution asymptotics, i.e. the asymptotic behavior of random coefficients at smaller and smaller scales in the same random realization of a spherical field. For such asymptotics, a crucial role can be played by terms which remain constant across different scales. In the case of usual needlets, which have bounded support over the multipoles, terms like these are simply dropped by construction. This is not so for Mexican needlets, which in any case include components at the lowest scales. These components are dominant when the angular power spectrum decays fast, and as such they prevent the possibility of asymptotic uncorrelation. In particular, we have correlation when the angular power spectrum is such that $\alpha > 4p + 2$.

Proposition 10.12   *Under condition 10.6, for* $\alpha > 4p + 2$, $\forall\ \varepsilon \in (0, 1)$, *there exists a positive* $\delta = \delta_\varepsilon$ *such that*

$$\lim_{j \to \infty} \inf Corr\left(\beta_{jk;p}, \beta_{jk';p}\right) > 1 - \varepsilon, \qquad (10.15)$$

for all $\left\{\xi_{jk}, \xi_{jk'}\right\}$ *such that* $d(\xi_{jk}, \xi_{jk'}) \le \delta$.

The proof of this result is provided in [124]. As mentioned earlier, the results in the previous two Propositions illustrate an interesting trade-off between the localization and correlation properties of spherical needlets. In particular, we can always achieve uncorrelation by choosing $p > (\alpha - 2)/4$; of course α is generally unknown and must be estimated from the data (in this sense standard needlets have better robustness properties). Introducing higher order terms implies lowering the weight of the lowest multipoles, i.e. improving the localization properties in frequency space. On the other hand, it may be expected that such an improvement of the localization properties in the frequency domain will lead to a worsening of the localization in pixel space, as a consequence of the "uncertainty principle" (*it is not possible for a function and its Fourier transform to be simultaneously very small*, see for instance [96]). Some numerical evidence on this phenomenon is provided in [179]; as mentioned earlier, overall in this paper Mexican needlets are shown to perform extremely well in circumstances of interest for CMB data analysis.

11

Needlets Estimation of Power Spectrum and Bispectrum

11.1 Introduction

Our purpose in the present chapter is to show how the asymptotic uncorrelation and localization results for spherical needlets (established in Chapter 10) may be used in the statistical analysis of spherical random fields, with a specific emphasis on CMB data analysis. We shall focus especially on the estimation of the angular power spectrum and bispectrum. For notational simplicity, we will stick throughout the chapter to the standard needlets case (i.e., compactly supported $b(.)$); however our arguments could be extended without efforts to Mexican needlets, provided of course that the correlation of random coefficients is ensured to decay at a suitable rate. The results in this chapter are largely taken by [13, 14], while the last section is mainly based on [123].

11.2 A general convergence result

Let us first fix some more notation on the set of cubature points $\{\xi_{jk}\}$. More precisely, and following the notation introduced in [14] we write

$$B^\circ(a, \alpha) = \{x : d(a, x) < \alpha\}, B(a, \alpha) = \{x : d(a, x) \leq \alpha\},$$

for the standard (open and closed) balls in S^2, where d stands for the usual spherical distance (that is, $d(x, y)$ indicates the distance between x and y along the great circle connecting them). We need also the following

Definition 11.1 For any $\varepsilon > 0$, we s ay that a set $\Xi_\varepsilon = \{x_1, ...x_N\}$ is a *maximal* ε-*net* if $x_1, ...x_N$ are in S^2, $\forall i \neq j, d(x_i, x_j) > \varepsilon$, and the set is maximal for this property, i.e.

$$\forall x \in S^2, d(x, \Xi_\varepsilon) \leq \varepsilon, \cup_{x_i \in \Xi_\varepsilon} B(x_i, \varepsilon) = S^2,$$
$$\text{and } \forall i \neq j, B(x_i, \varepsilon/2) \cap B(x_j, \varepsilon/2) = \emptyset.$$

Heuristically, an ε-net is a grid of point at distance at least ε from each other, and such that any extra point should be within a distance ε from a point in the grid. The number of points in a ε-net on the sphere can be bounded from above and from below, indeed it follows from Lemma 5 in [14] that $N \approx \varepsilon^{-2}$, or more precisely

Lemma 11.2   *Let* $\Xi_\epsilon = \{x_1, ... x_N\}$ *be a maximal ε-net. Then*

$$\frac{4}{\varepsilon^2} \leq N \leq \frac{4}{\varepsilon^2} \pi^2 . \tag{11.1}$$

Proof   Clearly, for all $x \in S^2$

$$\sigma(B(x, \eta)) = 2\pi \int_0^\eta \sin \vartheta d\vartheta = 2\pi(1 - \cos \eta) = 4\pi \sin^2 \frac{\eta}{2} .$$

Hence, using $\frac{2}{\pi} \leq \frac{\sin \eta/2}{\eta/2} \leq 1$ for $0 < \eta \leq \pi$ we have

$$\eta^2 \frac{4}{\pi} \leq 4\pi \frac{\eta^2}{4} \frac{\sin^2 \eta/2}{\eta^2/4} = \sigma(B(x, \eta)) \leq \pi \eta^2.$$

Now because $\cup_{x_i \in \Xi_\varepsilon} B(x_i, \varepsilon) = S^2$ we obtain

$$4\pi \leq \sum_{i=1}^N \sigma(B(x_i, \varepsilon)) \leq \pi N \varepsilon^2$$

and because $\cup_{x_i \in \Xi_\varepsilon} B(x_i, \varepsilon/2) \subset S^2$, $B(x_i, \varepsilon/2) \cap B(x_j, \varepsilon/2) = \emptyset$

$$4\pi \geq \sum_{i=1}^N \sigma(B(x_i, \varepsilon/2)) \geq N \frac{\varepsilon^2}{4} \frac{4}{\pi} = N \frac{\varepsilon^2}{\pi} ,$$

and the result follows immediately.   □

Given an ε-net, it is natural to partition the sphere into disjoint sets, each of them associated with a single point in the net. This task is accomplished by the well-known *Voronoi cells* construction.

Definition 11.3   Let Ξ_ϵ be a maximal ε-net. For all $x_i \in \Xi_\varepsilon$, the associated family of *Voronoi cells* is defined by:

$$\mathcal{V}(x_i) = \{x \in S^2 : \forall j \neq i, \ d(x, x_i) \leq d(x, x_j)\} .$$

We recall that $B(x_i, \varepsilon/2) \subset \mathcal{V}(x_i) \subset B(x_i, \varepsilon)$, hence $\sigma(\mathcal{V}(x_i)) \approx \varepsilon^2$. Also, if two Voronoi cells are adjacent, i.e. $\mathcal{V}(x_i) \cap \mathcal{V}(x_j) \neq \emptyset$, then by necessity $d(x_i, x_j) \leq 2\varepsilon$. It is proved in [14] that there are at most $6\pi^2$ adjacent cells to any given cell.

A remarkable property of the cubature points and weights $\{\xi_{jk}, \lambda_{jk}\}$ holds, namely (see [143], Corollary 4.4):

- The cubature points $\{\xi_{jk}\}$ can be taken to form a maximal ε_j-net, with $\varepsilon_j \approx B^{-j}$.
- The cubature weights $\{\lambda_{jk}\}$ are of order B^{-2j}, i.e. for all j, k we have $\lambda_{jk} \approx \sigma\big(\mathcal{V}(\xi_{jk})\big)$, the area of the associated Voronoi cell.

In what follows, we consider a centered strongly isotropic Gaussian field $T = \{T(x) : x \in S^2\}$, whose needlets coefficients $\{\beta_{jk}\}$ are defined via formula (10.4).

With this background, we are now ready to investigate the statistical properties of needlet-based estimators. Following [13], we can focus on polynomials functions of the normalized needlets coefficients; a general expression for such nonlinear statistics can be written as follows

$$h_{u,N_j} := \frac{1}{\sqrt{N_j}} \sum_{k=1}^{N_j} \sum_{q=1}^{Q} w_{uq} H_q(\widehat{\beta}_{jk}) \, , \widehat{\beta}_{jk} := \frac{\beta_{jk}}{\sqrt{E\beta_{jk}^2}} \, , u = 1, ..., U, \quad (11.2)$$

$$h_{N_j} = \big(h_{1,N_j}, ..., h_{U,N_j}\big)' \, . \tag{11.3}$$

where $H_q(.)$ denotes as before the qth order Hermite polynomials, N_j is the cardinality of coefficients corresponding to frequency j (we recall that we can take the cubature points $\{\xi_{jk}\}$ to form a B^{-j}-net, see [14], so that $N_j \approx B^{2j}$), and $\{w_{uq}\}$ is a set of deterministic weights that must ensure these statistics are asymptotically nondegenerate. In particular, the following condition is assumed to hold.

Condition 11.4 There exist j_0 such that for all $j > j_0$

$$\mathrm{rank}(\Omega_j) = U \, , \quad \Omega_j := Eh_{N_j}h'_{N_j} \, .$$

Note that Ω_j is a covariance matrix. Condition 11.4 is a standard invertibility assumption which will ensure our statistics are asymptotically nondegenerate (for instance, it rules out multicollinearity). Several examples of relevant polynomials are given in [13]; for instance, given a theoretical model for the angular power spectrum $\{C_l\}$, it is suggested in that reference that a goodness-of-fit statistic might be based upon

$$\frac{1}{\sqrt{N_j}} \sum_{k=1}^{N_j} H_2(\widehat{\beta}_{jk}) = \frac{1}{\sqrt{N_j}} \sum_{k=1}^{N_j} (\widehat{\beta}_{jk}^2 - 1)$$

$$= \frac{1}{\sqrt{N_j}} \sum_{k=1}^{N_j} \left(\frac{\beta_{jk}^2}{\lambda_{jk} \sum_{l \geq 1} b^2(\frac{l}{B^j}) \frac{2l+1}{4\pi} C_l} - 1 \right) \, .$$

The statistic

$$\widehat{\Gamma}_j = \frac{1}{\sqrt{N_j}} \sum_{k=1}^{N_j} \frac{\beta_{jk}^2}{\lambda_{jk}}$$

can then be viewed as an unbiased estimator for

$$\Gamma_j = E\widehat{\Gamma}_j = \sum_{l \geq 1} b^2 \left(\frac{l}{B^j}\right) \frac{2l+1}{4\pi} C_l \,.$$

We prove below consistency and asymptotic Gaussianity (for fully observed maps and without noise). As always in this framework, consistency has a non-standard meaning, as we do not have convergence to a fixed parameter, but rather convergence to unity of the ratio $\widehat{\Gamma}_j / \Gamma_j$.

Likewise, tests of Gaussianity could be implemented by focussing on the skewness and kurtosis of the wavelets coefficients (see for instance [141]), i.e. by focussing on

$$\frac{1}{\sqrt{N_j}} \sum_{k=1}^{N_j} \left\{ H_3(\widehat{\beta}_{jk}) + 3H_1(\widehat{\beta}_{jk}) \right\} = \frac{1}{\sqrt{N_j}} \sum_{k=1}^{N_j} \widehat{\beta}_{jk}^3 \text{ and}$$

$$\frac{1}{\sqrt{N_j}} \sum_{k=1}^{N_j} \left\{ H_4(\widehat{\beta}_{jk}) + 6H_2(\widehat{\beta}_{jk}) \right\} = \frac{1}{\sqrt{N_j}} \sum_{k=1}^{N_j} \left\{ \widehat{\beta}_{jk}^4 - 3 \right\} \,.$$

The joint distribution for these statistics is provided by the following results.

Theorem 11.5 *Assume T is a centered, Gaussian and isotropic random field; assume also that Conditions 10.6, 11.4 Then, as $N_j \to \infty$*

$$\Omega_j^{-1/2} h_{N_j} \overset{law}{\to} N_U(0, I_U),$$

where, as before, $N_U(0, I_U)$ indicates a U-dimensional i.i.d. centered Gaussian vector whose elements have unit variance.

Proof We prove the result in the simpler case where the components of the vector h_{N_j} are Hermite polynomials of different orders ≥ 2, that is: for every $u = 1, ..., U$, there exists an integer $q_u \geq 2$, as well as a constant $\widetilde{w}_{q_u}$ such that (i) $w_{uq} = \widetilde{w}_{q_u}$ for $q = q_u$ and $w_{uq} = 0$ otherwise, and (ii) $q_u \neq q_v$ for $u \neq v$. The general result can be deduced e.g. by using the so-called Cramer-Wold device for joint convergence. Owing to Theorem 4.20, we need only prove that

$$\lim_{N_j \to \infty} E\left\{ \left(\frac{h_{N_j, u}}{\sqrt{Var(h_{N_j, u})}} \right)^4 \right\} = 3,$$

or, equivalently, that the cumulants of order 4 of the previous sequence converge to zero. We have the following estimate:

$$
\left| Cum \left\{ \frac{1}{\sqrt{N_j}} \sum_{k_1=1}^{N_j} \widetilde{w}_{q_u} H_{q_u}(\beta_{jk_1}), ..., \frac{1}{\sqrt{N_j}} \sum_{k_4=1}^{N_j} \widetilde{w}_{q_u} H_{q_u}(\beta_{jk_4}) \right\} \right|
$$

$$
= \left| \frac{1}{N_j^2} \sum_{k_1=1}^{N_j} \cdots \sum_{k_4=1}^{N_j} \widetilde{w}_{q_u}^4 Cum \left\{ H_{q_u}(\beta_{jk_1}), ..., H_{q_u}(\beta_{jk_4}) \right\} \right|
$$

$$
= \left| \frac{1}{N_j^2} \sum_{k_1=1}^{N_j} \cdots \sum_{k_4=1}^{N_j} \widetilde{w}_{q_u}^4 \sum_G \prod_{1 \le u \le v \le 4} \gamma_{k_u k_v}^{\eta_{uv}(G)} \right|
$$

$$
\le \frac{\widetilde{w}_{q_u}^4}{N_j^2} \sum_G \left| \sum_{k_1,...,k_4=1}^{N_j} \prod_{1 \le u \le v \le 4} \gamma_{k_u k_v}^{\eta_{uv}(G)} \right| ,
$$

where we have used Proposition 4.15; in particular, in the above equations the sum in G runs over all non-flat and connected diagrams associated with a table with four rows of q_u elements, and $\eta_{uv}(G)$ counts the number of edges of g linking a vertex in the uth row with a vertex in the vth row. To conclude we just need to show that, as $N_j \to \infty$,

$$
\sum_{k_1,...,k_4=1}^{N_j} \prod_{1 \le u \le v \le 4} \gamma_{k_u k_v}^{\eta_{uv}(G)} = o(N_j^2).
$$

Note that $\left| \gamma_{k_u k_v}^{\eta_{uv}(G)} \right| \le |\gamma_{k_u k_v}|$ always, because $\gamma_{k_u k_v}$ are correlations and hence smaller than 1 in absolute value. Thus,

$$
\left| \sum_{k_1,k_2,k_3,k_4=1}^{N_j} \prod_{1 \le u \le v \le 4} \gamma_{k_u k_v}^{\eta_{uv}(G)} \right| \le \sum_{k_1,k_2,k_3,k_4=1}^{N_j} \left| \gamma_{k_1 k_2} \gamma_{k_2 k_3} \gamma_{k_3 k_4} \right|
$$

$$
\le \sum_{k_1}^{N_j} \sum_{k_2}^{N_j} |\gamma_{k_1 k_2}| \sum_{k_3}^{N_j} |\gamma_{k_2 k_3}| \sum_{k_4}^{N_j} |\gamma_{k_3 k_4}|
$$

$$
\le \{C_M'\}^3 N_j ,
$$

in view of Lemma 11.7 below, that we can combine with the inequality (10.9) to obtain that

$$
\sum_{k'}^{N_j} |\gamma_{kk'}| \le \sum_{k_1}^{N_j} \frac{C_M}{(1 + B^j d(\xi_{jk}, \xi_{jk'}))^M} \le C_M',
$$

thus finishing the proof. □

Remark 11.6   As discussed in Chapter 4, this result could be easily strengthened to provide an exact bound on the total variation distance d_{TV} for the speed of convergence to a Gaussian law.

Lemma 11.7   *If $M \geq 3$, for all $\xi_{jk} \in X_j$ there exist constant c_{1M}, c_{2M} such that*

$$\sum_{k'=1}^{N_j} \frac{1}{(1 + B^j d(\xi_{jk}, \xi_{jk'}))^M} \leq c_{1M}, \tag{11.4}$$

and

$$\sum_{k'} \frac{1}{\left\{1 + B^j d(\xi_{jk}, \xi_{jk'})\right\}^M} \frac{1}{\left\{1 + B^j d(\xi_{jk'}, \xi_{jk''})\right\}^M} \leq \frac{c_{2M}}{\left\{1 + B^j d(\xi_{jk}, \xi_{jk''})\right\}^M}. \tag{11.5}$$

Proof   The inequality (11.4) is proved in [13] and we follow their argument – an alternative proof is given in [143]. We have

$$\sum_{k=1}^{N_j} \frac{1}{(1 + B^j d(\xi_{jk}, \xi_{jk'}))^M}$$

$$= \frac{1}{\sigma(B(N_{j+1}^{-1}/2))} \sum_{k=1}^{N_j} \int_{B(\xi_{jk}, N_{j+1}^{-1/2}/2)} \frac{d\sigma(x)}{(1 + B^j d(\xi_{jk}, \xi_{jk'}))^M},$$

where $B(\xi_{jk}, N_{j+1}^{-1/2}/2)$ is the ball of radius $N_{j+1}^{-1/2}/2$ around ξ_{jk} and as before $\sigma(.)$ denotes the Lebesgue measure on the sphere. Clearly for $x \in B(\xi_{jk}, N_{j+1}^{-1/2}/2)$

$$d(\xi_{jk'}, x) \leq d(\xi_{jk}, \xi_{jk'}) + d(\xi_{jk}, x) \leq d(\xi_{jk}, \xi_{jk'}) + \frac{N_{j+1}^{-1/2}}{2}$$

$$\leq 2d(\xi_{jk}, \xi_{jk'}),$$

because $d(\xi_{jk}, \xi_{jk'}) \geq N_{j+1}^{-1/2}$, whence

$$\frac{1}{\sigma(B(N_{j+1}^{-1/2}/2))} \sum_{k=1}^{N_j} \int_{B(\xi_{jk}, N_{j+1}^{-1/2}/2)} \frac{d\sigma(x)}{(1 + B^j d(\xi_{jk}, \xi_{jk'}))^M}$$

$$\leq CB^{2j} \sum_{k=1}^{N_j} \int_{B(\xi_{jk}, N_{j+1}^{-1/2}/2)} \frac{2^M d\sigma(x)}{(1 + B^j d(\xi_{jk'}, x))^M}$$

$$\leq CB^{2j} \int_{S^2} \frac{2^M d\sigma(x)}{\left(1 + B^j \arccos(\langle \xi_{jk}, x \rangle)\right)^M}$$

$$= CB^{2j} \int_{S^2} \frac{2^M d\sigma(x)}{(1 + B^j \arccos(\langle \mathbf{e}_3, x \rangle))^M}$$

where (as before) we denoted by $\mathbf{e}_3 = (0, 0)$ the North Pole (in spherical coordinates), and we used spherical symmetry. Hence the previous quantity becomes

$$CB^{2j} \int_0^{2\pi} \int_0^\pi \frac{\sin\vartheta\, d\vartheta\, d\varphi}{(1 + B^j\vartheta)^M} = 2\pi CB^{2j} \int_0^\pi \frac{\sin\vartheta\, d\vartheta}{(1 + B^j\vartheta)^M}$$

$$\leq 2\pi CB^{2j} \int_0^\pi \frac{\vartheta\, d\vartheta}{(1 + B^j\vartheta)^M} \leq 2\pi CB^{2j} \left\{ \int_0^{B^{-j}} \vartheta\, d\vartheta + B^{-jM} \int_{B^{-j}}^\infty \vartheta^{1-M} d\vartheta \right\}$$

$$= 2\pi CB^{2j} \left\{ \frac{1}{2B^{2j}} + \frac{B^{-jM} B^{j(M-2)}}{(M-2)} \right\} \leq 2\pi C.$$

For (11.5), we follow [144], Lemma 4.8. Indeed, by an approximation argument as for (11.4) we can show that

$$\sum_{k'} \frac{1}{\left\{1 + B^j d(\xi_{jk}, \xi_{jk'})\right\}^M} \frac{1}{\left\{1 + B^j d(\xi_{jk'}, \xi_{jk''})\right\}^M}$$

$$\leq CB^{2j} \int_{S^2} \frac{1}{\left(1 + B^j \arccos(\langle \xi_{jk}, x \rangle)\right)^M} \frac{d\sigma(x)}{\left(1 + B^j \arccos(\langle \xi_{jk''}, x \rangle)\right)^M}.$$

Now write

$$S_{\xi_{jk}} := \left\{ x \in S^2 : d(x, \xi_{jk}) > d(\xi_{jk}, \xi_{jk''})/2 \right\},$$

$$S_{\xi_{jk''}} := \left\{ x \in S^2 : d(x, \xi_{jk''}) > d(\xi_{jk}, \xi_{jk''})/2 \right\}.$$

Clearly, $S^2 = S_{\xi_{jk}} \cup S_{\xi_{jk''}}$, and, by the triangle inequality

$$\int_{S_{\xi_{jk}}} \frac{1}{\left(1 + B^j \arccos(\langle \xi_{jk}, x \rangle)\right)^M} \frac{1}{\left(1 + B^j \arccos(\langle \xi_{jk''}, x \rangle)\right)^M} d\sigma(x)$$

$$\leq \frac{2^M}{\left(1 + B^j \arccos(\langle \xi_{jk}, \xi_{jk''} \rangle)\right)^M} \int_{S_{\xi_{jk}}} \frac{1}{\left(1 + B^j \arccos(\langle \xi_{jk''}, x \rangle)\right)^M} d\sigma(x)$$

$$\leq \frac{2^M}{\left(1 + B^j \arccos(\langle \xi_{jk}, \xi_{jk''} \rangle)\right)^M} \int_{S^2} \frac{1}{\left(1 + B^j \arccos(\langle \xi_{jk''}, x \rangle)\right)^M} d\sigma(x)$$

$$\leq \frac{c_M}{\left(1 + B^j \arccos(\langle \xi_{jk}, \xi_{jk''} \rangle)\right)^M},$$

and an analogous result holds for $\int_{S_{\xi_{jk''}}}$. The conclusion then follows easily from $\int_{S^2} \leq \int_{S_{\xi_{jk}}} + \int_{S_{\xi_{jk''}}}$. $\qquad\square$

11.3 Estimation of the angular power spectrum

The results in the previous Section yield as an immediate consequence consistency and weak convergence result for the estimator of the angular power spectrum. Indeed, let us consider again the spectral estimator

$$\widehat{\Gamma}_j := \frac{1}{N_j} \sum_k \frac{\beta_{jk}^2}{\lambda_{jk}} , \; E\widehat{\Gamma}_j = \sum_{l=B^{j-1}}^{B^{j+1}} b^2 \left(\frac{l}{B^j}\right) C_l (2l+1) . \tag{11.6}$$

Note that

$$E\beta_{jk}^2 = \lambda_{jk} E\widehat{\Gamma}_j ,$$

and Theorem 11.5 implies that

$$\left\{\frac{\widehat{\Gamma}_j}{E\widehat{\Gamma}_j} - 1\right\} = \frac{1}{N_j} \sum_k \left\{\frac{\beta_{jk}^2}{E\beta_{jk}^2} - 1\right\} = O_p(N_j^{-1/2}) = o_p(1) , \text{ as } j \to \infty ,$$

because

$$\begin{aligned}
Var\left\{\frac{\widehat{\Gamma}_j}{E\widehat{\Gamma}_j} - 1\right\} &= Var\left\{\frac{1}{N_j} \sum_k H_2\left(\frac{\beta_{jk}}{Var(\beta_{jk})}\right)\right\} \\
&\leq \frac{C}{N_j^2} \sum_{kk'} \frac{1}{\left\{1 + B^j d(\xi_{jk}, \xi_{jk'})\right\}^3} \\
&\leq \frac{C}{N_j^2} \sum_{kk'} \frac{1}{\left\{1 + B^j d(N, \xi_{jk'})\right\}^3} = O(N_j^{-1}) ,
\end{aligned}$$

in view of Lemma (11.7). Moreover

$$\left[Var\left\{\frac{\widehat{\Gamma}_j}{E\widehat{\Gamma}_j} - 1\right\}\right]^{-1/2} \left\{\frac{\widehat{\Gamma}_j}{E\widehat{\Gamma}_j} - 1\right\} \overset{law}{\to} N(0, 1) , \text{ as } j \to \infty .$$

These results make the implementation of confidence intervals and testing procedures viable (see again [13] for details).

Remark 11.8 A natural question for the practical implementation of these statistics is the estimation of their variance and normalizing factors. It is indeed shown by [14] how, under Gaussianity, this variance can be consistently estimated by subsampling techniques. For brevity's sake, we do not discuss these issues in detail here; intuitively, the idea is that under Gaussianity different subsets of the sphere can be considered asymptotically as independent realizations of the whole sphere, and hence be used for bootstrap-like evaluation of the sampling variability. Anologous considerations allow also the separate estimation of angular power spectra across different regions of the sky, for instance to test for symmetry/isotropy violations on real data. These ideas have indeed been applied already and quite successfully to data from the *WMAP* experiment, see for instance [163, 165, 176] and the references therein. More discussion on this, and the related issue of estimation in the presence of observational noise, are contained in the final chapter.

We have shown in the previous chapter how the coefficients $\{\beta_{jk}\}$ are asymptotically unaffected by the presence of missing regions as $j \to \infty$. The price for such robustness properties is clearly connected to the smoothing, i.e. in the presence of missing observations it turns out to be unfeasible to estimate each angular power spectrum mode C_l by itself, and one must stick to a slightly less ambitious goal, i.e. the estimation of joint values averaged over some subset of frequencies (chosen by the data analyst). There is, of course, a standard trade-off in the choice of the bandwidth parameter B, as values closer to unity entail a much better resolution, but yield worse localization properties on the sphere and therefore a possibly higher contamination from spurious observations; on the other hand, higher values of B yield more robust, but less informative estimates (numerical evidence on these phenomena is reported for instance in [139, 179]).

It is also possible to establish stronger results, i.e. a functional central limit theorem which provides uniform convergence of the sequence of estimators at hand. This is the issue to which we devote next section.

11.4 A functional central limit theorem

For most statistical applications, the convergence results provided earlier are not sufficient and it is natural to require uniform convergence results across the whole frequency band. For instance, when estimating the angular power spectrum from the data, physicists are clearly interested on the concordance between theory and observations across all multipoles l, and discussing tests on a single frequency j is not sufficient for their purposes. In this Section, we shall then show how to establish Functional Central Limit Theorems on general nonlinear transforms of the random coefficients.

In particular, following [13], we shall focus on the continuous-time vector process

$$W_J(r) := \frac{1}{\sqrt{J}} \sum_{j=2}^{[Jr]} \Omega_j^{-1/2} h_{N_j} \, , \, 0 \le r \le 1 \, .$$

In words, we take partial sums of our statistics, to be able to detect possible divergences with respect to the expected Gaussian behaviour and locate the frequency to which they correspond. For simplicity, we have taken the sum in steps of two, so the summands are independent. We have the following

Theorem 11.9   *Under Conditions 10.6, 11.4, as $J \to \infty$*

$$W_J(r) \Longrightarrow W(r),$$

where $W(r)$ denotes standard Brownian motion and $\Rightarrow$ denotes weak convergence in the Skorohod space $D[0, 1]^U$.

Proof   We recall that tightness in $D[0, 1]^U$ can be established by considering each component on its own. By a standard use of the Cramer-Wold device, the proof of convergence of the finite-dimensional distributions can also be reduced to the univariate case, so it is sufficient to focus on the case $U = 1$ (see e.g. [24] for a proof of these claims).

Write

$$W_J(r) := \frac{1}{\sqrt{J}} \sum_{j=2}^{[Jr]} \widetilde{h}_{N_j} , \, 0 \le r \le 1 , \widetilde{h}_{N_j} := \Omega_j^{-1/2} h_{N_j} .$$

Because the summands are independent and $E\widetilde{h}_{N_j}^2 = 1$ for all j, convergence of the finite-dimensional distributions is a simple consequence of the Lyapunov condition

$$\lim_{J\to\infty} \frac{\sum_{j=2,4,\dots}^{[Jr]} E\widetilde{h}_{N_j}^{2+\delta}}{\left\{Var\left(\sum_{j=2}^{[Jr]} \widetilde{h}_{N_j}\right)\right\}^2} = \lim_{J\to\infty} \frac{\sum_{j=2,4,\dots}^{[Jr]} E\widetilde{h}_{N_j}^{2+\delta}}{J^2 r^2} = 0 , \text{ some } \delta > 0 . \quad (11.7)$$

To establish (11.7), it is sufficient to take $\delta = 2$ and note that $E\widetilde{h}_{N_j}^4 < \infty$, uniformly for all j, from the same arguments as in the previous Section.

Likewise, to establish tightness we can focus on the criterion described in [24, p. 128], implying that a sequence $\{X_n\}$ is tight whenever an estimate of the type

$$E\left\{|X_n(r) - X_n(r_1)|^2 \, |X_n(r_2) - X_n(r)|^2\right\} \le C \, |r_2 - r_1|^2 \quad \text{for all } r_1 \le r \le r_2$$

is verified. In our case, we have

$$E\left\{|W_J(r) - W_J(r_1)|^2 \, |W_J(r_2) - W_J(r)|^2\right\}$$

$$= \frac{1}{J^2} E\left\{\left(\sum_{j=[Jr_1]+2}^{[Jr]} \widetilde{h}_{N_j}\right)^2 \left(\sum_{j=[Jr]+2}^{[Jr_2]} \widetilde{h}_{N_j}\right)^2\right\}$$

$$\le \frac{C}{J^2} \{[Jr] - [Jr_1]\} \{[Jr_2] - [Jr]\}$$

$$\leq 4C(r_2 - r_1)^4, \text{ for all } r_1 \leq r \leq r_2 .$$

$\square$

For practical applications, J should be such to cover the highest multipoles made available from a given experiment, i.e. $J \approx \log_B L$, where as in the previous chapters we denoted by L the maximum observed multipole (we recall $L \simeq 700$ for *WMAP*, while from *Planck* $L \simeq 2500/3000$ is expected).

Spherical wavelets in general, and needlets in particular, allow for many statistical applications, which go much beyond angular power spectrum estimation. A pioneering example was implemented on real data by [162]. These authors search for cross-correlation between CMB and Large Scale Structure (LSS) maps; the existence of such cross-correlation at the scales $l \simeq 30/50$ is indeed an effect of the so-called *Integrated Sachs-Wolfe effect,* a key prediction of many cosmological models in the presence of *Dark Energy,* see [51, 60] for details. Several other applications have been considered already, as mentioned in the introduction to Chapter 10. In the following Section, we devote our attention to applications to non-Gaussianity testing, which have already met quite a lot of success in the applied literature (see [123] and [164, 165, 175, 176]).

11.5 A central limit theorem for the needlets bispectrum

11.5.1 The needlets bispectrum

As we discussed in Chapter 9, there is by now a wide consensus in the literature on spherical random fields related to CMB that the most efficient procedures to probe non-Gaussianity are based upon the bispectrum, in the ideal circumstances where the spherical random field is fully observed. However, the properties of the bispectrum are also known to deteriorate dramatically in the presence of missing observations, see for instance [36] for some numerical evidence. Given the localization properties that we have previously discussed for needlets, it is natural to investigate whether they could be exploited to build a needlet version of the bispectrum which should be more robust in the presence of missing data.

Our aim in this chapter is then to combine ideas from the bispectrum and the needlets literature (i.e., the material of this and of the previous two chapters) to propose and analyze a needlets bispectrum, where the random coefficients in the needlets expansion are combined in a similar way to the bispectrum construction. The aim is to obtain a procedure which mimics the ability of the bispectrum to search for non-Gaussianity at the most efficient combination of frequencies, at the same time providing a much more robust construction in the presence of missing data, as typical of the needlets. This part is largely based on the recent paper [123]; applications to CMB data have been realized in [164, 165, 175, 176].

Again, we denote by T a centered Gaussian isotropic random field on the sphere.

We recall from Chapter 9 the (normalized) angular bispectrum, defined as

$$I_{l_1 l_2 l_3} = \sum_{m_1 m_2 m_3} \begin{pmatrix} l_1 & l_2 & l_3 \\ m_1 & m_2 & m_3 \end{pmatrix} \frac{a_{l_1 m_1} a_{l_2 m_2} a_{l_3 m_3}}{\sqrt{C_{l_1} C_{l_2} C_{l_3}}} .$$

As we mentioned in the same Chapter 9, an alternative formulation of the (normalized) bispectrum is the following (see also [117, 198]),

$$\widetilde{I}_{l_1 l_2 l_3} = \frac{1}{4\pi} \int_{S^2} \frac{T_{l_1}(x) T_{l_2}(x) T_{l_3}(x)}{\sqrt{Var(T_{l_1}(x)) Var(T_{l_2}(x)) Var(T_{l_3}(x))}} d\sigma(x) \tag{11.8}$$

$$= \int_{S^2} \sum_{m_1 m_2 m_3} \frac{\sqrt{4\pi} a_{l_1 m_1} a_{l_2 m_2} a_{l_3 m_3} Y_{l_1 m_1}(x) Y_{l_2 m_2}(x) Y_{l_3 m_3}(x)}{\sqrt{(2l_1 + 1)(2l_2 + 1)(2l_3 + 1) C_{l_1} C_{l_2} C_{l_3}}} d\sigma(x)$$

$$= \begin{pmatrix} l_1 & l_2 & l_3 \\ 0 & 0 & 0 \end{pmatrix} \sum_{m_1 m_2 m_3} \frac{a_{l_1 m_1} a_{l_2 m_2} a_{l_3 m_3}}{\sqrt{C_{l_1} C_{l_2} C_{l_3}}} \begin{pmatrix} l_1 & l_2 & l_3 \\ m_1 & m_2 & m_3 \end{pmatrix}$$

$$= \begin{pmatrix} l_1 & l_2 & l_3 \\ 0 & 0 & 0 \end{pmatrix} I_{l_1 l_2 l_3} .$$

In this chapter, to ease notation we shall assume that the cubature points are nested, i.e. $\mathcal{X}_j \subset \mathcal{X}_{j'}$, for $j' > j$, and that the cubature weights are constant over k, i.e. $\lambda_{jk} = 4\pi/N_j$, where as before $N_j = \text{card}\{\mathcal{X}_j\}$. Note that as we reported before there exist positive constants c_1, c_2 such that $4\pi c_1/N_j \le \lambda_{jk} \le 4\pi c_2/N_j$ for all j, k.

Under these conditions, we can introduce the (normalized) *needlets bispectrum* as

$$I_{j_1 j_2 j_3} := \sqrt{\frac{4\pi}{N_{j_3}}} \sum_{k_3} \widehat{\beta}_{j_1 k_1} \widehat{\beta}_{j_2 k_2} \widehat{\beta}_{j_3 k_3} \delta_{j_1 j_2 j_3}(k_1, k_2, k_3) , \quad j_1 \le j_2 \le j_3 , \tag{11.9}$$

where

$$\delta_{j_1 j_2 j_3}(k_1, k_2, k_3) = \mathbb{I}(\xi_{j_3 k_3} \in \mathcal{V}_{j_2 k_2}) \mathbb{I}(\xi_{j_2 k_2} \in \mathcal{V}_{j_1 k_1}) ,$$

where $\mathbb{I}(.)$ denotes the indicator function and $\mathcal{V}_{jk}$ is the Voronoi Cell that corresponds to ξ_{jk}. It is immediate to see that (11.9) can be seen as a natural development of (11.8), where the convolution with the orthonormal projector operator L_l is replaced by the (discretized) convolution with the frame operator projection $\sqrt{\lambda_{jk}} \sum_l b(l/B^j) Z_l$. Of course, in practice (11.8) is unfeasible and requires discretization to be implemented. In words, we are considering a version of the bispectrum where the exact identification of the multipoles is blurred by a form of suitable smoothing, with the purpose of a better robustness against missing observations.

The summation convention in (11.9) needs some further discussion. The role of $\delta_{j_1 j_2 j_3}(k_1, k_2, k_3)$ is to ensure that our sum runs over $k_2 = k_2(k_3)$ (the (unique) value of k_2 such that $\xi_{j_3 k_3} \in \mathcal{V}_{j_2 k_2}$), and $k_1 = k_1(k_3)$ (the (unique) value of k_1 such that $\xi_{j_2 k_2}, \xi_{j_3 k_3} \in \mathcal{V}_{j_1 k_1}$). In other words, the "finest grid" $\mathcal{X}_{j_3}$ is the one which leads the summation, whereas smaller frequency terms are identified with those *centres* whose corresponding Voronoi cells include the points being summed. In the sequel, for notational simplicity we write k_1, k_2 rather than $k_1(k_3), k_2(k_3)$, when no ambiguity is possible.

To investigate the asymptotic behaviour of the needlets bispectrum, once again an extensive use of the Diagram Formula and its extensions is needed. These results were proved in [123], and we do not provide full details here. We simply report the major statements, which is as follows.

Theorem 11.10 *Let $T(x)$ be a zero-mean, mean square continuous isotropic Gaussian random field, with angular power spectrum that satisfies Condition 10.6. As $j_1 \to \infty$, we have*

$$\frac{I_{j_1 j_2 j_3}}{\sqrt{E I_{j_1 j_2 j_3}^2}} \xrightarrow{law} N(0, 1).$$

We now consider the case where the variance of the needlets coefficients is unknown and estimated from the data. A natural estimator for σ_j is provided by

$$\widetilde{\sigma}_j^2 = \frac{1}{N_j} \sum_{\xi_{jk} \in \mathcal{X}_j} |\beta_{jk}|^2$$

where as before $N_j = \text{card}\{\mathcal{X}_j\} \approx B^{2j}$. We define our studentized statistics as $\widetilde{\beta}_{jk} := \beta_{jk}/\widetilde{\sigma}_j$ and we then consider

$$\widetilde{I}_{j_1 j_2 j_3} = \sqrt{\frac{4\pi}{N_{j_3}}} \sum_{k_3} \widetilde{\beta}_{j_1 k_1} \widetilde{\beta}_{j_2 k_2} \widetilde{\beta}_{j_3 k_3} \delta_{j_1 j_2 j_3}(k_1, k_2, k_3).$$

Our next result shows that this studentization procedure has no effect on asymptotic behaviour.

Theorem 11.11 *As $j_1 \to \infty$, we have $\{\sigma_{j_1}^2 \sigma_{j_2}^2 \sigma_{j_3}^2\}^{-1} \widetilde{\sigma}_{j_1}^2 \widetilde{\sigma}_{j_2}^2 \widetilde{\sigma}_{j_3}^2 \longrightarrow 1$ in* *probability, and hence*

$$\widetilde{I}_{j_1 j_2 j_3} \overset{law}{\to} N(0, 1) \,.$$

Remark 11.12 As above, the previous findings can be extended to uniform results, i.e., convergence to Brownian motion and Brownian sheet of partial sum processes constructed by averaging across different frequencies, see [123].

11.5.2 Behaviour under non-Gaussianity

To conclude, we shall provide some quick and informal discussion on the behaviour of our statistics under non-Gaussianity. We shall consider the same non-Gaussian models as in Chapters 7 and 9, namely a simple version of the highly popular Sachs-Wolfe model, the typical benchmark for applications to CMB data analysis. For brevity's sake, this section will be somewhat heuristic and we will use freely the approximations which are common in the physical literature.

We start from the expected value of the needlets bispectrum, which is provided by

$$EI_{j_1 j_2 j_3}$$

$$= \frac{1}{\sqrt{B^{2j_3}}} \sum_{k_3} E\widehat{\beta}_{j_1 k_1} \widehat{\beta}_{j_2 k_2} \widehat{\beta}_{j_3 k_3} \delta_{j_1 j_2 j_3}(k_1, k_2, k_3)$$

$$\approx \frac{1}{\sigma_{j_1} \sigma_{j_2} \sigma_{j_3}} \frac{1}{\sqrt{B^{2j_3}}} \sum_{k_3} \sum_{l_1, l_2, l_3} \sum_{m_1 m_2 m_3} b\left(\frac{l_1}{B^{j_1}}\right) b\left(\frac{l_2}{B^{j_2}}\right) b\left(\frac{l_3}{B^{j_3}}\right) E\left(a_{l_1 m_1} a_{l_2 m_2} a_{l_3 m_3}\right)$$

$$\times Y_{l_1 m_1}(\xi_{j_1 k_1}) Y_{l_2 m_2}(\xi_{j_2 k_2}) Y_{l_3 m_3}(\xi_{j_3 k_3}) \delta_{j_1 j_2 j_3}(k_1, k_2, k_3)$$

$$= \frac{1}{\sigma_{j_1} \sigma_{j_2} \sigma_{j_3}} B^{j_3} \sum_{l_1, l_2, l_3 = B^{j-1}}^{B^{j+1}} \sum_{m_1 m_2 m_3} b\left(\frac{l_1}{B^{j_1}}\right) b\left(\frac{l_2}{B^{j_2}}\right) b\left(\frac{l_3}{B^{j_3}}\right) b_{l_1 l_2 l_3}$$

$$\times \begin{pmatrix} l_1 & l_2 & l_3 \\ m_1 & m_2 & m_3 \end{pmatrix} \begin{pmatrix} l_1 & l_2 & l_3 \\ 0 & 0 & 0 \end{pmatrix} \sqrt{\frac{(2l_1 + 1)(2l_2 + 1)(2l_3 + 1)}{4\pi}}$$

$$\times \left\{ \frac{1}{B^{2j_3}} \sum_{k_1 k_2 k_3} Y_{l_1 m_1}(\xi_{j_1 k_1}) Y_{l_2 m_2}(\xi_{j_2 k_2}) Y_{l_3 m_3}(\xi_{j_3 k_3}) \delta_{j_1 j_2 j_3}(k_1, k_2, k_3) \right\} \,.$$

Here, we recall from Chapters 6 and 9 that $b_{l_1 l_2 l_3}$ is the so-called reduced bispectrum; for illustrative purposes, we assume here that the latter takes the approximative Sachs-Wolfe form which we discussed in Chapter 9, namely

$$b_{l_1 l_2 l_3} = -6 f_{NL} \left\{ C_{l_1} C_{l_2} + C_{l_1} C_{l_3} + C_{l_2} C_{l_3} \right\} \,,$$

compare also [18, 117, 198, 184, 67] for more discussion on this functional form. For our purposes below, we recall also that

$$
\begin{pmatrix} l & l & l \\ 0 & 0 & 0 \end{pmatrix} \approx \frac{(-1)^{-3l/2}}{l}, \begin{pmatrix} l_0 & l & l+l_0 \\ 0 & 0 & 0 \end{pmatrix} \approx \frac{(-1)^{-l_0+l}}{\sqrt{l}}, \quad (11.10)
$$

as can be checked from the analytic expression for Wigner's coefficients and Stirling's Formula. Now

$$
\left\{ \frac{1}{B^{2j_3}} \sum_{k_3} Y_{l_1 m_1}(\xi_{j_1 k_1}) Y_{l_2 m_2}(\xi_{j_2 k_2}) Y_{l_3 m_3}(\xi_{j_3 k_3}) \delta_{j_1 j_2 j_3}(k_1, k_2, k_3) \right\}
$$

$$
\approx \int_{S^2} Y_{l_1 m_1}(\xi) Y_{l_2 m_2}(\xi) Y_{l_3 m_3}(\xi) d\xi
$$

$$
= \begin{pmatrix} l_1 & l_2 & l_3 \\ m_1 & m_2 & m_3 \end{pmatrix} \begin{pmatrix} l_1 & l_2 & l_3 \\ 0 & 0 & 0 \end{pmatrix} \sqrt{\frac{(2l_1+1)(2l_2+1)(2l_3+1)}{4\pi}},
$$

whence

$$
EI_{j_1 j_2 j_3} = \frac{1}{\sigma_{j_1} \sigma_{j_2} \sigma_{j_3}} B^{j_3} \sum_{l_1, l_2, l_3 = B^{j-1}}^{B^{j+1}} b\left(\frac{l_1}{B^{j_1}}\right) b\left(\frac{l_2}{B^{j_2}}\right) b\left(\frac{l_3}{B^{j_3}}\right) b_{l_1 l_2 l_3} \begin{pmatrix} l_1 & l_2 & l_3 \\ 0 & 0 & 0 \end{pmatrix}^2
$$

$$
\times \frac{(2l_1+1)(2l_2+1)(2l_3+1)}{4\pi} \sum_{m_1 m_2 m_3} \begin{pmatrix} l_1 & l_2 & l_3 \\ m_1 & m_2 & m_3 \end{pmatrix}^2
$$

$$
= \frac{B^{j_3}}{\sigma_{j_1} \sigma_{j_2} \sigma_{j_3}} \sum_{l_1, l_2, l_3 = B^{j-1}}^{B^{j+1}} b\left(\frac{l_1}{B^{j_1}}\right) b\left(\frac{l_2}{B^{j_2}}\right) b\left(\frac{l_3}{B^{j_3}}\right) b_{l_1 l_2 l_3}
$$

$$
\times \begin{pmatrix} l_1 & l_2 & l_3 \\ 0 & 0 & 0 \end{pmatrix}^2 \frac{(2l_1+1)(2l_2+1)(2l_3+1)}{4\pi},
$$

in view of the orthonormality properties of the Wigner's $3j$ coefficients. To keep the analogy with the cosmological literature, we shall focus on "equilateral" and "squeezed" configurations, see Chapter 9 and [8, 123, 131, 184]. In the equilateral case $j_1 = j_2 = j_3 = j$ we have

$$
EI_{jjj} = \frac{B^j}{\sigma_{j_1} \sigma_{j_2} \sigma_{j_3}} \sum_{l_1, l_2, l_3 = B^{j-1}}^{B^{j+1}} b\left(\frac{l_1}{B^j}\right) b\left(\frac{l_2}{B^j}\right) b\left(\frac{l_3}{B^j}\right) b_{l_1 l_2 l_3}
$$

$$
\times \begin{pmatrix} l_1 & l_2 & l_3 \\ 0 & 0 & 0 \end{pmatrix}^2 \frac{(2l_1+1)(2l_2+1)(2l_3+1)}{4\pi}.
$$

Now recall $l \approx B^j$, $\sigma_j^3 \approx C_{B^j}^{3/2} B^{3j}$, $b_{l_1 l_2 l_3} \approx f_{NL} l_1^{-\alpha} l_2^{-\alpha}$ so that, after some manipulations,

$$
B^j \sum_{l_1,l_2,l_3=B^{j-1}}^{B^{j+1}} b\left(\frac{l_1}{B^j}\right) b\left(\frac{l_2}{B^j}\right) b\left(\frac{l_3}{B^j}\right) \frac{b_{l_1 l_2 l_3}}{\sigma_j^3} \left(\begin{array}{ccc} l_1 & l_2 & l_3 \\ 0 & 0 & 0 \end{array}\right)^2
$$
$$
\times \frac{(2l_1+1)(2l_2+1)(2l_3+1)}{4\pi}
$$
$$
\simeq B^j \sum_{l_1,l_2,l_3=B^{j-1}}^{B^{j+1}} b\left(\frac{l_1}{B^j}\right) b\left(\frac{l_2}{B^j}\right) b\left(\frac{l_3}{B^j}\right) \frac{f_{NL} l_1^{-\alpha} l_2^{-\alpha}}{C_{B^j}^{3/2} B^{3j}} \left(\begin{array}{ccc} l_1 & l_2 & l_3 \\ 0 & 0 & 0 \end{array}\right)^2 B^{3j}
$$
$$
\simeq B^j \sum_{l_1,l_2,l_3=B^{j-1}}^{B^{j+1}} b\left(\frac{l_1}{B^j}\right) b\left(\frac{l_2}{B^j}\right) b\left(\frac{l_3}{B^j}\right) \frac{f_{NL} B^{-2j\alpha}}{B^{-3j\alpha/2} B^{3j}} B^j \simeq f_{NL} B^{2j} B^{-j\alpha/2}.
$$

This suggests the expected value of the needlets bispectrum can either diverge or converge to zero, according to the asymptotic behaviour of the angular power spectrum; in particular, it does diverge for all $\alpha < 4$. On the other hand, for $j_1 \ll j_2 = j_3$ by an analogous argument we obtain

$$
EI_{j_1 j_2 j_2} \simeq \frac{f_{NL} B^{j_2}}{B^{-j_1(\alpha/2-1)} B^{-j_2(\alpha-2)}} \sum_{l_1,l_2,l_3=B^{j-1}}^{B^{j+1}} b\left(\frac{l_1}{B^j}\right) b\left(\frac{l_2}{B^j}\right) b\left(\frac{l_3}{B^j}\right) l_1^{-\alpha} l_2^{-\alpha}
$$
$$
\times \left(\begin{array}{ccc} l_1 & l_2 & l_3 \\ 0 & 0 & 0 \end{array}\right)^2 B^{j_1} B^{2j_2}
$$
$$
\simeq \frac{f_{NL} B^{j_2}}{B^{-j_1(\alpha/2-1)} B^{-j_2(\alpha-2)}} \sum_{l_1,l_2,l_3=B^{j-1}}^{B^{j+1}} b\left(\frac{l_1}{B^j}\right) b\left(\frac{l_2}{B^j}\right) b\left(\frac{l_3}{B^j}\right) B^{-j_1\alpha} B^{-j_2\alpha} B^{j_2}
$$
$$
\simeq f_{NL} B^{-j_1\alpha/2} B^{2j_2}.
$$

As for the usual bispectrum, the previous computations suggest that the power is maximized by "squeezing" frequencies, i.e. by maximizing the differences between the "side lengths" j_1 and j_2. This is the same sort of qualitative result which was discussed in Chapter 9 (compare also with [131] and [36]). Our heuristic calculations in this Section suggest very clearly that the needlets bispectrum may enjoy the same good power properties, at the same time healing the difficulties which follow from the presence of missing observations.

These claims have been confirmed by many applications of the needlets bispectrum to data from the *WMAP* experiment. In particular, [164, 175] consider test for non-Gaussianity and estimation of the non-linearity parameter f_{NL}. They are both able to confirm that the performance of the needlet bispectrum is comparable to the most efficient statistical techniques currently available. In physical units, as discussed earlier the fluctuations in the CMB temperature value are in the order of 10^{-4} degrees; the quadratic term in the potential is then in the order of 10^{-8}. At the *WMAP* resolution level and under realistic assumptions concerning noise and missing data, the needlet bispectrum based estimator of f_{NL} provides an estimated standard error in the order of $25 - 30$, with some fluctuations depending on different assumptions on noise and missing observations. In practice, analyzing *WMAP* data the value $f_{NL} = 0$, which corresponds to exactly Gaussian circumstances, seems in a boundary region where no definite conclusions can be drawn; more precise conclusions are expected to be possible when *Planck* data will be publicly available (early 2013). It should be noted that standard inflationary models predict a value for $f_{NL} \simeq 1$; much higher values can still be compatible with inflation, but would require more elaborate scenarios, such as those known as multiple fields inflation, see [18, 19, 20, 184] for more discussions and details.

Other recent applications of the needlets bispectrum include the results of [176], where the bispectrum is evaluated on subregions of the full CMB sky as a test of possible asymmetries (violations of statistical isotropy), and those of [165], where variations on the dependency across different frequencies are considered, again as a test for asymmetries. The literature is evolving extremely fast, however, so it is impossible at this stage to provide a fully updated overview.

12

Spin Random Fields

12.1 Introduction

In the previous chapters of this book, we provided a systematic analysis of scalar-valued spherical random fields, with special emphasis on applications in astrophysics and cosmology. These same fields of applications are now prompting stochastic models which are more sophisticated (and more intriguing) than ordinary, scalar-valued random fields. For instance, in areas as diverse as brain imaging and astrophysics more and more examples are emerging of random fields which do not take ordinary (scalar) values, but rather have a domain given by more sophisticated mathematical structures. In this final chapter, we shall be again concerned with astrophysical and cosmological applications, but several related issues can be found in other disciplines, see for instance [178] for related mathematical models in the field of brain mapping.

In particular, almost all mathematical statistics papers in the CMB area have been concerned with the temperature component of this radiation; however, most recent and forthcoming experiments (such as *Planck*, which was launched on May 14, 2009, or the projected mission *CMBPOL*) are focussing on a more complex and elusive feature, namely the so-called polarization of the Cosmic Microwave Background. The physical significance of the latter is explained for instance in [33, 181] and [108]; we do not enter into these motivations here, but we do stress how the analysis of such a feature is supposed to provide extremely rewarding physical informations. For instance, detection of a nonzero angular power spectrum for the so-called *B*-modes of polarization data (to be defined later) would provide the first experimental evidence of primordial gravitational waves; this would result in an impressive window into the General Relativity picture of the primordial Big Bang dynamics and as such it is certainly one of the most interesting perspective of current physical research.

Here, however, we shall not go deeper into these physical perspectives, as we prefer to focus instead on the new mathematical ideas which are forced in by the analysis of these datasets. A rigorous understanding requires some technicalities which are postponed to the next section; however we hope to convey the general idea as follows. We can imagine that experiments recording CMB radiation are measuring in each direction $\xi \in S^2$ a random ellipse on $T_\xi S^2$, the tangent plane at that point. The "width" of this ellipse, which is a standard random variable, corresponds to temperature data, on which the mathematical statistics literature has so far concentrated. The other identifying features of this ellipse (elongation and orientation) are collected in polarization data, which are then to be viewed as a random field taking values in a space of algebraic curves. In more formal terms, this can be summarized by saying that we shall be concerned with random sections of fiber bundles over the sphere; from a more group theoretical point of view, we shall show that polarization random fields are related to so-called *spin-weighted representations* of the group of rotations $SO(3)$. A further mathematical interpretation, which is entirely equivalent but shall not be pursued here, is to view these data as realizations of random matrices fields (compare with [108, 204]). Quite interestingly, there are other, unrelated physical issues where the mathematical and statistical formalism turns out to be identical. In particular gravitational lensing data, which have currently drawn much interest in astrophysics and will certainly make up a core issue for research in the next two decades, can be shown to have the same (spin 2, see below) mathematical structure, see for instance [28, 115] . More generally, similar issues may arise when dealing with random deformations of shapes, as discussed for instance in [4].

The construction of a wavelet system for spin functions is addressed by Geller and Marinucci in [74]; the idea in that paper is to extend the needlet approach initiated by [143, 144] to this new broader geometrical setting, and investigate the stochastic properties of the resulting spin needlets coefficients, thus generalizing results from [13, 14]. A wide range of possible applications to the analysis of polarization data is discussed in [72]. The possibility to use spin needlets for angular power spectrum estimation for spin fields, largely along the same lines as in the scalar case (see Chapter 11), was first discussed by [73]. More recently, an alternative wavelet construction in the spin case was provided by [75], who introduced so-called *mixed needlets;* the main difference with the approach advocated here is that the latter allows for needlet coefficients which are scalar valued, and hence to some extent simpler to handle. However, we decided to focus here on the *pure spin* construction as the theory is somewhat closer to the one presented in the previous chapters. Our presentation will follow closely Geller, Lan and Marinucci [73]; we refer also to Geller et al. [75, 78, 80], Malyarenko [129], Leonenko and Sakhno [126], for related developments.

The plan of this chapter is as follows: in Section 2 we present the motivations for our analysis, i.e. some minimal physical background on polarization. In Section 3 and 4 we introduce the geometrical formalism on spin line bundles and spin needlets, respectively. In Section 5 we define spin random fields, while Sections 6 and 7 are devoted to the derivation of their asymptotic properties in the presence of missing observations and noise, and to related statistical tests. We stress that the material of this final chapter relates to research which is definitely at an earlier stage than the rest of this monograph; as such, we hope it will provide a stimulus for further developments.

12.2 Motivations

The classical theory of electromagnetic radiation entails that the latter can be characterized in terms of the so-called *Stokes' parameters* Q and U, which are defined as follows. An electromagnetic wave propagating in direction z has components

$$E_x(z,t) = E_{0x}\cos(\tau + \delta_x), E_y(z,t) = E_{0y}\cos(\tau + \delta_y), \qquad (12.1)$$

where $\tau := \omega t - kz$ is the so-called propagator, E_{0x}, E_{0y} and δ_x, δ_y are amplitude and phase factors and $\nu = 2\pi\omega/k$ is the frequency of the wave. Relations (12.1) can be viewed as the parametric equations of an ellipse which is drawn by the incoming radiation on the plane perpendicular to the direction of motion: indeed, some elementary algebra yields

$$\frac{E_x^2(z,t)}{E_{0x}^2} + \frac{E_y^2(z,t)}{E_{0y}^2} - 2\frac{E_x(z,t)}{E_{0x}}\frac{E_y(z,t)}{E_{0y}}\cos\delta = \sin^2\delta, \delta := \delta_y - \delta_x.$$

The magnitude of the ellipse is given by

$$T = E_{0x}^2 + E_{0y}^2;$$

here, T has the nature of a scalar quantity, that is to say, it is readily seen to be invariant under rotation of the local coordinate axes x and y. It can hence be viewed as an intrinsic quantity measuring the total intensity of radiation; from the physical point of view, this is exactly the nature of CMB temperature observations which have been the focus of so much research over the last decade. It should be noted that, despite the non-negativity constraint, in the physical literature on CMB experiments T is usually taken to be Gaussian around its mean, in excellent agreement with observations. This apparent paradox is explained by the fact that the variance of T is several order of magnitudes smaller than its mean, so the Gaussian approximation is justifiable.

The characterization of the polarization ellipse is completed introducing Stokes' parameters, which are defined as

$$Q = E_{0x}^2 - E_{0y}^2 \,, U = 2E_{0x}E_{0y}\cos\delta \,. \tag{12.2}$$

To provide a flavour of their geometrical meaning, we recall from elementary geometry that the parametric equations of a circle are obtained in the special case $E_{0x} = E_{0y}$, $\delta_x = \delta_y + \pi/2$, whence the circle corresponds to $Q = U = 0$. On the other hand, it is not difficult to see that a segment aligned on the x axis is characterized by $Q = T$, a segment aligned on the y axis by $Q = -T$, for a segment on the line $y = \pm x$ we have $\delta_x - \delta_y = 0, \pi$, and hence $Q = 0$, $U = \pm T$, respectively. The key feature to note, however, is the following: while T does not depend on any choice of coordinates, this is not the case for Q and U, i.e. the latter are not geometrically intrinsic quantities. However, as these parameters identify an ellipse, it is natural to expect that they will be invariant to rotations by 180° degrees and multiples thereof. This is the first step to understand the introduction of spin random fields below.

Indeed, it is convenient to identify $\mathbb{R}^2$ with the complex plane $\mathbb{C}$ by the standard identification $w = x + iy$; a change of coordinates corresponding to a rotation by γ radians can then be expressed as $w' = \exp(i\gamma)w$, and some elementary algebra shows that the induced transform on (Q, U) can be written as

$$\begin{pmatrix} Q' \\ U' \end{pmatrix} = \begin{pmatrix} \cos 2\gamma & \sin 2\gamma \\ -\sin 2\gamma & \cos 2\gamma \end{pmatrix} \begin{pmatrix} Q' \\ U' \end{pmatrix},$$

or more compactly

$$Q' + iU' = \exp(i2\gamma)(Q + iU) \,. \tag{12.3}$$

In the physicists' terminology, (12.3) identifies the Stokes' parameters as spin 2 objects, that is, a rotation by an angle γ changes these parameters by 2γ. As mentioned before, this can be intuitively visualized by focussing on an ellipse, which is clearly invariant by rotations of $180°$. To compare with other cases, standard (scalar) random fields are clearly invariant (or better covariant) with respect to the choice of any coordinate axis in the local tangent plan, and hence they are spin zero fields; a vector field is spin 1, while we can envisage random fields taking values in higher order algebraic curves and thus having any integer spin $s \geq 2$.

As remarked earlier, it is very important to notice that polarization is not the only possible motivations for the analysis of spin random fields. For instance, an identical formalism is derived when dealing with gravitational lensing, i.e. the deformation of images induced by gravity according to general relativity equations. Gravitational lensing is now the object of an enormous interest, which has led to a lively debate on suitable statistical methods (see for instance [28, 115]). Some discussion on the statistical procedures which are made possible by the application of spin needlets to lensing data is provided by [59], where thresholding techniques for spin nonparametric regression are introduced and (nearly) adaptive properties over L^p losses are established for the corresponding estimators.

12.3 Geometric background

In this section, we will provide a more rigorous background on spin functions. Despite the fact that our motivating applications are limited to the case $s = 2$, we will discuss here the case of a general integer $s \in \mathbb{N}$, which does not entail any extra difficulty. Our discussion requires some background in differential geometry, for which we refer for instance to [3] and [25]. The construction of spin functions is discussed in more details in [74], building upon a well-established physical literature described for instance in [33, 146, 181].

We recall first the tangent plane $T_\xi S^2$, where $\xi \in S^2$ and $T_\xi S^2$ is defined as usual as the linear space generated by the collection of tangent vectors in ξ. To proceed further to spin random fields, we need to recall from geometry the notion of a fibre bundle. The latter consists of the data (E, B, π, F), where E, B, and F are topological spaces and $\pi : E \to B$ is a continuous surjection satisfying a local triviality condition outlined below. The space B is called the base space of the bundle, E the total space, and F the fiber; the map π is called the projection map (or bundle projection). In our case, the base space is simply the unit sphere $B = S^2$; it is tempting to view the fibers as ellipses, (or vectors, for $s = 1$, or more general algebraic curves, for $s \geq 3$) lying in $T_\xi S^2$; however one must bear in mind that to characterize the ellipses we would need the triple T, Q, U, while we shall only focus on the Stoke's parameters Q and U.

The basic intuition behind fibre bundles is that they behave locally as simple Cartesian products $B \times F$; in the special case where F is a vector space the term vector bundles is used. The former intuition is implemented by requiring that for all $\xi \in S^2$ there exists a neighbourhood $U = U(\xi)$ such that $\pi^{-1}(U)$ is homeomorphic to $U \times F$, in such a way that π carries over to the projection onto the first factor. In particular, the diagram in Fig. 12.1 should commute (where

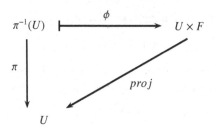

Figure 12.1   A commuting diagram

ϕ is a homeomorphism and *proj* is the natural projection). The set $\pi^{-1}(\xi)$ is homeomorphic to F and is called the fiber over ξ. The fiber bundles we shall consider are smooth, that is, E, B, and F are required to be smooth manifolds and all the projections above are required to be smooth maps.

In our case, we shall be dealing with a complex line bundle which is uniquely identified by fixing transition functions to explicit the transformation laws under changes of coordinates of these smooth manifolds. Following [74] (see also [84, 146]), we define $U_I := S^2 \setminus \{N, S\}$ as the chart covering the sphere with the exception of the North and South Poles, with the usual coordinates (ϑ, φ). We define also the rotated charts $U_R = RU_I$; in this new charts, we will use the natural coordinates (ϑ_R, φ_R) (it can be shown that this definition can be extended by continuity to cover also $\{N, S\}$). On each point ξ, we consider the tangent plane $T_\xi S^2$, with the orthonormal basis which is obtained by taking as an element $\partial/\partial\varphi_R$ for fixed values of ϑ_R, in the direction of an increasing φ. Again as in [74], we label $\psi_{\xi R_1 R_2}$ the angle between $\partial/\partial\varphi_{R_1}$ and $\partial/\partial\varphi_{R_2}$, for any $\xi \in U_{R_1} \cap U_{R_2}$; this angle is independent of any choice of coordinates. Now we say that $f \in C_s^\infty(S^2)$, i.e. f is a smooth spin s function, if for all $R_1, R_2 \in SO(3)$ we have that

$$f_{R_2}(\xi) = \exp(is\psi_{\xi R_2 R_1})f_{R_1}(\xi) ,$$

so that $\exp(is\psi_{\xi R_2 R_1})$ provides the transition functions of the line bundles from the chart U_{R_1} to U_{R_2}.

An alternative, group theoretic point of view can be motivated as follows. Consider the group of rotations $SO(3)$; as discussed in Chapter 3, it is a well-known fact that each element $g \in SO(3)$ can be expressed as

$$g = R_{\mathbf{e}_3}(\varphi)R_{\mathbf{e}_2}(\vartheta)R_{\mathbf{e}_3}(\psi) , 0 \le \vartheta \le \pi , 0 \le \varphi, \psi \le 2\pi , \qquad (12.4)$$

where $R_{\mathbf{e}_3}(.)$ and $R_{\mathbf{e}_2}(.)$ represent rotations around the $\mathbf{e}_3$ and $\mathbf{e}_2$ axis, respectively. From that same chapter, we recall also that the elements of Wigner's matrices can be expressed in terms of the Euler angles as

$$D^l_{m_1 m_2}(g) = \exp(-im_1\varphi)d^l_{m_1 m_2}(\vartheta)\exp(-im_2\psi).$$

We know that, as a consequence of the Peter-Weyl Theorem, all square integrable functions on $SO(3)$ can be expanded, in the mean-square sense, as

$$f(g) = \sum_l \sum_{m_1 m_2} b^l_{m_1 m_2} D^l_{m_1 m_2}(g),$$

where the coefficients $\{b^l_{m_1 m_2}\}$ can be recovered from the inverse Fourier transform

$$b^l_{m_1 m_2} = \int_{SO(3)} f(g)\overline{D}^l_{m_1 m_2}(g)dg.$$

As detailed in Chapter 3, we can loosely say that square-integrable, scalar-valued function on the sphere live in the space generated by the column (or equivalently row) $s = 0$ of the Wigner's D matrices of irreducible representations. Now from Peter-Weyl Theorem we know that each of the columns $s = -l, ..., l$ spans a space of irreducible representations, and these spaces are mutually orthogonal; it is then a natural question to ask what is the physical significance of these further spaces. It turns out that these are strictly related to spin functions, indeed we can expand the fiber bundle of spin s functions as

$$f_s(\vartheta, \varphi) = \sum_{l \geq s} \sum_m b^l_{ms} D^l_{ms}(\varphi, \vartheta, \psi)\Big|_{\psi=0}. \tag{12.5}$$

Spin s functions can then be related to the so-called *spin weighted representations* of $SO(3)$, see for instance [31]. Note that throughout this chapter l is always defined to be such that $l \geq s$, although we shall omit to write it for brevity's sake.

The analogy with the scalar case can actually be pursued further than that. Again in Chapter 3, we proved that the elements D^l_{m0}, $m = -l, ..., l$ of the Wigner's D matrices are proportional to the spherical harmonics Y_{lm}, which are the eigenfunctions of the spherical Laplacian operator $\Delta_{S^2} Y_{lm} = -l(l+1)Y_{lm}$. This equivalence holds in much greater generality and for all integers s and $l \geq s$ there exist a differential operator $\eth\bar{\eth}$ such that $-\eth\bar{\eth}D^l_{ms} = (l-s)(l+s+1)D^l_{ms}$. The operators $\eth, \bar{\eth}$ are defined as follows, in terms of their action on any spin s function $f_s(.)$,

$$\eth f_s(\vartheta, \varphi) = -(\sin\vartheta)^s \left[\frac{\partial}{\partial\vartheta} + \frac{i}{\sin\vartheta}\frac{\partial}{\partial\varphi}\right](\sin\vartheta)^{-s} f_s(\vartheta, \varphi),$$

$$\bar{\eth} f_s(\vartheta, \varphi) = -(\sin\vartheta)^{-s} \left[\frac{\partial}{\partial\vartheta} - \frac{i}{\sin\vartheta}\frac{\partial}{\partial\varphi}\right](\sin\vartheta)^s f_s(\vartheta, \varphi).$$

The previous expressions should be written more rigorously as $\eth_{U_R} f_{U_{RS}}, \bar{\eth}_{U_R} f_{U_{RS}}$ as both the left- and right-hand side depend on the choice of coordinates, although the line bundle clearly doesn't. The spin s spherical harmonics can then be identified as

$$Y_{l;ms}(\vartheta, \varphi) = \sqrt{\frac{2l+1}{4\pi}} \overline{D}^l_{m-s}(\varphi, \vartheta, 0) = (-1)^s \sqrt{\frac{2l+1}{4\pi}} \exp(im\varphi) d^l_{m-s}(\vartheta).$$

$$(12.6)$$

We note that we are using here the same phase convention as in [60] and [129], while in [73] a different phase factor is used. The spin spherical harmonics can be shown to satisfy

$$Y_{l;ms} = \begin{cases} [(l-s)(l+s+1)]^{-1/2} \eth Y_{l;m,s-1}, \\ -[(l+s)(l-s+1)]^{-1/2} \bar{\eth} Y_{l;m,s+1}, \end{cases} \tag{12.7}$$

which motivates the name of *spin raising* and *spin lowering* operators for $\eth, \bar{\eth}$. Further properties of the spin spherical harmonics follow easily from their proportionality to elements of Wigner's D matrices; indeed we have (orthonormality)

$$\int_{S^2} Y_{l;ms}(x) \overline{Y}_{l';m's}(x) d\sigma(x) = \int_0^{2\pi} \int_0^\pi Y_{lms}(\vartheta, \varphi) \overline{Y}_{l'm'}(\vartheta, \varphi) \sin \vartheta d\vartheta d\varphi = \delta^{l'}_l \delta^{m'}_m,$$

and

$$\sum_{m=-l}^l Y_{l;ms}(x) \overline{Y}_{l;ms}(x) = \frac{2l+1}{4\pi}.$$

By combining (12.5) and (12.6), the spectral representation of spin functions is derived, as always in the L^2 sense:

$$f_s(\vartheta, \varphi) = \sum_l \sum_m a_{lms} Y_{lms}(\vartheta, \varphi). \tag{12.8}$$

From (12.8), a further, extremely important characterization of spin functions was first introduced in [146], see [74] for a much more complete discussion. In particular, it can be shown that there exist a scalar complex-valued function $f(\vartheta, \varphi) = f_E(\vartheta, \varphi) + i f_B(\vartheta, \varphi)$, such that

$$f(\vartheta, \varphi) = \sum_{lm} a_{lm;E} Y_{lm}(\vartheta, \varphi) + i \sum_{lm} a_{lm;B} Y_{lm}(\vartheta, \varphi), \tag{12.9}$$

where $f_E(\vartheta, \varphi), f_B(\vartheta, \varphi)$ are real-valued scalar functions. The components

$$\{a_{lm;E}, a_{lm;B}\}$$

are known in the physical literature as the E (electric) and B (magnetic) modes of CMB polarization, see again ([33, 108, 181, 204]).

Remark 12.1 The previous discussion can be recast in a slightly different form, as suggested in [129]. View $SO(2)$ as a closed subgroup of $SO(3)$, with elements generically denoted by k. This is an Abelian subgroup, and assuming k is parametrized by the Euler angle $\psi \in [0, 2\pi]$ the irreducible representations of $SO(2)$ are well-known to be one-dimensional and given by

$$W_s(k) : \mathbb{C} \rightarrow \mathbb{C} : x \mapsto \exp(is\psi)x,$$

where $s \in \mathbb{N}$. Let $g \in SO(3)$, and consider the action

$$k : \{SO(3) \times \mathbb{C}\} \rightarrow \left\{S^2, \mathbb{C}\right\} , k(g, x) = (gk, \exp(is\psi)x) .$$

We denote by $\mathcal{E}_s$ the quotient space of orbits of the above action; that is, two elements (g_1, x_1) and (g_2, x_2) belong to the same equivalence class if there exist $k \in SO(2)$ such that $(g_2, x_2) = (g_1 k, W_s(k)x_1)$. For $s = 0$, this is clearly isomorphic to $\left\{S^2, \mathbb{C}\right\}$, i.e. the space of complex-valued functions on the sphere. For $s \neq 0$, we obtain indeed the same spin fiber bundle we defined before $\left\{\mathcal{E}_s, \pi, S^2\right\}$, by taking the projection

$$\pi : \mathcal{E}_s \rightarrow S^2 , \pi(g, x) = gSO(2) ,$$

where we denoted as usual $gSO(2)$ the equivalence class $\{gk : k \in SO(2)\}$; for $g \in SO(3)$, this is isomorphic to $SO(3)/SO(2) \simeq S^2$.

12.4 Spin needlets and spin random fields

We are now in the position to recall the construction of spin needlets, as provided by [74]. As recalled in Chapter 10, needlets have been defined in [143, 144] as

$$\psi_{jk}(\xi) = \sqrt{\lambda_{jk}} \sum_l b\left(\frac{l}{B^j}\right) \sum_{m=-l}^{l} Y_{lm}(\xi) \overline{Y}_{lm}\left(\xi_{jk}\right) , \xi \in S^2, \qquad (12.10)$$

where $\left\{\xi_{jk}, \lambda_{jk}\right\}$ are a set of cubature points and weights ensuring that

$$\sum_{jk} \lambda_{jk} Y_{lm}\left(\xi_{jk}\right) \overline{Y_{l'm'}}\left(\xi_{jk}\right) = \int_{S^2} Y_{l_1 m_1}(\xi) \overline{Y}_{l_2 m_2}(\xi) d\sigma(\xi) = \delta_l^{l'} \delta_m^{m'} ,$$

$b(.)$ is a compactly supported, C^∞ function, and $B > 1$ is a user-chosen "bandwidth" parameter.

We keep here the same notations as in the previous two chapters; more precisely, for a fixed $B > 1$, we shall denote by $\{X_j\}_{j=0}^\infty$ the nested sequence of cubature points corresponding to the space $\mathcal{K}_{[2B^{j+1}]}$, where $[.]$ represents as usual integer part and $\mathcal{K}_L = \oplus_{l=0}^L \mathcal{H}_l$ is the space spanned by spherical harmonics up to order L. As discussed earlier, it is known that $\{X_j\}_{j=0}^\infty$ can be taken such that the cubature points for each j are almost uniformly ε_j–distributed with $\varepsilon_j := \kappa B^{-j}$, the coefficients $\{\lambda_{jk}\}$ are such that $c_1 B^{-2j} \leq \lambda_{jk} \leq c_2 B^{-2j}$, where c_1, c_2 are finite real numbers, and card$\{X_j\} \approx B^{2j}$. Spin needlets are then defined as (see [74])

$$\psi_{jk;s}(\xi) = \sqrt{\lambda_{jk}} \sum_l b\left(\frac{l}{B^j}\right) \sum_{m=-l}^l Y_{l;ms}(\xi) \overline{Y}_{l;ms}(\xi_{jk}) , \qquad (12.11)$$

$$\beta_{jk;s} = \int_{S^2} f_s(\xi)\overline{\psi}_{jk;s}(\xi)d\sigma(\xi) = \sqrt{\lambda_{jk}} \sum_l b\left(\frac{l}{B^j}\right) \sum_{m=-l}^l a_{l;ms} Y_{l;ms}(\xi_{jk}) . \qquad (12.12)$$

As before, $\{\lambda_{jk}, \xi_{jk}\}$ are cubature points and weights, $b(\cdot) \in C^\infty$ is non-negative, and has a compact support in $[1/B, B]$. Exact cubature points can be defined for the spin as for the scalar case. Indeed, we noted in Remark (3.9) of Chapter 3 that the squares $|D_{mn}^l(\varphi, \vartheta, \psi)|^2$ are polynomials of degree $2l$ in $x_3 = \cos \vartheta$ for all l, m, n. More generally, the product $Y_{l_1;m_1n}(\vartheta, \varphi)\overline{Y}_{l_2;m_2n}(\vartheta, \varphi)$ is itself a polynomial of degree at most $l_1 + l_2$, as shown by [16, 75].

Hence, because $Y_{l_1;m_1n}(\vartheta, \varphi)\overline{Y}_{l_2;m_2n}(\vartheta, \varphi)$ is a standard polynomial on the sphere, it is straightforward to see that for cubature points $\xi_{jk} \in X_j$ such that $B^j > l_1 + l_2$ we have

$$\int_{S^2} Y_{l_1;m_1n}(x)\overline{Y}_{l_2;m_2n}(x)d\sigma(x) = \sum_{\xi_{jk}} \lambda_{jk} Y_{l_1;m_1n}(\xi_{jk})\overline{Y}_{l_2;m_2n}(\lambda_{jk}) .$$

In other words, we can use for spin needlets the same cubature points and weights as for the scalar case, to obtain for instance the reconstruction formula

$$\sum_{jk} \beta_{jk;s}\psi_{jk;s}(x)$$

$$= \sum_{jk} \lambda_{jk} \sum_{l_1,l_2} b\left(\frac{l_1}{B^j}\right) b\left(\frac{l_2}{B^j}\right) \sum_{m_1,m_2} a_{l_1;ms} Y_{l_1;m_1s}(\xi_{jk}) \overline{Y}_{l_2;m_2s}(\xi_{jk}) Y_{l_2;m_2s}(x)$$

$$= \sum_j \sum_{l_1,l_2} b\left(\frac{l_1}{B^j}\right) b\left(\frac{l_2}{B^j}\right) \sum_{m_1,m_2} a_{l_1;m_1s} \left\{ \sum_k \lambda_{jk} Y_{l_1;m_1s}(\xi_{jk}) \overline{Y}_{l_2;m_2s}(\xi_{jk}) \right\} Y_{l;m_2s}(x)$$

$$= \sum_j \sum_{l_1,l_2} b\left(\frac{l_1}{B^j}\right) b\left(\frac{l_2}{B^j}\right) \sum_{m_1,m_2} a_{l_1;m_1s} \delta_{l_1}^{l_2} \delta_{m_1}^{m_2} Y_{l_2;m_2s}(x)$$

$$= \sum_l \sum_j b^2\left(\frac{l}{B^j}\right) \sum_{m=-l}^l a_{l;ms} Y_{l;ms}(x) = \sum_l \sum_{m=-l}^l a_{l;ms} Y_{l;ms}(x) = f_s(x) .$$

For practical CMB data analysis, these cubature points can be identified for instance with the centres of the pixels provided by Healpix (see [85]), with only a minor approximation. Note that in [74] and [75] in the argument of $b(.)$ the multipole l is actually replaced by the square root of the spin eigenvalues $\sqrt{(l-s)(l+s+1)}$. This formulation is instrumental for the derivation of the main properties of spin needlets by means of differential arguments as in [76, 77, 78]; we stress, however, that this is actually a minor difference, as all our results are asymptotic and of course

$$\lim_{l\to\infty} \frac{\sqrt{(l-s)(l+s+1)}}{l} = 1, \text{ for all fixed } s .$$

Indeed, as detailed in the Appendix, even in the proof of the localization of standard needlets it is slightly more convenient to take $l + \frac{1}{2}$ rather than l for the argument of $b(.)$.

The expression (12.11) bears an obvious resemblance with (12.10), but it is also important to point out some crucial differences. An important feature is as follows: (12.11) cannot be viewed as a well-defined scalar or spin function, because $Y_{l;ms}(\xi)$, $\overline{Y}_{l;ms}(\xi_{jk})$ are spin (s and $-s$) functions defined on different point of S^2, and as such they cannot be multiplied in any meaningful way (their product depends on the local choice of coordinates). Hence, (12.11) should be written more rigorously as

$$\psi_{jk;s}(\xi) = \sqrt{\lambda_{jk}} \sum_l b\left(\frac{l}{B^j}\right) \sum_{m=-l}^{l} \left\{Y_{l;ms}(\xi) \otimes \overline{Y}_{l;ms}(\xi_{jk})\right\} ,$$

$$\overline{\psi}_{jk;s}(\xi) = \sqrt{\lambda_{jk}} \sum_l b\left(\frac{l}{B^j}\right) \sum_{m=-l}^{l} \left\{\overline{Y}_{l;ms}(\xi) \otimes Y_{l;ms}(\xi_{jk})\right\} ,$$

where we denoted by $\otimes$ the tensor product of spin functions; spin needlets can the be viewed as spin $\{-s, s\}$ operators (written $Z_{-s,s}$), which act on a space of spin s functions square integrable functions to produce a sequence of spin s square-summable coefficients, i.e. $Z_{-s,s} : L_s^2 \to \ell_s^2$. This action is actually an isometry, as a consequence of the tight frame property, see also [16] and [79].

Remark 12.2 Using a notation closer to the physical literature, we could view spin s quantities as "bra" entities, i.e. write $\langle T(\xi)$, $\langle \beta_{jk;s}$, and spin $-s$ as "ket" quantities, i.e. write for instance $\overline{Y}_{l;ms}(\xi)\rangle$, compare [127] and [23]. With this notation we can write

$$\int_{S^2} f_s(\xi)\overline{\psi}_{jk;s}(\xi)d\sigma(\xi)$$

$$= \sqrt{\lambda_{jk}} \sum_{l} b\left(\frac{l}{B^j}\right) \sum_{m=-l}^{l} \int_{S^2} \left\langle f_s(\xi) \, \overline{Y}_{l;ms}(\xi) \right\rangle \left\langle Y_{l;ms}(\xi_{jk}) \, d\sigma(\xi) \right.$$

$$= \sqrt{\lambda_{jk}} \sum_{l} b\left(\frac{l}{B^j}\right) \sum_{m=-l}^{l} a_{l;ms} \left\langle Y_{l;ms}(\xi_{jk}) \right\rangle ,$$

which is a well-defined spin quantity, as the inner product $\left\langle f_s(\xi), \overline{Y}_{l;ms}(\xi) \right\rangle$ yields a well-defined, complex-valued scalar.

The absolute value of spin needlets is indeed a well-defined scalar function, and this allows to discuss localization properties. In this framework, the main result is established in [74], where it is shown that for any $M \in \mathbb{N}$ there exists a constant $c_M > 0$ such that, for every $\xi \in S^2$:

$$\left| \psi_{jk;s}(\xi) \right| \leq \frac{c_M B^j}{(1 + B^j \arccos(\langle \xi_{jk}, \xi \rangle))^M} \text{ uniformly in } (j,k), \qquad (12.13)$$

i.e. the tails decay quasi-exponentially.

We are now able to focus on the core of this chapter, which is related to the analysis of spin random fields. As mentioned in the previous discussion, we have in mind circumstances where stochastic analysis must be developed on polarization random fields $\{Q \pm iU\}$, which are spin ± 2 random functions.

Hence we shall now assume we deal with random isotropic spin functions f_s, by which we mean that there exist a probability space $(\Omega, \mathfrak{I}, P)$, such that for all choices of charts U_R, the ordinary random function $(f_s)_R$, defined on $\Omega \times S^2$, is jointly $\mathfrak{I} \times \mathcal{B}(U_R)$ measurable, where $\mathcal{B}(U_R)$ denotes the Borel sigma-algebra on U_R. In particular, for the spin 2 random function $(Q + iU)(\xi)$ as for the scalar case, the following representation holds, in the mean square sense ([74])

$$\{Q + iU\} = \sum_{lm} a_{l;m2} Y_{l;m2} ,$$

i.e.

$$\lim_{L \to \infty} E \int_{S^2} \left| \{Q + iU\}(\xi) - \sum_{l=2}^{L} \sum_{m=-l}^{l} a_{l;m2} Y_{l;m2}(\xi) \right|^2 d\sigma(\xi) = 0 .$$

Note that the quantity on the left-hand side is a well-defined scalar, for all L. The sequence $\{a_{l;m2} = a_{lm;E} + i a_{lm;B}\}$ is complex-valued and is such that, for all l_1, l_2, m_1, m_2,

$$E a_{l_1 m_1;E} a_{l_2 m_2;E} = E a_{l_1 m_1;B} a_{l_2 m_2;E} = E a_{l_1 m_1;E} a_{l_2 m_2;B} = E a_{l_1 m_1;E} \overline{a}_{l_2 m_2;B} = 0 ,$$

and

$$E a_{l_1 m_1;E} \overline{a}_{l_2 m_2;E} = C_{l_1 E} \delta_{l_1}^{l_2} \delta_{m_1}^{m_2} , \quad E a_{l_1 m_1;B} \overline{a}_{l_2 m_2;B} = C_{l_1 B} \delta_{l_1}^{l_2} \delta_{m_1}^{m_2},$$

where

$$\sum_l \frac{2l+1}{4\pi} C_{lE} \, , \; \sum_l \frac{2l+1}{4\pi} C_{lB} < \infty \, .$$

The spin (or total) angular power spectrum is then defined as

$$E|a_{l;m2}|^2 =: C_l = \{C_{lE} + C_{lB}\} \, , \, l = 2, 3, \dots \, .$$

12.5 Spin needlets spectral estimator

In this section, we shall establish an asymptotic result for the spectral estimator of spin needlets in the high-resolution sense, i.e. we will investigate the
asymptotic behaviour of our statistics as the frequency band diverges. We note
first, however, one very important issue. As we mentioned earlier, spin needlet
coefficients are not in general scalar quantities. It is possible to choose a single
chart to cover all points other than the North and South Pole; these two points
can be clearly neglected without any effect on asymptotic results. The resulting spin coefficients will in general depend on the chart, and should hence be
written as $\{\beta_{R;jks}\}$; however the choice of the chart will only produce an arbitrary phase factor $\exp(is\psi_k)$. The point is that, because in this chapter we are
only concerned with quadratic statistics, the phase factor is automatically lost
and our statistics for the spin spectral estimator will be invariant with respect
to the choice of coordinates. In view of this, from now on we can neglect the
issues relative to the choice of charts; we will deal with needlet coefficients as
scalar-valued complex quantities, i.e. we will take the chart as fixed, and for
notational simplicity we write $\{\beta_{jks}\}$ rather than $\{\beta_{R;jks}\}$.

We begin by introducing some regularity conditions on the polarization angular power spectrum Γ_l, which are basically the same as (10.6) in Chapter
10 and [73, 74], see also [13], [14] and [123, 64, 124, 140] for closely related
assumptions.

Condition 12.3   The random field $\{Q + iU\}(\xi)$ is Gaussian and isotropic with
angular power spectrum such that

$$C_l = l^{-\alpha} g(l) > 0 \, , \text{ where } c_0^{-1} \le g(l) \le c_0 \, , \, \alpha > 2 \, , \text{ for all } l \in \mathbb{N} \, ,$$

and for every $r \in \mathbb{N}$ there exist $c_r > 0$ such that

$$|\frac{d^r}{du^r} g(u)| \le c_r u^{-r} \, , \, u \in (|s|, \infty) \, .$$

Remark 12.4 As for (10.6), the condition is fulfilled for instance by angular power spectra of the form

$$C_l = \frac{F_1(l)}{l^\beta F_2(l)} \, ,$$

where $F_1(l), F_2(l) > 0$ are polynomials of degree $q_1, q_2 > 0, \beta + q_2 - q_1 = \alpha$.

By (12.12), it is readily seen that

$$E\beta_{jk;s}\beta_{j'k';s} = \sqrt{\lambda_{jk}\lambda_{j'k'}} \sum_{l,l'} b\left(\frac{l}{B^j}\right) b\left(\frac{l'}{B^{j'}}\right) \sum_{m,m'} Ea_{l;ms}a_{l';m's}Y_{l;ms}\left(\xi_{jk}\right) Y_{l';m's}\left(\xi_{j'k'}\right)$$

$$= 0$$

because

$$Ea_{l;ms}a_{l';m's} = Ea_{l_1 m_1;E}a_{l_2 m_2;E} + 2Ea_{l_1 m_1;B}a_{l_2 m_2;E} + Ea_{l_1 m_1;E}a_{l_2 m_2;B} = 0 \, .$$

On the other hand, the covariance $Cov\left(\beta_{jk;s}, \overline{\beta}_{jk';s}\right) = E\beta_{jk;s}\overline{\beta}_{jk';s}$ is in general non-zero. In view of (12.12, it is immediate to see that

$$\left|Cov\left(\beta_{jk;s}, \overline{\beta}_{jk';s}\right)\right| = \left|\sqrt{\lambda_{jk}}\sqrt{\lambda_{jk'}} \sum_l b^2\left(\frac{l}{B^j}\right) C_l \frac{(2l+1)}{4\pi} K^{ls}\left(\xi_{jk}, \xi_{jk'}\right)\right| \, ; \tag{12.14}$$

where

$$K^{ls}(x, x') := \sum_{m=-l}^{l} Y_{lms}(x) \overline{Y}_{lms}(x') \, . \tag{12.15}$$

For $k = k'$ we obtain as a special case from (12.14) that

$$E\left|\beta_{jk;s}\right|^2 = \lambda_{jk} \sum_l b^2\left(\frac{l}{B^j}\right) C_l \frac{(2l+1)}{4\pi} \approx B^{-\alpha j} \, . \tag{12.16}$$

From (12.14) and (6.20) we obtain

$$\left|Corr\left(\beta_{jk;s}, \overline{\beta}_{jk';s}\right)\right| = \frac{\left|\sum_l b^2\left(\frac{l}{B^j}\right) C_l \frac{(2l+1)}{4\pi} K^{ls}\left(\xi_{jk}, \xi_{jk'}\right)\right|}{\sum_l b^2\left(\frac{l}{B^j}\right) C_l \frac{(2l+1)}{4\pi}} \, . \tag{12.17}$$

The key result for the development of the high-frequency asymptotic theory in the next sections is the following uncorrelation result, which was provided by [74]; under Condition 12.3,

$$\left|Corr\left(\beta_{jk;s}, \overline{\beta}_{jk';s}\right)\right| \le \frac{C_M}{\left\{1 + B^j d(\xi_{jk}, \xi_{jk'})\right\}^M} \, , \text{ for all } M \in \mathbb{N} \, , \text{ some } C_M > 0 \, . \tag{12.18}$$

The proof is close to analogous results for the scalar case reported in Chapter 10, and hence omitted.

In view of (12.16), let us now denote

$$\Gamma_{j;s} := \sum_k \left|\beta_{jk;s}\right|^2 = \sum_k \lambda_{jk} \sum_l b^2 \left(\frac{l}{B^j}\right) C_l \frac{(2l+1)}{4\pi}$$

$$= \sum_l b^2 \left(\frac{l}{B^j}\right) C_l (2l+1) \ .$$

Under Condition 12.3, it is immediate to see that

$$C_0 B^{(2-\alpha)j} \le \Gamma_{j;s} \le C_1 B^{(2-\alpha)j}. \tag{12.19}$$

A question of great practical relevance is the asymptotic behaviour of $\sum_k \left|\beta_{jk;s}\right|^2$ as an estimator for $\Gamma_{j;s}$, as done for the scalar case in Chapter 11. In the spin case, angular power spectrum estimation was first considered by [74], under the unrealistic assumptions that the spin random field $P = Q + iU$ is observed on the whole sphere and without noise; our analysis here follows closely [73], where noise and gaps are accounted for. Indeed, here we shall be concerned with the much more realistic case where some parts of the domain S^2 are "masked" by the presence of foreground contamination; more precisely, we assume data are collected only on a subset $S^2 \setminus G$, G denoting the masked region. In this section, we do not consider the presence of observational noise, which shall be dealt with in the following section. In the sequel, for some (arbitrary small) constant $\varepsilon > 0$, we define as before $G_\varepsilon = \left\{x \in S^2 : d(x,G) \le \varepsilon\right\}$. Consider

$$\widehat{\Gamma}^*_{j;sG} := \left\{\sum_{k:\xi_{jk}\in S^2\setminus G_\varepsilon} \lambda_k\right\}^{-1} \sum_{k:\xi_{jk}\in S^2\setminus G_\varepsilon} \left|\beta^*_{jk;s}\right|^2 \tag{12.20}$$

where

$$\beta^*_{jk;s} = \int_{S^2\setminus G} P(x)\overline{\psi}_{jk;s}(x)dx \ .$$

Our aim will be to prove the following

Theorem 12.5 *Under condition (12.3), we have*

$$\frac{\widehat{\Gamma}^*_{j;sG} - \Gamma_{j;s}}{\sqrt{Var\left\{\widehat{\Gamma}^*_{j;sG}\right\}}} \xrightarrow{law} N(0,1) \ , \ as \ j \to \infty \ .$$

Proof The proof will be basically in two steps; define

$$\widehat{\Gamma}_{j;sG} := \left\{\sum_{k:\xi_{jk}\in S^2\setminus G_\varepsilon} \lambda_k\right\}^{-1} \sum_{k:\xi_{jk}\in S^2\setminus G_\varepsilon} \left|\beta_{jk;s}\right|^2 \ , \tag{12.21}$$

which is clearly an unfeasible version of (12.20), where the $\beta^*_{jk;s}$ have been replaced by the coefficients (in the observed region) evaluated without gaps. The idea will be to show that

$$\frac{\widehat{\Gamma}_{j;sG} - \Gamma_{j;s}}{\sqrt{Var\{\widehat{\Gamma}_{j;sG}\}}} \overset{law}{\to} N(0,1)\,, \quad \frac{\sqrt{Var\{\widehat{\Gamma}_{j;sG}\}}}{\sqrt{Var\{\widehat{\Gamma}^*_{j;sG}\}}} \to 1$$

$$\text{and}\quad \frac{\widehat{\Gamma}^*_{j;sG} - \widehat{\Gamma}_{j;sG}}{\sqrt{Var\{\widehat{\Gamma}^*_{j;sG}\}}} \overset{p}{\to} 0\,, \text{ as } j \to \infty\,.$$

The proof of these three statements is provided in separate Propositions below.

□

Proposition 12.6 *As $j \to \infty$, under Condition 12.3 we have*

$$\frac{\widehat{\Gamma}_{j;sG} - \Gamma_{j;s}}{\sqrt{Var\{\widehat{\Gamma}_{j;sG}\}}} \overset{law}{\to} N(0,1)\,.$$

Proof Notice that

$$\left(\sum_{k:\xi_{jk}\in S^2\backslash G_\varepsilon} \lambda_k\right)^2 Var\left(\widehat{\Gamma}_{j;sG}\right) = Var\left[\sum_k |\beta_{jk;s}|^2\right] = \sum_{k,k'} |E\beta_{jk;s}\bar{\beta}_{jk';s}|^2$$

$$= \sum_{k,k'} \lambda_{jk}\lambda_{jk'} \left|\sum_l b^2\left(\frac{l}{B^j}\right) C_l \frac{(2l+1)}{4\pi} K^{ls}\left(\xi_{jk}, \xi_{jk'}\right)\right|^2\,.$$

By standard manipulations we obtain the upper bound

$$Var\left[\sum_k |\beta_{jk;s}|^2\right]$$

$$\leq C_M B^{2(2-\alpha)j} \sum_{k,k'} \lambda_{jk}\lambda_{jk'} \frac{1}{\left[1 + B^j d(\xi_{jk}, \xi_{jk'})\right]^{2M}}$$

$$\leq C_M B^{2(2-\alpha)j} \left[\sup_{k'} \lambda_{jk'}\right] \sum_k \lambda_{jk} \sum_{k'} \frac{1}{\left[1 + d(\xi_{jk}, \xi_{jk'})\right]^{2M}}$$

$$= \sum_k \lambda_{jk} O(B^{2(1-\alpha)j})\,,$$

in view of (12.18) (12.19) and $\lambda_{jk} \approx B^{-2j}$. On the other hand, we also have the trivial lower bound

$$\sum_{k,k'} |E\beta_{jk;s}\bar{\beta}_{jk';s}|^2 \geq \sum_k |E\beta_{jk;s}\bar{\beta}_{jk;s}|^2 = \Gamma^2_{j;s} \sum_k \lambda^2_{jk} \geq c \sum_k \lambda_{jk} B^{2(1-\alpha)j}\,,$$

whence we have

$$Var\left\{\sum_k |\beta_{jk;s}|^2\right\} \approx \left(\sum_k \lambda_{jk}\right) B^{2(1-\alpha)j} \approx B^{2(1-\alpha)j}. \qquad (12.22)$$

It is immediate to check that these statistics belong to the second-order Wiener chaos; hence, by the simplified method of moments presented in Chapter 4, to establish the Central Limit Theorem it suffices to focus on the fourth-order cumulant; the proof that

$$Cum_4\left\{\frac{\sum_k |\beta_{jk;s}|^2 - \left(\sum_k \lambda_{jk}\right)\Gamma_{j;s}}{\sqrt{Var\left\{\sum_k |\beta_{jk;s}|^2\right\}}}\right\} \to 0 \text{ as } j \to \infty,$$

is a standard application of the diagram formulae presented in Chapter 4: indeed from (12.16) and (12.18) we have

$$Cum_4\left\{\sum_k |\beta_{jk;s}|^2 - \left(\sum_k \lambda_{jk}\right)\Gamma_{j;s}\right\}$$

$$= 6 \sum_{k_1,k_2,k_3,k_4} E\beta_{jk_1;s}\bar{\beta}_{jk_2;s} E\beta_{jk_2;s}\bar{\beta}_{jk_3;s} E\beta_{jk_3;s}\bar{\beta}_{jk_4;s} E\beta_{jk_4;s}\bar{\beta}_{jk_1;s}$$

$$\leq CB^{-4\alpha_j} \sum_{k_1,k_2,k_3,k_4} \frac{1}{\left\{1 + B^j d(\xi_{jk_1}, \xi_{jk_2})\right\}^M} \frac{1}{\left\{1 + B^j d(\xi_{jk_2}, \xi_{jk_3})\right\}^M}$$

$$\times \frac{1}{\left\{1 + B^j d(\xi_{jk_3}, \xi_{jk_4})\right\}^M} \frac{1}{\left\{1 + B^j d(\xi_{jk_4}, \xi_{jk_1})\right\}^M}.$$

By iterated application of (11.5), this quantity is bounded by

$$\leq CB^{-4\alpha_j} \sum_{k_1,k_2,k_3} \frac{1}{\left\{1 + B^j d(\xi_{jk_1}, \xi_{jk_2})\right\}^M} \frac{1}{\left\{1 + B^j d(\xi_{jk_2}, \xi_{jk_3})\right\}^M} \frac{1}{\left\{1 + B^j d(\xi_{jk_3}, \xi_{jk_1})\right\}^M}$$

$$\leq CB^{-4\alpha_j} \sum_{k_1,k_2} \frac{1}{\left\{1 + B^j d(\xi_{jk_1}, \xi_{jk_2})\right\}^M} \frac{1}{\left\{1 + B^j d(\xi_{jk_2}, \xi_{jk_1})\right\}^M}$$

$$\leq CB^{-4\alpha_j} \sum_{k_1,k_2} \frac{1}{\left\{1 + B^j d(\xi_{jk_1}, \xi_{jk_2})\right\}^M} \frac{1}{\left\{1 + B^j d(\xi_{jk_2}, \xi_{jk_1})\right\}^M}$$

$$\leq CB^{-4\alpha_j} \sum_{k_1,k_2} 1 = O\left(B^{(2-4\alpha)j}\right),$$

with C varying from line to line. Thus from (12.22)

$$Cum_4\left\{\frac{\sum_k |\beta_{jk;s}|^2 - \left(\sum_k \lambda_{jk}\right)\Gamma_{j;s}}{\sqrt{Var\left\{\sum_k |\beta_{jk;s}|^2\right\}}}\right\} = O\left(B^{(2-4\alpha)j}\right)O\left(B^{-(4-4\alpha)j}\right) = O\left(B^{-2j}\right),$$

and the Proposition is established. □

Remark 12.7 As discussed in Chapter 4, it would be immediate to strengthen this result providing an exact bound on the total variation distance d_{TV} for the speed of convergence to a Gaussian law, i.e.

$$d_{TV}\left(\frac{\widehat{\Gamma}_{j;sG} - \Gamma_{j;s}}{\sqrt{Var\left\{\widehat{\Gamma}_{j;sG}\right\}}}, z\right) = O\left(B^{-j}\right), Z \overset{law}{=} N(0, 1).$$

Next we turn to the following

Proposition 12.8 *As $j \to \infty$, under Condition 12.3 we have*

$$\frac{\sqrt{Var\left\{\widehat{\Gamma}_{j;sG}\right\}}}{\sqrt{Var\left\{\widehat{\Gamma}_{j;sG}^*\right\}}} \to 1$$

Proof Again in view of the Diagram Formula, it is enough to focus on

$$Var\left(\sum_{k:\xi_{jk}\in S^2\backslash G_\varepsilon} |\beta_{jk;s}|^2\right) - Var\left(\sum_{k:\xi_{jk}\in S^2\backslash G_\varepsilon} |\beta_{jk;s}^*|^2\right)$$

$$= O\left(\sum_{k,k'} \left|E\beta_{jk;s}\overline{\beta}_{jk';s}\right|^2 - \sum_{k,k'} \left|E\beta_{jk;s}^*\overline{\beta}_{jk';s}^*\right|^2\right).$$

Now notice that

$$\left|E\beta_{jk;s}\overline{\beta}_{jk';s}\right|^2 - \left|E\beta_{jk;s}^*\overline{\beta}_{jk';s}^*\right|^2 = E\beta_{jk;s}\overline{\beta}_{jk';s}\left(E\overline{\beta}_{jk;s}\beta_{jk';s} - E\overline{\beta}_{jk;s}^*\beta_{jk';s}^*\right)$$

$$+E\overline{\beta}_{jk;s}^*\beta_{jk';s}^*\left(E\beta_{jk;s}\overline{\beta}_{jk';s} - E\beta_{jk;s}^*\overline{\beta}_{jk';s}^*\right),$$

and

$$E\overline{\beta}_{jk;s}\beta_{jk';s} - E\overline{\beta}_{jk;s}^*\beta_{jk';s}^* = E\overline{\beta}_{jk;s}\left(\beta_{jk';s} - \beta_{jk';s}^*\right) + E\beta_{jk';s}^*\left(\overline{\beta}_{jk;s} - \overline{\beta}_{jk;s}^*\right)$$

$$\leq \left\{E\left|\overline{\beta}_{jk;s}\right|^2\right\}^{1/2}\left\{E\left|\beta_{jk';s} - \beta_{jk';s}^*\right|^2\right\}^{1/2} + \left\{E\left|\beta_{jk';s}^*\right|^2\right\}^{1/2}\left\{E\left|\overline{\beta}_{jk;s} - \overline{\beta}_{jk;s}^*\right|^2\right\}^{1/2}.$$

$$(12.23)$$

Hence

$$E|\beta_{jk;s} - \beta_{jk;s}^*|^2$$

$$\leq E\left\{\int_G P(x)\overline{\psi}_{jk;s}(x)dx\right\}^2 \leq E\left\{\sup_{x\in G}\left\{\overline{\psi}_{jk;s}(x)\right\}\int_{G^\varepsilon}|P(x)|dx\right\}^2$$

$$\leq \left[\sup_{x\in G}\left\{\overline{\psi}_{jk;s}(x)\right\}\right]^2 E\left\{\int_G |P(x)|dx\right\}^2$$

$$\leq \left[\sup_{x\in G}\left\{\overline{\psi}_{jk;s}(x)\right\}\right]^2 E\left\{\left[\int_G 1dx\right]\left[\int_G |P(x)|^2 dx\right]\right\}$$

$$\leq 4\pi\left[\sup_{x\in G}\left\{\overline{\psi}_{jk;s}(x)\right\}\right]^2 E\left\{\left[\int_G |P(x)|^2 dx\right]\right\} = O\left(\frac{B^{2j}}{[1 + B^j\varepsilon]^{2M}}\right).$$

Now recall that

$$E|\beta_{jk;s}|^2 = O\left(B^{-\alpha j}\right),$$

whence $E|\beta_{jk;s}^*|^2 = O\left(B^{-\alpha j}\right)$, if $M > \alpha/2 + 1$. Hence, in view of (12.23)

$$\left|E\overline{\beta}_{jk;s}^*\beta_{jk';s}^* - E\overline{\beta}_{jk;s}\beta_{jk';s}\right| \leq \frac{CB^{(1-\alpha/2)j}}{[1 + B^j\varepsilon]^M}, \qquad (12.24)$$

for some constant $C > 0$. Also, from (12.23) and (12.24) we obtain that

$$\sum_{k,k'}\left(\left|E\beta_{jk;s}\overline{\beta}_{jk';s}\right|^2 - \left|E\beta_{jk;s}^*\overline{\beta}_{jk';s}^*\right|^2\right)$$

$$\leq \sum_{k,k'}\left(\left|E\beta_{jk;s}\overline{\beta}_{jk';s}\right| + \left|E\beta_{jk;s}^*\overline{\beta}_{jk';s}^*\right|\right)O\left(\frac{B^{-j\alpha/2}}{[1 + B^j\varepsilon]^M}\right)$$

$$\leq O\left(\frac{B^{(1-\alpha/2)j}}{[1 + B^j\varepsilon]^M}\right)\Gamma_{j;s}\sum_{k,k'}\frac{C_M\sqrt{\lambda_{jk}\lambda_{jk'}}}{\left\{1 + B^j d(\xi_{jk}, \xi_{jk'})\right\}^M}$$

$$= O\left(\frac{B^{3(1-\alpha/2)j}}{[1 + B^j\varepsilon]^M}\right).$$

Recall from (12.22) that $Var\left(\sum_{k:\xi_{jk}\in S^2\backslash G}|\beta_{jk;s}|^2\right) = O\left(B^{2(1-\alpha)j}\right)$. Hence for large enough M, that is $M > 1 + \alpha/2$, the statement of the Proposition is established. □

Proposition 12.9 *As $j \to \infty$, under Condition 12.3 we have*

$$\frac{\widehat{\Gamma}_{j;sG}^* - \widehat{\Gamma}_{j;sG}}{\sqrt{Var\left\{\widehat{\Gamma}_{j;sG}^*\right\}}} \xrightarrow{p} 0.$$

Proof We have

$$
E\left\{\left[\sum_{k:\xi_{jk}\in S^2\backslash G_\varepsilon} \lambda_k\right]\left(\widehat{\Gamma}^*_{j;sG} - \widehat{\Gamma}_{j;sG}\right)\right\}^2 = E\left\{\sum_k |\beta_{jk;s}|^2 - |\beta^*_{jk;s}|^2\right\}^2,
$$

which we can expand as follows

$$
E\left\{\sum_k \overline{\beta}_{jk;s}\left(\beta_{jk;s} - \beta^*_{jk;s}\right) + \sum_k \beta^*_{jk;s}\left(\overline{\beta}_{jk;s} - \overline{\beta}^*_{jk;s}\right)\right\}^2
$$

$$
= E\left\{\sum_k \overline{\beta}_{jk;s}\left(\beta_{jk;s} - \beta^*_{jk;s}\right)\right\}^2 + E\left\{\sum_k \beta^*_{jk;s}\left(\overline{\beta}_{jk;s} - \overline{\beta}^*_{jk;s}\right)\right\}^2
$$

$$
+2E\left\{\sum_k \overline{\beta}_{jk;s}\left(\beta_{jk;s} - \beta^*_{jk;s}\right)\right\}\left\{\sum_k \beta^*_{jk;s}\left(\overline{\beta}_{jk;s} - \overline{\beta}^*_{jk;s}\right)\right\}
$$

$$
= \sum_{k,k'}\left[E\overline{\beta}_{jk;s}\left(\beta_{jk';s} - \beta^*_{jk';s}\right)E\overline{\beta}_{jk';s}\left(\beta_{jk;s} - \beta^*_{jk;s}\right)\right.
$$

$$
\left.+E\beta^*_{jk;s}\left(\overline{\beta}_{jk';s} - \overline{\beta}^*_{jk';s}\right)E\beta^*_{jk';s}\left(\overline{\beta}_{jk;s} - \overline{\beta}^*_{jk;s}\right)\right]
$$

$$
+\left\{\sum_k E\overline{\beta}_{jk;s}\left(\beta_{jk;s} - \beta^*_{jk;s}\right)\right\}^2 + \left\{\sum_k E\beta^*_{jk;s}\left(\overline{\beta}_{jk;s} - \overline{\beta}^*_{jk;s}\right)\right\}^2
$$

$$
+2\left\{\sum_k E\overline{\beta}_{jk;s}\left(\beta_{jk;s} - \beta^*_{jk;s}\right)\right\}\left\{\sum_k E\beta^*_{jk;s}\left(\overline{\beta}_{jk;s} - \overline{\beta}^*_{jk;s}\right)\right\}
$$

$$
+2\left\{\sum_{k,k'} E\overline{\beta}_{jk;s}\left(\beta_{jk';s} - \beta^*_{jk';s}\right)E\beta^*_{jk';s}\left(\overline{\beta}_{jk;s} - \overline{\beta}^*_{jk;s}\right)\right\}
$$

$$
+2\left\{\sum_{k,k'} E\overline{\beta}_{jk;s}\beta^*_{jk';s}E\left(\overline{\beta}_{jk;s} - \overline{\beta}^*_{jk;s}\right)\left(\beta_{jk';s} - \beta^*_{jk';s}\right)\right\}.
$$

Now recall again

$$
E\left|\beta_{jk;s}\right|^2, E\left|\beta^*_{jk;s}\right|^2 \le CB^{-\alpha j}, \text{ and } E\left|\beta_{jk;s} - \beta^*_{jk;s}\right|^2 \le \frac{C'B^{2j}}{[1 + B^j\varepsilon]^M},
$$

whence from the same steps as in the previous Proposition, we have

$$
E\overline{\beta}_{jk;s}\left(\beta_{jk';s} - \beta^*_{jk';s}\right), \ E\overline{\beta}^*_{jk;s}\left(\beta_{jk';s} - \beta^*_{jk';s}\right) \le \frac{CB^{(1-\alpha/2)j}}{[1 + B^j\varepsilon]^M}.
$$

It follows that

$$
E\left\{\sum_k |\beta_{jk;s}|^2 - |\beta^*_{jk;s}|^2\right\}^2 \le \frac{CB^{(6-\alpha)j}}{[1 + B^j\varepsilon]^{2M}}.
$$

By arguments in the previous Propositions, we know that

$$Var\left\{\left[\sum_{k:\xi_{jk}\in S^2\setminus G_\varepsilon}\lambda_k\right]\widehat{\Gamma^*_{j;sG}}\right\}\approx\left(\sum_{k:\xi_{jk}\in S^2\setminus G_\varepsilon}\lambda_{jk}\right)B^{2(1-\alpha)j};$$

thus the statement is established, provided we take $M > 2 + \alpha/2$.   □

Remark 12.10   In general the expression for $Var\left\{\widehat{\Gamma}_{j;sG}\right\}$, $Var\left\{\widehat{\Gamma^*}_{j;sG}\right\}$ depends on the unknown angular power spectrum. However, the normalizing factors can be consistently estimated by subsampling techniques, following the same steps as in [14].

12.6 Detection of asymmetries

In this Section, we shall consider one more possible application of spin needlets to problems of interest for cosmology. In particular, a highly debated issue in modern cosmology relates to the existence of "features", i.e. asymmetries in the distribution of CMB radiation (for instance between the Northern and the Southern hemispheres, in Galactic coordinates). These issues have been the subject of dozens of physical papers, in the last few years, some of them exploiting scalar needlets, see [163].

In order to investigate this issue, we shall employ a similar technique as [14] for the scalar case. More precisely, we shall focus on the difference between the estimated angular power spectrum over two different regions of the sky. Let us consider A_1, A_2, two subsets of S^2 such that $A_1 \cap A_2 = \emptyset$; we do not assume that $A_1 \cup A_2 = S^2$, i.e. we admit the presence of missing observations. For practical applications, A_1 and A_2 can be visualized as the spherical caps centered at the north and south pole N, S (i.e. $A_1 = \left\{x \in S^2 : d(x, N) \leq \pi/2\right\}$, $A_2 = \left\{x \in S^2 : d(x, S) \leq \pi/2\right\}$) but the results would hold without any modification for general subsets and could be easily generalized to a higher number of regions. We shall then focus on the statistic

$$\frac{\widehat{\Gamma^*}_{j;sA_1} - \widehat{\Gamma^*}_{j;sA_2}}{\sqrt{Var\left\{\widehat{\Gamma^*}_{j;sA_1}\right\} + Var\left\{\widehat{\Gamma^*}_{j;sA_2}\right\}}},$$

where

$$\widehat{\Gamma^*}_{j;sA_1} := \left\{\sum_{k:\xi_{jk}\in A_1^\varepsilon}\lambda_k\right\}^{-1}\sum_{k:\xi_{jk}\in A_1^\varepsilon}\left|\beta^*_{jk;s}\right|^2,$$

$$\widehat{\Gamma}^*_{j;sA_2} := \left\{ \sum_{k:\xi_{jk}\in A_2^\varepsilon} \lambda_k \right\}^{-1} \sum_{k:\xi_{jk}\in A_2^\varepsilon} \left| \beta^*_{jk;s} \right|^2 , \text{ some } \varepsilon > 0 ,$$

and

$$A_1^\varepsilon = A_1 \cap \left\{ S^2 \backslash A_{2\varepsilon} \right\} = \{x \in A_1 : d(x, A_2) \geq \varepsilon\} ,$$

$$A_2^\varepsilon = A_2 \cap \left\{ S^2 \backslash A_{1\varepsilon} \right\} = \{x \in A_2 : d(x, A_1) \geq \varepsilon\} .$$

We are here able to establish the following

Proposition 12.11 *As $j \to \infty$, we have*

$$\left(\begin{array}{c} \left[Var\left\{\widehat{\Gamma}^*_{j;sA_1}\right\} \right]^{-1/2} \left(\widehat{\Gamma}^*_{j;sA_1} - \Gamma_{j;s} \right) \\ \left[Var\left\{\widehat{\Gamma}^*_{j;sA_2}\right\} \right]^{-1/2} \left(\widehat{\Gamma}^*_{j;sA_2} - \Gamma_{j;s} \right) \end{array} \right) \overset{law}{\to} N(0_2, I_2) ,$$

where $(0_2, I_2)$ are, respectively, the 2×1 vector of zeros and the 2×2 identity matrix.

Proof By the Cramer-Wold device, the proof can follow very much the same steps as for the univariate case. We first establish the asymptotic uncorrelation of the two components, i.e. we show that

$$\lim_{j\to\infty} \left[Var\left\{\widehat{\Gamma}^*_{j;sA_1}\right\} Var\left\{\widehat{\Gamma}^*_{j;sA_2}\right\} \right]^{-1/2} E\left\{ \left(\widehat{\Gamma}^*_{j;sA_1} - \Gamma_{j;s}\right)\left(\widehat{\Gamma}^*_{j;sA_2} - \Gamma_{j;s}\right) \right\} = 0 . \tag{12.25}$$

Now

$$E\left(\widehat{\Gamma}^*_{j;sA_1} - \Gamma_{j;s}\right)\left(\widehat{\Gamma}^*_{j;sA_2} - \Gamma_{j;s}\right) = E\left(\widehat{\Gamma}^*_{j;sA_1} - \Gamma_{j;s}\right) E\left(\widehat{\Gamma}^*_{j;sA_2} - \Gamma_{j;s}\right)$$

$$+ 2 \left\{ \sum_{k:\xi_{jk}\in A_1^\varepsilon} \lambda_k \sum_{k:\xi_{jk}\in A_2^\varepsilon} \lambda_k \right\}^{-1} \sum_{k:\xi_{jk}\in A_1^\varepsilon} \sum_{k':\xi_{jk'}\in A_2^\varepsilon} \left| E\beta^*_{jk;s}\overline{\beta^*_{jk';s}} \right|^2 . \tag{12.26}$$

In view of (12.18) and Proposition 12.9, we have

$$|(12.25)| \leq \left(\Gamma_{j;s}\right)^2 \sum_{k:\xi_{jk}\in S^2\backslash A_1^\varepsilon} \sum_{k':\xi_{jk'}\in S^2\backslash A_2^\varepsilon} \frac{C\lambda_{jk}\lambda_{jk'}}{\left[1 + B^j d(\xi_{jk}, \xi_{jk'})\right]^{2M}}$$

$$\leq \frac{C\left(\Gamma_{j;s}\right)^2 \left[\sup_k \lambda_{jk}\right]^2}{\left[1 + 2B^j\varepsilon\right]^{2(M-1)}} = O\left(B^{2(1-\alpha-M)j}\right) .$$

Thus (12.25) is established, in view of (12.22) and Propositions (12.8), (12.9). For the fourth order cumulant, we shall write

$$X = \left[Var\left\{\widehat{\Gamma}^*_{j;sA_1}\right\} \right]^{-1/2} \left(\widehat{\Gamma}^*_{j;sA_1} - \Gamma_{j;s} \right) , \tag{12.27}$$

and

$$Y = \left[Var \left\{ \widehat{\Gamma}^*_{j;sA_2} \right\} \right]^{-1/2} \left(\widehat{\Gamma}^*_{j;sA_2} - \Gamma_{j;s} \right) . \tag{12.28}$$

Recall that

$$Cum_4 (X + Y) = Cum_4 (X) + Cum_4 (Y) + 4Cum(X, Y, Y, Y)$$
$$+ 6Cum(X, X, Y, Y) + 4Cum(X, X, X, Y) ;$$

by results in the previous section, we have immediately $Cum_4 (X)$, $Cum_4 (Y) \to$ 0, as $j \to \infty$. On the other hand, in view of Proposition 12.9 and the equivalence between convergence in probability and in L^p for Gaussian subordinated processes (see [105]), we can replace $\widehat{\Gamma}^*_{j;sA_i}$ by $\widehat{\Gamma}_{j;sA_i}$ in (12.27) and (12.28), and we have easily

$$Cum(X, Y, Y, Y)$$
$$\le CB^{4(\alpha-1)j} \left(\Gamma_{j;s} \right)^2 \sum_{k:\xi_{jk} \in A_1^\varepsilon} \sum_{\xi_{jk_1}, \dots, \xi_{jk_3} \in A_2^\varepsilon} \frac{\lambda_{jk} \lambda_{jk_1} \lambda_{jk_3} \lambda_{jk_3}}{\left[1 + B^j d(\xi_{jk}, \xi_{jk_1}) \right]^M}$$
$$\times \frac{1}{\left[1 + B^j d(\xi_{jk_2}, \xi_{jk_1}) \right]^M \left[1 + B^j d(\xi_{jk_3}, \xi_{jk_2}) \right]^M \left[1 + B^j d(\xi_{jk}, \xi_{jk_3}) \right]^M}$$
$$\le \frac{CB^{4(\alpha-1)j} \left(\Gamma_{j;s} \right)^2 \left[\sup_k \lambda_{jk} \right]^4}{\left[1 + 2B^j \varepsilon \right]^{2(M-1)}} = O \left(B^{-2(M+1)j} \right).$$

Similarly, we have

$$Cum(X, X, X, Y), Cum(X, X, Y, Y) \le CB^{-2(M+1)j}.$$

Thus the Proposition is established, provided we choose $M > 2 + \alpha$.   □

Remark 12.12   An obvious consequence of Proposition 12.11 is

$$\frac{\widehat{\Gamma}^*_{j;sA_1} - \widehat{\Gamma}^*_{j;sA_2}}{\sqrt{Var \left\{ \widehat{\Gamma}^*_{j;sA_1} \right\} + Var \left\{ \widehat{\Gamma}^*_{j;sA_2} \right\}}} \overset{law}{\to} N(0, 1) .$$

This result provides the asymptotic justification to implement on polarization data the same testing procedures as those considered for instance by [163] to search for features and asymmetries in CMB scalar data; i.e., it is possible to estimate for instance the angular power spectrum on the Northern and Southern hemisphere and test whether they are statistically different, as suggested by some empirical findings of the recent cosmological literature (see e.g. [163]).

12.7 Estimation with noise

In the previous sections, we worked under a simplifying assumption, i.e. we figured that although observations on some parts of the sphere were completely unattainable, data on the remaining part were available free of noise. In this Section, we aim at relaxing this assumption; in particular, we shall consider the more realistic circumstances where, while we still take some regions of the sky to be completely unobservable, even for those where observations are available the latter are partially contaminated by noise.

To understand our model for noise, we recall a few basic facts on the underlying physics, as in Chapter 8. As discussed, a key issue about (scalar and polarized) CMB radiation experiments is that they actually measure radiation across a set of different electromagnetic frequencies, ranging from 30 GHz to nearly 900 GHz. We shall hence assume that D detectors are available at frequencies $\nu_1, ..., \nu_D$, so that the following vector random field is observed:

$$P_{\nu_r}(x) = P(x) + N_{\nu_r}(x) \ ;$$

here, both $P(x), N_\nu(x)$ are taken to be Gaussian spin random fields, independent among them and such that, while the signal $P(x)$ is identical across all frequencies, the noise $N_\nu(x)$ is not. More precisely, we shall assume for noise the same regularity conditions as for the signal P, again under the justification that they seem mild and general:

Condition 12.13 The (spin) random field $N_\nu(x)$ is Gaussian and isotropic, independent from $P(x)$ and with total angular power spectrum $\{C_{lN}\}$ such that

$$C_{lN} = l^{-\gamma} g_N(l) > 0 \ , \text{ where } c_{0N}^{-1} \le g_N(l) \le c_{0N} \ , \gamma > 2 \ , l \in \mathbb{N} \ ,$$

and for every $r \in \mathbb{N}$ there exist $c_r > 0$ such that

$$|\frac{d^r}{du^r} g_N(u)| \le c_{rN} u^{-r} \ , u \in (|s|, \infty) \ .$$

It follows from our previous assumptions that for each frequency ν_r we shall be able to evaluate

$$\int_{S^2} P_{\nu_r}(x) \overline{\psi}_{jk;s}(x) d\sigma(x) =: \beta_{jk;sr} = \beta_{jk;sP} + \beta_{jk;sN_r}$$

where clearly

$$\beta_{jk;sP} = \int_{S^2} P(x) \overline{\psi}_{jk;s}(x) d\sigma(x) \ , \beta_{jk;sN_r} = \int_{S^2} N_{\nu_r}(x) \overline{\psi}_{jk;s}(x) d\sigma(x) \ .$$

Now it is immediate to note that

$$E|\beta_{jk;sr}|^2 = E|\beta_{jk;sP} + \beta_{jk;sN_r}|^2$$

$$= E\beta_{jk;sP}\overline{\beta}_{jk;sP} + E\beta_{jk;sN_r}\overline{\beta}_{jk;sN_r} + E\beta_{jk;sN_r}\overline{\beta}_{jk;sP} + E\beta_{jk;sP}\overline{\beta}_{jk;sN_r}$$
$$= E|\beta_{jk;sP}|^2 + E|\beta_{jk;sN_r}|^2 \,,$$

so that the estimator $\sum_k |\beta_{jk;sr}|^2$ will now be upward biased. In the next subsections we shall discuss two possible solutions for dealing with this bias terms, along the lines of [167] and Chapter 8, and we will provide statistical procedures to test for estimation bias. We note first that correlation of needlet coefficients across different channels are provided by

$$E\beta_{jk;sr}\overline{\beta}_{jk';sr} = E\beta_{jk;sP}\overline{\beta}_{jk';sP} + E\beta_{jk;sN_r}\beta_{jk';sN_r} \,.$$

Denote

$$\Gamma^N_{j;s} = \sum_k E|\beta_{jk;sN_r}|^2 = \sum_l b^2\left(\frac{l}{B^j}\right)\frac{2l+1}{4\pi}C_{lN} \,;$$

as before, it is easy to obtain that $C_1 B^{(2-\gamma)j} \le \Gamma^N_{j;s} \le C_2 B^{(2-\gamma)j}$, and likewise under Condition 12.13,

$$\left|Corr\left(\beta_{jk;sr},\overline{\beta}_{jk';sr}\right)\right| \le \frac{C_M}{\left\{1 + B^j d(\xi_{jk},\xi_{jk'})\right\}^M} \,, \text{ for all } M \in \mathbb{N} \,. \tag{12.29}$$

12.7.1 The needlet auto-power spectrum estimator

In many circumstances, it can be reasonable to assume that the angular power spectrum of the noise component, C_{lN}, is known in advance to the experimenter. For instance, if noise is primarily dominated by instrumental components, then its behaviour may possibly be calibrated before the experimental devices are actually sent in orbit, or otherwise by observing a peculiar region where the signal has been very tightly measured by previous experiments. Assuming the angular power spectrum of noise to be known, the expected value for the bias term is immediately derived:

$$E|\beta_{jk;sN_r}|^2 = \sum_l b^2\left(\frac{l}{B^j}\right)\frac{2l+1}{4\pi}C_{lN_r} \,,$$

whence it is natural to propose the bias-corrected estimator

$$\widetilde{\Gamma}^{AP}_j := \frac{1}{D}\sum_k\sum_r\left\{|\beta_{jk;sr}|^2 - E|\beta_{jk;sN_r}|^2\right\}$$
$$= \frac{1}{D}\sum_k\sum_r\left\{\left(\beta_{jk;sP} + \beta_{jk;sN_r}\right)\left(\overline{\beta}_{jk;sP} + \overline{\beta}_{jk;sN_r}\right) - E|\beta_{jk;sN_r}|^2\right\}$$
$$= \sum_k|\beta_{jk;sP}|^2 +$$

$$\frac{1}{D}\left\{\sum_k \sum_r \left(\beta_{jk;sP}\overline{\beta}_{jk;sN_r} + \beta_{jk;sN_r}\overline{\beta}_{jk;sP} + \left[\left|\beta_{jk;sN_r}\right|^2 - E\left|\beta_{jk;sN_r}\right|^2\right]\right)\right\}.$$

We label the previous statistics the needlet auto-power spectrum estimator (*AP*, compare Chapter 8 and [73, 167]). The derivation of the following Proposition is rather standard, and hence omitted for brevity's sake.

Proposition 12.14 *As $j \to \infty$, we have*

$$\frac{\widetilde{\Gamma}_j^{AP} - \Gamma_j}{\sqrt{Var\left\{\widetilde{\Gamma}_j^{AP}\right\}}} \overset{law}{\to} N(0,1),$$

where

$$Var\left\{\widetilde{\Gamma}_j^{AP}\right\} = O(B^{2(1-\min(\alpha,\gamma))j}).$$

As before, the normalizing variance in the denominator can be consistently estimated by subsampling techniques, along the lines of [14]. It should be noticed that the rate of convergence for $\left\{\widetilde{\Gamma}_j^{AP} - \Gamma_j\right\} = O(B^{(1-\min(\alpha,\gamma))j})$ is the same as in the noiseless case for $\gamma \geq \alpha$, whereas it is slower otherwise, when the noise is asymptotically dominating. The "signal-to-noise" ratio $\Gamma_j / \sqrt{Var\left\{\widetilde{\Gamma}_j^{AP}\right\}}$ is easily seen to be in the order of $B^{2j-\alpha j}/B^{(1-\min(\alpha,\gamma))j} = B^{j(1+\min(\alpha,\gamma)-\alpha)}$, whence it decays to zero unless $\alpha \leq \gamma + 1$.

12.7.2 The needlet cross-power spectrum estimator

To handle the bias term, we shall pursue here a different strategy than the previous subsection, dispensing with any prior knowledge of the spectrum of the noise component. The idea is the same as in Chapter 8, i.e. to exploit the fact that, while the signal is perfectly correlated among the different frequency components, noise is by assumption independent. We shall hence focus on the needlets cross-angular power spectrum estimator (*CP*), defined as

$$\widetilde{\Gamma}_j^{CP} := \frac{1}{D(D-1)} \sum_k \sum_{r_1 \neq r_2} \beta_{jk;sr_1}\overline{\beta}_{jk;sr_2}$$

$$= \frac{1}{D(D-1)} \sum_k \sum_{r_1 \neq r_2} \left(\beta_{jk;sP} + \beta_{jk;sN_{r_1}}\right)\left(\overline{\beta}_{jk;sP} + \overline{\beta}_{jk;sN_{r_2}}\right)$$

$$= \sum_k \left|\beta_{jk;sP}\right|^2$$

$$+ \frac{1}{D(D-1)}\left\{\sum_k \sum_{r_1 \neq r_2}\left(\beta_{jk;sP}\overline{\beta}_{jk;sN_{r_2}} + \beta_{jk;sN_{r_1}}\overline{\beta}_{jk;sP} + \beta_{jk;sN_{r_1}}\overline{\beta}_{jk;sN_{r_2}}\right)\right\}.$$

In view of the previous independence assumptions, it is then immediately seen that the above estimator is unbiased for Γ_j, i.e.

$$E\widetilde{\Gamma}_j^{CP} = \sum_k E\left|\beta_{jk;sP}\right|^2 = \sum_l b^2 \left(\frac{l}{B^j}\right) \frac{2l+1}{4\pi} C_l .$$

We are actually able to establish a stronger result, namely

Proposition 12.15 *As $j \to \infty$, we have*

$$\frac{\widetilde{\Gamma}_j^{CP} - \Gamma_j}{\sqrt{Var\left\{\widetilde{\Gamma}_j^{CP}\right\}}} \overset{law}{\to} N(0,1) , \ Var\left\{\widetilde{\Gamma}_j^{CP}\right\} = O(B^{2(1-\min(\alpha,\gamma))j}) .$$

We omit also this (standard) proof for brevity's sake. We can repeat here the same comments as in the previous subsection, concerning the possibility of estimating the normalizing variance by subsampling techniques, along the lines of [14], and the roles of α, γ for the rate of convergence $\left\{\widetilde{\Gamma}_j^{CP} - \Gamma_j\right\} = O(B^{(1-\min(\alpha,\gamma))j})$.

12.7.3 Hausman test for noise misspecification

In the previous two subsections, we have considered two alternate estimators for the angular power spectrum, in the presence of observational noise. As we observed in Chapter 8, it is a standard result (compare [167]) that the auto-power spectrum estimator enjoys a smaller variance, provided of course that the model for noise is correct. Loosely speaking, we can hence conclude that the auto-power spectrum estimator is more efficient when noise is correctly specified, while the cross-power spectrum estimator is more robust, as it does not depend on any previous knowledge on the noise angular power spectrum. Once again, an obvious question at this stage is whether the previous results can be exploited to implement a procedure to search consistently for noise misspecification. The answer is indeed positive, as it was shown in [73] along the lines of the procedure suggested by [167] for the scalar case.

Proposition 12.16 *Under Assumptions 12.3 and 12.13 , we have*

$$\frac{\widetilde{\Gamma}_{j;s}^{CP} - \widetilde{\Gamma}_{j;s}^{AP}}{\sqrt{Var\left\{\widetilde{\Gamma}_{j;s}^{CP} - \widetilde{\Gamma}_{j;s}^{AP}\right\}}} \overset{law}{\to} N(0,1),$$

where

$$Var\left\{\widetilde{\Gamma}_{j;s}^{CP} - \widetilde{\Gamma}_{j;s}^{AP}\right\} = O(B^{2(1-\gamma)j})$$

Proof The proof is again quite standard, and we only need to provide the main details. Notice first that

$$\widetilde{\Gamma}_{j;s}^{CP} - \widetilde{\Gamma}_{j;s}^{AP} = \frac{1}{D(D-1)} \sum_k \sum_{r_1 \neq r_2} \beta_{jk;sr_1} \bar{\beta}_{jk;sr_2} - \frac{1}{D} \sum_k \sum_r \left\{ |\beta_{jk;sr}|^2 - E|\beta_{jk;sN_r}|^2 \right\}$$

$$= \frac{1}{D(D-1)} \sum_k \left\{ (D-1) \sum_r E|\beta_{jk;sN_r}|^2 - \sum_{r_1 \neq r_2} |\beta_{jk;sr_1} - \beta_{jk;sr_2}|^2 \right\},$$

and applying again the Diagram Formula, we have that

$$Var\left(\widetilde{\Gamma}_{j;s}^{CP} - \widetilde{\Gamma}_{j;s}^{AP}\right)$$

$$= \frac{1}{D^2(D-1)^2} \sum_{k_1,k_2} \sum_{r_1 \neq r_2, r_3 \neq r_4,} \left| E\left(\beta_{jk_1;sr_1} - \beta_{jk_1;sr_2}\right)\left(\bar{\beta}_{jk_2;sr_3} - \bar{\beta}_{jk_2;sr_4}\right)\right|^2.$$

Similarly to the discussion for (12.22), we can show that

$$Var\left(\widetilde{\Gamma}_{j;s}^{CP} - \widetilde{\Gamma}_{j;s}^{AP}\right) = O\left(D^2 B^{2(1-\gamma))j}\right).$$

Once again, the next step is to consider the fourth order cumulants,

$$Cum_4 \left\{ \sum_k \left(\sum_{r_1 \neq r_2} |\beta_{jk;sr_1} - \beta_{jk;sr_2}|^2 - (D-1) \sum_r E|\beta_{jk;sN_r}|^2 \right) \right\}$$

$$= 6 \sum_{k_1,\ldots,k_4} \sum_{r_{2n} \neq r_{2n-1}, n=1,\ldots,4} E\left(\beta_{jk_1;sr_1} - \beta_{jk_1;sr_2}\right)\left(\bar{\beta}_{jk_2;sr_3} - \bar{\beta}_{jk_2;sr_4}\right)$$

$$\times E\left(\beta_{jk_2;sr_3} - \beta_{jk_2;sr_4}\right)\left(\bar{\beta}_{jk_3;sr_5} - \bar{\beta}_{jk_3;sr_6}\right) E\left(\beta_{jk_3;sr_5} - \beta_{jk_3;sr_6}\right)\left(\bar{\beta}_{jk_4;sr_7} - \bar{\beta}_{jk_4;sr_8}\right)$$

$$\times E\left(\beta_{jk_4;sr_7} - \beta_{jk_4;sr_8}\right)\left(\bar{\beta}_{jk_1;sr_1} - \bar{\beta}_{jk_1;sr_2}\right)$$

$$\leq C_M D^4 \left(\Gamma_{j;s}^N\right)^4 \sum_{k_1,\ldots,k_4} \frac{\lambda_{jk_1} \lambda_{jk_2} \lambda_{jk_3} \lambda_{jk_4}}{\left[1 + d(\xi_{jk_1}, \xi_{jk_2})\right]^M \left[1 + d(\xi_{jk_2}, \xi_{jk_3})\right]^M}$$

$$\times \frac{1}{\left[1 + d(\xi_{jk_3}, \xi_{jk_4})\right]^M \left[1 + d(\xi_{jk_4}, \xi_{jk_1})\right]^M}$$

$$\leq CD^4 B^{(2-4\gamma)j}.$$

in view of (12.29), choosing $M \geq 3$. Now it is easy to see that

$$Cum_4 \left\{ \frac{\widetilde{\Gamma}_{j;s}^{CP} - \widetilde{\Gamma}_{j;s}^{AP}}{\sqrt{Var\{\widetilde{\Gamma}_{j;s}^{CP} - \widetilde{\Gamma}_{j;s}^{AP}\}}} \right\} \to 0,$$

whence the Proposition is established, again using the results of Chapter 4. $\square$

Remark 12.17 Note that $Var\left\{\widetilde{\Gamma}_{j;s}^{CP}\right\}, Var\left\{\widetilde{\Gamma}_{j;s}^{AP}\right\}, 2Cov\left\{\widetilde{\Gamma}_{j;s}^{CP}, \widetilde{\Gamma}_{j;s}^{AP}\right\}$ are robust to misspecification of the noise, because variance and covariance are translation invariant. It follows that the denominator can (once again) be consistently estimated by subsampling techniques, as in [14].

Under the alternative of noise misspecification, we have easily

$$\frac{\widetilde{\Gamma}_{j;s}^{CP} - \widetilde{\Gamma}_{j;s}^{AP}}{\sqrt{Var\left\{\widetilde{\Gamma}_{j;s}^{CP} - \widetilde{\Gamma}_{j;s}^{AP}\right\}}} \xrightarrow{law} N(\delta_j, 1)$$

where

$$\delta_j := \frac{E|\beta_{jk;sN_r}|^2 - \Gamma_{j;sN_r}}{\sqrt{Var\left\{\widetilde{\Gamma}_{j;s}^{CP} - \widetilde{\Gamma}_{j;s}^{AP}\right\}}}$$

where $\Gamma_{j;sN_r}$ is the bias-correction term which is wrongly adopted. The derivation of the power properties of this testing procedure is then immediate.

Remark 12.18 (On Mixed Needlets.) The pure spin needlets we considered here are not feasible for some statistical applications: for instance, throughout this paper we have only been considering estimation and testing for the total angular power spectrum $C_l = C_{lE} + C_{lB}$. The separate estimation of the two components (E and B modes) is of great interest for physical applications, and can be achieved by focussing on the *mixed needlets* construction of [75]. In short, mixed needlets are defined as

$$\psi_{jk;sM}(x) = \sqrt{\lambda_{jk}} \sum_{l=|s|}^{\infty} b\left(\frac{l}{B^j}\right) \sum_m Y_{lm,s}(x)\overline{Y}_{lm}(\xi_{jk}),$$

with *mixed spin needlet coefficients given by*

$$\beta_{jk,sM} = \int_{S^2} f_s(x)\overline{\psi}_{jk;sM}(x)dx = \sqrt{\lambda_{jk}} \sum_l b\left(\frac{l}{B^j}\right) \sum_m a_{lm,s}Y_{lm}(\xi_{jk}) . \quad (12.30)$$

It is readily seen that

$$\beta_{jk;M} = \sqrt{\lambda_{jk}} \sum_{lm} b\left(\frac{l}{B^j}\right) \{a_{lm;E} + ia_{lm;B}\} Y_{lm}(\xi_{jk})$$

$$= \beta_{jk;E} + i\beta_{jk;B}$$

where $\beta_{jk;E}, \beta_{jk;B}$ are real, i.e. $\beta_{jk;E} = Re(\beta_{jk;B}), \beta_{jk;M} = Im(\beta_{jk;M})$. It is also possible to verify that $\beta_{jk;E}, \beta_{jk,B}$ are exactly the coefficients we would obtain by evaluating directly a scalar needlet transform on the scalar functions f_E, f_B. Moreover, mixed needlets enjoy analogous types of localization and uncorrelation properties as for the scalar and pure spin cases, and hence they can be exploited, analogously to what we did here, for angular power spectrum estimation, Gaussianity testing and several other applications which have been already detailed earlier in this monograph. More generally, several applications can be envisaged to adaptive estimation, for instance nonparametric regression with a view to applications for polarization and weak gravitational lensing, see [59]. These developments, however, are beyond the scope of this monograph.

13

Appendix

In this Appendix, we report for completeness some explicit characterizations of orthogonal polynomials and spherical harmonics. These expressions are included to make the book as self-contained as possible, and as a reference for applied researchers involved in numerical work. Most of the results below are taken from [1, 193, 195].

13.1 Orthogonal polynomials

13.1.1 Jacobi polynomials

For $x \in [1, -1]$, we introduce the weight function

$$w_{\alpha,\beta}(x) := (1 - x)^{\alpha}(1 + x)^{\beta} , \ \alpha,\beta > -1 .$$

The *Jacobi polynomials* are defined as the family $\left\{ P_l^{(\alpha,\beta)}(x) : l = 0, 1, 2, ... \right\}$, such that

$$w_{\alpha,\beta}(x) \times P_l^{(\alpha,\beta)}(x) = \frac{1}{2^l l!} \frac{d^l}{dx^l} \left\{ w_{\alpha,\beta}(x) \left(x^2 - 1 \right)^l \right\} , \qquad (13.1)$$

which is known as Rodrigues' formula, see for instance [193, p. 67]. Given an arbitrary polynomial $Q_{l'}$, $l' < l$, by iterating integration by parts it is simple to see that

$$\int_{-1}^{1} P_l^{(\alpha,\beta)}(x) Q_{l'}(x) w_{\alpha,\beta}(x) dx = 0 ,$$

i.e. $P_l^{(\alpha,\beta)}(x)$ is orthogonal to all polynomials of order smaller than l, with respect to the measure $w_{\alpha,\beta}(x)dx$. The scaling factor is chosen to ensure the standard normalization

$$P_l^{(\alpha,\beta)}(1) = \left(\begin{array}{c} l + \alpha \\ l \end{array} \right) .$$

In [193, formula (4.3.3)] it is also shown that

$$\int_{-1}^{1} P_{l_1}^{(\alpha,\beta)}(x) P_{l_2}^{(\alpha,\beta)}(x) w_{\alpha,\beta}(x) dx$$

$$= \delta_{l_1}^{l_2} \left\{ \frac{2^{\alpha+\beta+1}}{2l_1+\alpha+\beta+1} \frac{\Gamma(l_1+\alpha+1)\Gamma(l_1+\beta+1)}{\Gamma(l_1+\alpha+\beta+1)l_1!} \right\}. \tag{13.2}$$

Jacobi polynomials satisfy the symmetry relations

$$P_l^{(\alpha,\beta)}(x) = (-1)^l P_l^{(\beta,\alpha)}(-x) \,, \; P_l^{(\alpha,\beta)}(-1) = (-1)^l \binom{l+\beta}{l} \,.$$

We have also the analytic expressions (see [193, formula (4.3.2)])

$$P_l^{(\alpha,\beta)}(x) = \sum_{s=0}^{l} \binom{l+\alpha}{s} \binom{l+\beta}{l-s} \left(\frac{x-1}{2}\right)^{l-s} \left(\frac{x+1}{2}\right)^{s},$$

while for $l+\alpha, l+\beta, l+\alpha+\beta$ nonnegative integers we can write

$$P_l^{(\alpha,\beta)}(x) = (l+\alpha)!(l+\beta)! \sum_{s} \frac{1}{s!(l+\alpha-s)!(\beta+s)!(l-s)!} \left(\frac{x-1}{2}\right)^{l-s} \left(\frac{x+1}{2}\right)^{s},$$

where the sum extends over all integer s such that the factorials are nonnegative. Setting $\alpha = \beta$, we obtain the *Gegenbauer*, or *ultraspherical* polynomials , defined as

$$G_l^{\alpha}(x) = \frac{\Gamma(2\alpha+l)\Gamma(\alpha+\frac{1}{2})}{\Gamma(2\alpha)\Gamma(\alpha+l+\frac{1}{2})} P_l^{(\alpha-\frac{1}{2},\alpha-\frac{1}{2})}(x) \,.$$

Let us now provide an example of applications of Jacobi's polynomials to the analysis of orthogonality of Wigner's $D(.)$ functions. Recall first from (3.15) that

$$d_{mn}^{l}(\vartheta) = \left[\frac{(l-n)!(l+n)!}{(l-m)!(l+m)!} \right]^{\frac{1}{2}} \left(\sin\frac{\vartheta}{2}\right)^{m-n} \left(\cos\frac{\vartheta}{2}\right)^{m+n} P_{l-m}^{(m-n,m+n)}(\cos\vartheta) \,.$$

In (3.29) we proved that

$$\int_{SU(2)} D_{mn}^{l}(g) \overline{D}_{m'n'}^{l'}(g) dg = \frac{1}{2l+1} \delta_l^{l'} \delta_m^{m'} \delta_n^{n'} \,.$$

This is a special case of Schur's orthogonality relationships. However, it can be instructive to provide an analytic proof, as follows. Note that

$$\frac{1}{16\pi^2} \int_0^{2\pi} \int_0^{\pi} \int_{-2\pi}^{2\pi} D_{m_1 n_1}^{l_1}(\varphi,\vartheta,\psi) \overline{D}_{m_2 n_2}^{l_2}(\varphi,\vartheta,\psi) \sin\vartheta \, d\psi d\vartheta d\varphi$$

$$= \frac{1}{16\pi^2} \int_0^{2\pi} \int_0^{\pi} \int_{-2\pi}^{2\pi} \exp\{i(m_2 - m_1)\varphi\} d^{l_1}_{m_1 n_1}(\vartheta) d^{l_2}_{m_2 n_2}(\vartheta)$$

$$\times \exp\{i(n_2 - n_1)\psi\} \sin\vartheta \, d\psi \, d\vartheta \, d\varphi$$

$$= \frac{\delta^{m_2}_{m_1} \delta^{n_2}_{n_1}}{2} \int_0^{\pi} d^{l_1}_{m_1 n_1}(\vartheta) d^{l_2}_{m_2 n_2}(\vartheta) d\cos\vartheta \, .$$

It is hence sufficient to focus on $m_1 = m_2 = m$, $n_1 = n_2 = n$: using (3.21), we have

$$\frac{1}{2} \int_0^{\pi} d^{l_1}_{mn}(\vartheta) d^{l_2}_{mn}(\vartheta) d\cos\vartheta \tag{13.3}$$

$$= 2^{-2m-1} \left[\frac{(l_1 - m)!(l_1 + m)!}{(l_1 - n)!(l_1 + n)!} \right]^{\frac{1}{2}} \left[\frac{(l_2 - m)!(l_2 + m)!}{(l_2 - n)!(l_2 + n)!} \right]^{\frac{1}{2}} \tag{13.4}$$

$$\times \int_{-1}^{1} w_{m-n,m+n}(x) P^{(m-n,m+n)}_{l_1-m}(x) P^{(m-n,m+n)}_{l_2-m}(x) dx \, . \tag{13.5}$$

Now from (3.19) (or, equivalently, (13.2)), we deduce that this integral is identically zero unless $l_1 = l_2 = l$, and in this case

$$(13.5) = 2^{-2m-1} \left[\frac{(l - m)!(l + m)!}{(l - n)!(l + n)!} \right] \frac{2^{2m+1}}{2l + 1} \frac{\Gamma(l - n + 1)\Gamma(l + n + 1)}{\Gamma(l + m + 1)(l - m)!} = \frac{1}{2l + 1} \, ,$$

as expected. Note that for all $l \geq 1$

$$\int_{SO(3)} D^l_{mn}(g) dg = \int_{SO(3)} D^l_{mn}(g) \overline{D^0_{00}(g)} dg = 0 \, .$$

The special case $P^{(0,0)}_l(x) = G^{\frac{1}{2}}_l(x)$ produces Legendre polynomials, to which the next subsection is devoted.

13.1.2 Legendre polynomials

As noted before, Jacobi polynomials for $\alpha = \beta = 0$ provide by definition the so-called *Legendre polynomials* $\{P_l : l = 0, 1, ...\}$, which hence satisfy

$$P_l(x) = \frac{1}{2^l l!} \frac{d^l}{dx^l} (x^2 - 1)^l \, ,$$

while (13.2) becomes

$$\int_{-1}^{1} P_l(x) P_{l'}(x) dx = \frac{2}{2l + 1} \delta^{l'}_l \, .$$

Legendre polynomials can also be defined for $-1 \leq x \leq 1$, $l = 0, 1, 2, ...$ by the generating formula (see [1, Section 22.9])

$$\frac{1}{\sqrt{(1 - 2xs + s^2)}} = \sum_{l=0}^{\infty} P_l(x)s^l ,$$

or by means of *Legendre's equation*, (see [1, Section 22.6]), i.e.

$$(1 - x^2)\frac{d^2 P_l(x)}{dx^2} - 2x\frac{dP_l(x)}{dx} + l(l + 1)P_l(x) = 0 .$$

They also satisfy the following recurrence relationship

$$(l + 1)P_{l+1}(x) = (2l + 1)xP_l(x) - lP_{l-1}(x) , \forall l \in N , x \in [-1, 1] ,$$

which can be rewritten as

$$(2l + 1)(x - 1)P_l(x) = (l + 1)P_{l+1}(x) - (2l + 1)P_l(x) + lP_{l-1}(x) . \qquad (13.6)$$

Also, we have the differential recursive relationship (see [1, Section 22.8])

$$(1 - x^2)\frac{dP_l(x)}{dx} = -lxP_l(x) + lP_{l-1}(x) .$$

Note also that $P_l(x) = (-1)^l P_l(-x)$, for all l, x. The first few Legendre polynomials are given by

$$P_0(x) = 1 , P_1(x) = x ,$$

$$P_2(x) = \frac{1}{2}(3x^2 - 1) , P_3(x) = \frac{1}{2}(5x^2 - 3x) ,$$

$$P_4(x) = \frac{1}{8}(35x^4 - 30x^2 + 3) , P_5(x) = \frac{1}{8}(63x^5 - 70x^3 + 15x) .$$

13.1.3 Associated Legendre functions

The content of this section can be completely deduced from [1, Ch. 8]. For an integer $l \geq 0$, the *associated Legendre functions* are defined as

$$P_l^m(x) = \frac{(-1)^m}{2^l l!}(1 - x^2)^{m/2}\frac{d^{l+m}}{dx^{l+m}}(x^2 - 1)^l , \text{ for } m = -l, ..., l$$

or also, for $m \geq 0$

$$P_l^m(x) = (-1)^m(1 - x^2)^{m/2}\frac{d^m}{dx^m}P_l(x) , P_l^{-m} = (-1)^m\frac{(l - m)!}{(l + m)!}P_l^m . \qquad (13.7)$$

Associated Legendre functions also satisfy following recurrence relation

$$(2l + 1) xP_l^m = (l - m + 1) P_{l+1}^m + (l + m) P_{l-1}^m ,$$

$$\left(1 - x^2\right)\frac{dP_l^m}{dx} = (l + 1)\,xP_l^m - (l - m + 1)\,P_{l+1}^m \;.$$

It should be noted that $P_l^m(x)$ is actually a polynomial only for even values of m. The relationship with Jacobi polynomials is given by (for $m \geq 0$)

$$P_l^m(x) = \frac{(-1)^m}{2^m l!}\,(l + m)!(1 - x)^{m/2}(1 + x)^{m/2}P_{l-m}^{(m,m)}(x)$$

i.e., setting $x = \cos\vartheta$

$$P_l^m(\cos\vartheta) = \frac{(-1)^m}{2^m l!}\,(l + m)!\sin^m\vartheta P_{l-m}^{(m,m)}(\cos\vartheta) \;.$$

Expressions for negative values of m can be found by symmetry arguments, i.e. exploting (13.7) to obtain ($m \leq 0$)

$$P_l^m = P_l^{-|m|} = (-1)^m\frac{(l - |m|)!}{(l + |m|)!}P_l^{|m|} = \frac{1}{2^{|m|}l!}\,(l + m)!\sin^{|m|}\vartheta P_{l-|m|}^{(|m|,|m|)}(\cos\vartheta) \;.$$

The first few associated Legendre functions are provided by

$$P_0^0(x) = 1 \;,\; P_1^{-1}(x) = \frac{1}{2}(1 - x^2)^{\frac{1}{2}} \;,\; P_1^0(x) = x \;,\; P_1^1(x) = -(1 - x^2)^{\frac{1}{2}} \;,$$

$$P_2^{-2}(x) = \frac{1}{8}(1 - x^2) \;,\; P_2^{-1}(x) = \frac{1}{2}(1 - x^2)^{\frac{1}{2}}x \;,\; P_2^0(x) = \frac{1}{2}(3x^2 - 1) \;,$$

$$P_2^1(x) = -3(1 - x^2)^{\frac{1}{2}}x \;,\; P_2^2(x) = 3(1 - x^2) \;,$$

or after the identification $x = \cos\vartheta$

$$P_0^0(\cos\vartheta) = 1 \;,\; P_1^{-1}(\cos\vartheta) = \frac{1}{2}\sin\vartheta \;,$$

$$P_1^0(\cos\vartheta) = \cos\vartheta \;,\; P_1^1(\cos\vartheta) = -\sin\vartheta \;,$$

$$P_2^{-2}(\cos\vartheta) = \frac{1}{8}\sin^2\vartheta \;,\; P_2^{-1}(\cos\vartheta) = \frac{1}{2}\sin\vartheta\cos\vartheta \;,$$

$$P_2^0(\cos\vartheta) = \frac{1}{2}(3\cos^2\vartheta - 1) \;,$$

$$P_2^1(\cos\vartheta) = -3\sin\vartheta\cos\vartheta \;,\; P_2^2(\cos\vartheta) = 3\sin^2\vartheta \;.$$

The associated Legendre functions form an orthogonal set of functions on the interval $[-1, 1]$. Thus

$$\int_{-1}^1 P_l^m(x)P_{l'}^m(x)dx = \frac{2}{2l + 1}\frac{(l + m)!}{(l - m)!}\delta_{l'}^l \;.$$

The following alternative notation is often found in the literature:

$$P_{lm}(x) = (-1)^m P_l^m(x) \;.$$

13.1.4 Chebyshev polynomials

Define the Chebyshev polynomials of the first and second kind as, respectively

$$T_n(\cos\varphi) = \cos n\varphi \, , \, U_n(\cos\varphi) = \frac{\sin(n+1)\varphi}{\sin\varphi} \, .$$

It is simple to show that $T_n(x)$ and $U_n(x)$ are indeed polynomials of degree n in $x = \cos\varphi$; indeed from standard trigonometric identities we have ([197], Chapter 6.9)

$$\cos(n+1)\varphi = \cos n\varphi\cos\varphi - \sin n\varphi\sin\varphi \, ,$$
$$\cos(n-1)\varphi = \cos n\varphi\cos\varphi + \sin n\varphi\sin\varphi \, ,$$

whence

$$T_{n+1}(x) = 2xT_n(x) - T_{n-1}(x) \, ,$$

i.e.

$$T_0(x) = 1 \, , \, T_1(x) = x \, , \, T_2(x) = 2x^2 - 1 \, , \, T_3(x) = 4x^3 - 2x^2 - 2x + 1 \, , \ldots$$

Also

$$\frac{d}{dx}T_n(x) = nU_{n-1}(x) \, ,$$

so that $\{U_n\}$ is also a family of polynomials such that U_n has degree n. We have moreover the relations

$$\int_{-1}^{1} T_n(x)T_m(x)(1-x^2)^{-\frac{1}{2}}dx = \int_0^\pi \cos n\varphi\cos m\varphi d\varphi = \begin{cases} 0 & \text{if } m \neq n \\ \pi & \text{if } m = n = 0 \\ \pi/2 & \text{if } m = n \geq 1 \end{cases}$$

$$\int_{-1}^{1} U_n(x)U_m(x)(1-x^2)^{\frac{1}{2}}dx = \int_0^\pi \sin(n+1)\varphi\sin(m+1)\varphi d\varphi = \frac{\pi}{2}\delta_m^n \, ,$$

so that (up to a normalization factor) Chebyshev polynomials are actually Jacobi polynomials of parameters $\alpha = \beta = -\frac{1}{2}$ and $\alpha = \beta = \frac{1}{2}$, respectively (see [193, formula (4.17)]).

13.2 Spherical harmonics and their analytic properties

The relations discussed in this section can be all retrieved from [195, Chapter 5]. We recall first that the spherical harmonics are defined explicitly as follows:

for $0 \le \vartheta \le \pi, 0 \le \varphi < 2\pi$,

$$Y_{lm}(\vartheta, \varphi) = \exp(im\varphi) \sqrt{\frac{2l+1}{4\pi} \frac{(l-m)!}{(l+m)!}} P_l^m(\cos \vartheta)$$

$$= (-1)^m \exp(im\varphi) \sqrt{\frac{2l+1}{4\pi} \frac{(l-m)!}{(l+m)!}} P_{lm}(\cos \vartheta) , m = -l, ..., l .$$

In terms of the Legendre polynomials spherical harmonics can be expressed as

$$Y_{lm}(\vartheta, \varphi) = (-1)^m \exp(im\varphi) \sqrt{\frac{2l+1}{4\pi} \frac{(l-m)!}{(l+m)!}} \sin^m \vartheta \frac{d^m}{(d\cos\vartheta)^m} P_l(\cos \vartheta) , (m \ge 0)$$

(see [195, formula (5.2.1.6)]) while in terms of Jacobi polynomials we get, for $m \ge 0$

$$Y_{lm}(\vartheta, \varphi) = (-1)^m \frac{\exp(im\varphi)}{2^m l!} \sqrt{\frac{2l+1}{4\pi}} (l+m)!(l-m)! \sin^m \vartheta P_{l-m}^{(m,m)}(\cos \vartheta) .$$

(see [195, formula (5.2.7.38)]). The following symmetry properties hold:

$$\overline{Y}_{lm}(\vartheta, \varphi) = (-1)^m Y_{l-m}(\vartheta, \varphi) = Y_{lm}(\vartheta, -\varphi)$$

and (see [195, Section 5.5])

$$Y_{lm}(\pi - \vartheta, \varphi) = (-1)^{l+m} Y_{lm}(\vartheta, \varphi) ,$$
$$Y_{lm}(\vartheta, \pi + \varphi) = (-1)^m Y_{lm}(\vartheta, \varphi) ,$$
$$Y_{lm}(\pi - \vartheta, \pi + \varphi) = (-1)^l Y_{lm}(\vartheta, \varphi) ,$$
$$Y_{lm}(-\vartheta, \varphi) = (-1)^m Y_{lm}(\vartheta, \varphi) ,$$
$$Y_{lm}(\vartheta, -\varphi) = (-1)^m Y_{l-m}(\vartheta, \varphi) ,$$
$$Y_{lm}(-\vartheta, -\varphi) = Y_{l-m}(\vartheta, \varphi) .$$

13.2.1 Power series representations

In the sequel, we report a few representations of the spherical harmonics as power series of trigonometric functions of ϑ or $\vartheta/2$ – see [195, Chapter 5] for much more results in this area. It should be noted that each of these expressions includes only a finite number of non-zero summands, so that they can be implemented on a computer without numerical approximations. More explicitly, all the sums below are defined over arguments such that the arguments of the factorials are nonnegative. Thus, $Y_{lm}(\vartheta, \varphi)$ can be written as (see [195, Sections 5.2.2 - 5.2.3])

$$(-1)^m e^{im\varphi} \sqrt{\frac{2l+1}{4\pi} \frac{(l+m)!}{(l-m)!}} \left(\tan\frac{\vartheta}{2}\right)^m \sum_s (-1)^s \frac{(l+s)!}{(l-s)!} \frac{\left(\sin\frac{\vartheta}{2}\right)^{2s}}{s!(s+m)!}$$

$$= (-1)^m e^{im\varphi} \sqrt{\frac{2l+1}{4\pi} \frac{(l-m)!}{(l+m)!}} \left(\sin\frac{\vartheta}{2}\cos\frac{\vartheta}{2}\right)^m \sum_s (-1)^s \frac{(l+m+s)!}{(l-m-s)!} \frac{\left(\sin\frac{\vartheta}{2}\right)^{2s}}{s!(s+m)!}$$

$$= (-1)^l e^{im\varphi} \sqrt{\frac{2l+1}{4\pi} \frac{(l+m)!}{(l-m)!}} \left(\cot\frac{\vartheta}{2}\right)^m \sum_s (-1)^s \frac{(l+s)!}{(l-s)!} \frac{\left(\cos\frac{\vartheta}{2}\right)^{2s}}{s!(s+m)!}$$

$$= (-1)^l e^{im\varphi} \sqrt{\frac{2l+1}{4\pi} \frac{(l-m)!}{(l+m)!}} \left(\sin\frac{\vartheta}{2}\cos\frac{\vartheta}{2}\right)^m \sum_s (-1)^s \frac{(l+m+s)!}{(l-m-s)!} \frac{\left(\cos\frac{\vartheta}{2}\right)^{2s}}{s!(s+m)!}.$$

13.2.2 Differential and integral relationships

Spherical harmonics satisfy the differential equation

$$\Delta_{S^2} Y_{lm} + l(l+1)Y_{lm} = \left\{\frac{1}{\sin\vartheta}\frac{\partial}{\partial\vartheta}\left(\sin\vartheta\frac{\partial}{\partial\vartheta}\right) + \frac{1}{\sin^2\vartheta}\frac{\partial^2}{\partial\varphi^2}\right\} Y_{lm} + l(l+1)Y_{lm} = 0.$$

Repeated substitutions show that the following expression provides $2l+1$ linearly independent solutions: for $m = -l, ..., l$,

$$Y_{lm}(\vartheta, \varphi) = \frac{\exp(im\varphi)}{2^l l!} \sqrt{\frac{2l+1}{4\pi}\frac{(l+m)!}{(l-m)!}} (\sin\vartheta)^{-m} \frac{d^{l-m}}{(d\cos\vartheta)^{l-m}} (\cos^2\vartheta - 1)^l.$$

In terms of definite integrals, particularly important are the so-called Mehler-Dirichlet formulae (see [195, Section 5.3.1])

$$Y_{lm}(\vartheta, \varphi) = (-1)^m \frac{\sqrt{2}}{\pi} \exp(im\varphi) \sqrt{\frac{2l+1}{4\pi}\frac{(l-m)!}{(l+m)!}} (2m-1)!! \left(\frac{\sin\vartheta}{2}\right)^m$$

$$\times \int_0^\vartheta \frac{\cos\left[(2l+1)\psi/2\right]d\psi}{(\cos\psi - \cos\vartheta)^{m+1/2}}, \ m \geq 0, \tag{13.8}$$

$$Y_{lm}(\vartheta, \varphi) = \frac{\sqrt{2}}{\pi} \exp(im\varphi) \sqrt{\frac{2l+1}{4\pi}\frac{(l-m)!}{(l+m)!}} (2m-1)!! \left(\frac{\sin\vartheta}{2}\right)^m$$

$$\times \int_\vartheta^\pi \frac{\sin\left[(2l+1)\psi/2\right]d\psi}{(\cos\psi - \cos\vartheta)^{m+1/2}}, \ m \geq 0. \tag{13.9}$$

For $m = 0$, we set $(2m - 1)!! = 1$ and these equations provide a representation for Legendre polynomials which (following [143]) will be used below to establish localization properties of spherical needlets.

13.3 The proof of needlets' localization.

In this section, the representation properties of Legendre polynomials are used to provide the proof of the fundamental localization result established by Narcowich, Petrushev and Ward in [143]. This is a slight reformulation of the localization result provided in Proposition 10.5. In particular, we consider the following kernel:

$$\Psi_\varepsilon(\cos \vartheta) := \sum_l \kappa\left(\varepsilon\left(l + \frac{1}{2}\right)\right) \frac{2l + 1}{4\pi} P_l(\cos \vartheta) \, , \, 0 < \varepsilon < 1 \, ,$$

where $\kappa(.)$ is an even, positive function, such that for some $M \in \mathbb{N}$, all $r = 0, 1, 2, ..., M$, there exist positive constants C_r such

$$\left|\kappa^{(r)}(x)\right| \leq \frac{C_r}{(1 + |x|)^{M+2-r}} \, .$$

We note that this assumption are trivially satisfied by $b(.)$ in Proposition 10.5, which is compactly supported, belongs to C^M and can be extended to the negative axis simply taking $b(-|x|) := b(|x|)$. In the definition of the needlets, ε should be viewed as $\varepsilon = B^{-j}$ and a normalizing factor of order $\sqrt{\lambda_{jk}} \approx \varepsilon$ is included; a minor difference which is easily accounted for is the presence of an additive $\frac{1}{2}$ term in the argument of $\kappa(.)$. Note that for the argument below the function $\kappa(.)$ is not requred to have compact support, nor to be zero at the origin.

The proof below follows the argument in [143] and will require some well-known auxiliary results from standard Fourier analysis; we refer to [190] for background material. To this aim, for any function $f : \mathbb{R} \to \mathbb{R}$, $f \in L^1(\mathbb{R})$, we define its Fourier trasform as usual as

$$\mathcal{F}[f](\omega) = \widehat{f}(\omega) := \int_\mathbb{R} f(x) \exp(-ix\omega) dx \, .$$

We recall the following standard properties: assuming that

$$\int_\mathbb{R} |x|^r |f(x)| \, dx, \, \int_\mathbb{R} |\omega|^r \left|\widehat{f}(\omega)\right| d\omega < \infty \, ,$$

we have

$$\frac{d^r}{d\omega^r} \widehat{f}(\omega) = \int_\mathbb{R} f(x) \frac{d^r}{d\omega^r} \exp(-ix\omega) dx$$

$$= (-i)^r \int_{\mathbb{R}} x^r f(x) \exp(-ix\omega) dx$$
$$= (-i)^r \mathcal{F}[x^r f(x)](\omega) \,,$$

and

$$\mathcal{F}[(f^r)](\omega) = i^r \omega^r \mathcal{F}[f](\omega) \,.$$

Moreover we shall also exploit the well-known *Poisson Summation formula*, (see for instance [190, Chapter 7] or [94, Chapter 4]). Assume that

$$|f(x)| + \left|\widehat{f}(x)\right| \le \frac{C}{1 + |x|^{\alpha+1}} \,, \alpha > 0 \,. \tag{13.10}$$

We have that, for all $\omega \in [0, 2\pi]$,

$$\sum_{\tau=-\infty}^{\infty} f(\tau) \exp(-i\omega\tau) = \sum_{\mu=-\infty}^{\infty} \widehat{f}(\omega + 2\mu\pi) \,.$$

A simple proof can be provided as follows; define

$$\widehat{S}(\omega) := \sum_{\mu=-\infty}^{\infty} \widehat{f}(\omega + 2\mu\pi) \,;$$

clearly $\widehat{S}(\omega)$ is a 2π periodic function, with Fourier coefficients

$$\int_{-\pi}^{\pi} \widehat{S}(\omega) \exp(-i\omega\tau) d\omega = \int_{-\pi}^{\pi} \sum_{\mu=-\infty}^{\infty} \widehat{f}(\omega + 2\mu\pi) \exp(-i\omega\tau) d\omega$$

$$= \sum_{\mu=-\infty}^{\infty} \int_{-\pi}^{\pi} \widehat{f}(\omega + 2\mu\pi) \exp(-i\omega\tau) d\omega$$

$$= \sum_{\mu=-\infty}^{\infty} \int_{(2\mu-1)\pi}^{(2\mu+1)\pi} \widehat{f}(\omega) \exp(-i\omega\tau) d\omega$$

$$= \int_{-\infty}^{\infty} \widehat{f}(\omega) \exp(-i\omega\tau) d\omega$$

$$= 2\pi \mathcal{F}^{-1}[\widehat{f}](-\tau)$$

$$= 2\pi f(-\tau) \,.$$

Hence we have pointwise, under (13.10)

$$\widehat{S}(\omega) = \sum_{\tau} f(-\tau) \exp(i\omega\tau) \,,$$

and the result follows. The localization result we shall now provide is the following

Theorem 13.1 *There exist some positive constant C_M such that, for all $0 <$ $\varepsilon < 1$, $0 \leq \vartheta \leq \pi$,*

$$\Psi_\varepsilon(\cos \vartheta) \leq \frac{C_M \varepsilon^{-2}}{1 + \left(\frac{\vartheta}{\varepsilon}\right)^M} .$$

Proof We recall first the Mehler-Dirichlet representation formula for Legendre polynomials (see 13.9 or [193], page 85-86), stating that

$$\begin{aligned}
P_l(\cos \vartheta) &= \frac{\sqrt{2}}{\sqrt{\pi}\Gamma(\frac{1}{2})} \int_\vartheta^\pi \frac{\cos\left((2l+1)\phi/2 - \pi/2\right)}{\sqrt{\cos \vartheta - \cos \phi}} d\phi \\
&= \frac{\sqrt{2}}{\pi} \int_\vartheta^\pi \frac{\sin\left(\frac{(2l+1)\phi}{2}\right)}{\sqrt{\cos \vartheta - \cos \phi}} d\phi .
\end{aligned}$$

We can hence write

$$\Psi_\varepsilon(\cos \vartheta) = \frac{\sqrt{2}}{2\pi^2} \int_\vartheta^\pi \frac{K_\varepsilon(\phi)}{\sqrt{\cos \vartheta - \cos \phi}} d\phi ,$$

where

$$\begin{aligned}
K_\varepsilon(\phi) &:= \frac{1}{2\pi} \sum_{l=0}^\infty \kappa\left(\varepsilon\left(l + \frac{1}{2}\right)\right) \frac{2l+1}{2} \sin\left(\frac{(2l+1)\phi}{2}\right) \\
&= \frac{1}{4\pi} \sum_{l=-\infty}^\infty \kappa\left(\varepsilon\left(l + \frac{1}{2}\right)\right) \frac{2l+1}{2} \sin\left(\frac{(2l+1)\phi}{2}\right) .
\end{aligned}$$

Define also

$$g(u) = \frac{1}{2\pi} \kappa(\varepsilon u) u \sin(u\phi) ,$$

a function which is even in u and has a zero in $u = 0$. We have immediately

$$K_\varepsilon(\phi) = \sum_{l=0}^\infty g\left(l + \frac{1}{2}\right) = \frac{1}{2} \sum_{\mu=-\infty}^\infty g\left(\mu + \frac{1}{2}\right) .$$

Now

$$\begin{aligned}
\int_{\mathbb{R}} g\left(\mu + \frac{1}{2}\right) \exp(-i\mu\omega) d\mu &= \widehat{g}(\omega) \exp\left(i\frac{\omega}{2}\right) \\
&= \frac{1}{2\pi i} \frac{1}{\varepsilon} \exp\left(i\frac{\omega}{2}\right) \left(i\frac{d}{d\omega}\right) \widehat{\kappa}\left(\frac{\phi + \omega}{\varepsilon}\right) ,
\end{aligned}$$

and hence by Poisson summation formula

$$|K_\varepsilon(\phi)| = \frac{1}{4\pi\varepsilon} \left| \sum_{\nu=-\infty}^\infty \exp(i\nu\pi) \frac{d}{d\omega} \widehat{\kappa}\left(\frac{\phi + \omega}{\varepsilon}\right) \right|_{\omega=2\pi\nu} \right|$$

$$= \frac{1}{4\pi\varepsilon} \left| \sum_{\nu=-\infty}^{\infty} (-1)^{\nu} \frac{d}{d\phi} \widehat{\kappa} \left(\frac{\phi + 2\pi\nu}{\varepsilon} \right) \right| .$$

Now write

$$\frac{d}{d\phi} \widehat{\kappa} \left(\frac{\phi + 2\pi\nu}{\varepsilon} \right) = \widehat{\kappa}^{(1)} \left(\frac{\phi + 2\pi\nu}{\varepsilon} \right) .$$

Standard properties of the Fourier transform yield

$$\mathcal{F} \left(\frac{d^r}{dt^r} \{t\kappa(t)\} \right) (\phi) = i^{r+1} \phi^r \widehat{\kappa}^{(1)}(\phi) ,$$

whence

$$\left| \frac{\phi + 2\pi\nu}{\varepsilon} \right|^r \left| \widehat{\kappa}^{(1)} \left(\frac{\phi + 2\pi\nu}{\varepsilon} \right) \right| \leq \frac{1}{\varepsilon} \left\| \frac{d^r}{dt^r} \{t\kappa(t)\} \right\|_{L^1} .$$

Call

$$V_{M,\kappa} := \max_{r \leq M} \left\| \frac{d^r}{dt^r} \{t\kappa(t)\} \right\|_{L^1} ;$$

choosing $r = 0$ and $r = M$ we have hence proved that

$$\left| \widehat{\kappa}^{(1)} \left(\frac{\phi + \omega}{\varepsilon} \right) \right| \leq \frac{2V_{M,\kappa}/\varepsilon}{1 + \left| \frac{\phi+\omega}{\varepsilon} \right|^M}$$

and

$$|K_{\varepsilon}(\phi)| \leq \frac{1}{4\pi\varepsilon} \sum_{\nu=-\infty}^{\infty} \frac{2V_{M,\kappa}/\varepsilon}{1 + \left| \frac{\phi+2\pi\nu}{\varepsilon} \right|^M} \leq \frac{1}{2\pi} \sum_{\nu=-\infty}^{\infty} \frac{V_{M,\kappa}\varepsilon^{-2}}{1 + \left| \frac{\phi+2\pi\nu}{\varepsilon} \right|^M} ,$$

which holds for all $\phi \in \mathbb{R}$. In the sequel, c denotes a positive constant which may vary from line to line. For our range of interest $\phi \in [0, 2\pi]$, it can be easily verified that the dominant term corresponds to $\nu = 0$, and estimating the remaining series as an integral yields, after some manipulations

$$|K_{\varepsilon}(\phi)| \leq \frac{cV_{M,\kappa}\varepsilon^{-2}}{1 + \left| \frac{\phi}{\varepsilon} \right|^M} .$$

Hence we have

$$\Psi_{\varepsilon}(\cos\vartheta) \leq \int_{\vartheta}^{\pi} \frac{cV_{M,\kappa}\varepsilon^{-2}}{1 + \left| \frac{\phi}{\varepsilon} \right|^M} \frac{1}{\sqrt{\cos\vartheta - \cos\phi}} d\phi .$$

Consider first the case $\varepsilon \leq \vartheta \leq \frac{\pi}{2}$; we use the trigonometric identity

$$\cos\vartheta - \cos\phi = 2 \sin\frac{\phi + \vartheta}{2} \sin\frac{\phi - \vartheta}{2} .$$

Using the fact that $\frac{\sin u}{u}$ is decreasing for $u \in [0, \pi]$, it is simple to check that

$$\frac{\partial}{\partial \vartheta} \left\{ \frac{\sin \frac{\phi+\vartheta}{2}}{\frac{\phi+\vartheta}{2}} \times \frac{\sin \frac{\phi-\vartheta}{2}}{\frac{\phi-\vartheta}{2}} \right\}, \frac{\partial}{\partial \varphi} \left\{ \frac{\sin \frac{\phi+\vartheta}{2}}{\frac{\phi+\vartheta}{2}} \times \frac{\sin \frac{\phi-\vartheta}{2}}{\frac{\phi-\vartheta}{2}} \right\} < 0$$

whence we obtain,

$$\cos \vartheta - \cos \phi = 2(\phi^2 - \vartheta^2) \frac{\sin \frac{\phi+\vartheta}{2}}{\frac{\phi+\vartheta}{2}} \frac{\sin \frac{\phi-\vartheta}{2}}{\frac{\phi-\vartheta}{2}}$$

$$\geq 2(\phi^2 - \vartheta^2) \frac{\sin \frac{3}{4}\pi}{\frac{3}{4}\pi} \frac{\sin \frac{\pi}{4}}{\frac{\pi}{4}}$$

$$\geq 2(\phi^2 - \vartheta^2) \frac{2\sqrt{2}}{3\pi} \times \frac{2\sqrt{2}}{\pi} .$$

We have

$$\Psi_\varepsilon(\cos \vartheta) \leq c V_{M,\kappa} \varepsilon^{-2} \int_\vartheta^\pi \frac{1}{1 + \left|\frac{\phi}{\varepsilon}\right|^M} \frac{1}{\sqrt{\phi^2 - \vartheta^2}} d\phi$$

$$= \frac{c V_{M,\kappa} \varepsilon^{-2}}{\left|\frac{\vartheta}{\varepsilon}\right|^M} \int_1^{\pi/\vartheta} \frac{1}{1 + u^M} \frac{1}{\sqrt{u^2 - 1}} du$$

$$\leq \frac{c V_{M,\kappa} \varepsilon^{-2}}{\left|\frac{\vartheta}{\varepsilon}\right|^M} \int_1^\infty \frac{1}{1 + u^M} \frac{1}{\sqrt{u^2 - 1}} du$$

$$\leq \frac{c V_{M,\kappa} \varepsilon^{-2}}{\left|\frac{\vartheta}{\varepsilon}\right|^M} \int_0^1 \frac{1}{\sqrt{1 - u^2}} du \leq \frac{c V_{M,\kappa} \varepsilon^{-2}}{\left|\frac{\vartheta}{\varepsilon}\right|^M} \frac{\pi}{2} .$$

On the other hand, for $\pi \geq \vartheta \geq \frac{\pi}{2}$ we set $\widetilde{\phi} = \pi - \phi$, $\widetilde{\vartheta} = \pi - \vartheta$, which yields

$$\Psi_\varepsilon(\cos \vartheta) \leq \int_0^{\widetilde{\vartheta}} \frac{c V_{M,\kappa} \varepsilon^{-2}}{1 + \left|\frac{\pi-\widetilde{\phi}}{\varepsilon}\right|^M} \frac{1}{\sqrt{\cos \widetilde{\phi} - \cos \widetilde{\vartheta}}} d\widetilde{\phi}$$

where, in our domain of interest

$$0 \leq \frac{\widetilde{\vartheta} + \widetilde{\phi}}{2} \leq \frac{\pi}{2} , 0 \leq \frac{\widetilde{\vartheta} - \widetilde{\phi}}{2} \leq \frac{\pi}{4} .$$

Thus

$$\cos \widetilde{\phi} - \cos \widetilde{\vartheta} = 2(\widetilde{\vartheta}^2 - \widetilde{\phi}^2) \frac{\sin \frac{\widetilde{\vartheta}-\widetilde{\phi}}{2}}{\frac{\widetilde{\vartheta}-\widetilde{\phi}}{2}} \frac{\sin \frac{\widetilde{\vartheta}+\widetilde{\phi}}{2}}{\times \frac{\widetilde{\vartheta}+\widetilde{\phi}}{2}}$$

$$\geq 2(\widetilde{\vartheta}^2 - \widetilde{\phi}^2) \frac{2}{\pi} \frac{2 \times \sqrt{2}}{\pi} ,$$

and

$$\Psi_\varepsilon(\cos\vartheta) \le cV_{M,\kappa}\varepsilon^{-2} \int_0^{\widetilde{\vartheta}} \frac{1}{1+\left|\frac{\pi-\widetilde{\phi}}{\varepsilon}\right|^M} \frac{1}{\sqrt{(\widetilde{\vartheta}^2-\widetilde{\phi}^2)}} d\widetilde{\phi}$$

$$\le \frac{cV_{M,\kappa}\varepsilon^{-2}}{1+\left|\frac{\vartheta}{\varepsilon}\right|^M} \int_0^{\widetilde{\vartheta}} \frac{1}{\sqrt{(\widetilde{\vartheta}^2-\widetilde{\phi}^2)}} d\widetilde{\phi}$$

$$= \frac{cV_{M,\kappa}\varepsilon^{-2}}{1+\left|\frac{\vartheta}{\varepsilon}\right|^M} \int_0^1 \frac{1}{\sqrt{1-u^2}} du = \frac{cV_{M,\kappa}\varepsilon^{-2}}{1+\left|\frac{\vartheta}{\varepsilon}\right|^M} \frac{\pi}{2}.$$

Finally, for $0 \le \vartheta \le \varepsilon$ it is enough to show that

$$|K_\varepsilon(\phi)| \le \varepsilon^{-2} \sum_{l=0}^{\infty} \kappa\left(\varepsilon\left(l+\frac{1}{2}\right)\right) \varepsilon^2\left(l+\frac{1}{2}\right)$$

$$\le c\varepsilon^{-2} \sum_{l=0}^{\infty} \frac{C_r}{(1+\varepsilon l)^{M+2}} \varepsilon l \int_{\varepsilon l}^{\varepsilon(l+1)} dx$$

$$\le c\varepsilon^{-2} \int_0^{\infty} \frac{1}{(1+y)^{M+1}} dy \le c\varepsilon^{-2}.$$

Hence for all $0 \le \vartheta \le \pi$, some $C_M > 0$ we have the bound

$$\Psi_\varepsilon(\cos\vartheta) \le \frac{C_M\varepsilon^{-2}}{1+\left(\frac{\vartheta}{\varepsilon}\right)^M},$$

as claimed. $\quad\square$

References

[1] Abramowitz, M., Stegun, I. (1964) *Handbook of Mathematical Functions*, Dover.

[2] Adler, R.J. (1981) *The Geometry of Random Fields*, J. Wiley.

[3] Adler, R.J., Taylor, J.E. (2007) *Random Fields and Geometry*, Springer-Verlag.

[4] Anderes, E., Chatterjee, S. (2009) Consistent estimates of deformed isotropic Gaussian random fields on the plane, *Annals of Statistics*, 37, No. 5A, 2324-2350.

[5] Antoine, J.-P., Vandergheynst, P. (1999) Wavelets on the sphere: a group-theoretic approach, *Applied and Computational Harmonic Analysis*, 7, 262-291.

[6] Antoine, J.-P., Vandergheynst, P. (2007) Wavelets on the sphere and other conic sections, *Journal of Fourier Analysis and its Applications*, 13, 369-386.

[7] Arjunwadkar, M., Genovese, C.R., Miller, C.J., Nichol, R.C., Wasserman, L. (2004) Nonparametric inference for the Cosmic Microwave Background, *Statistical Science*, 19, 308-321.

[8] Babich, D., Creminelli, P., Zaldarriaga, M. (2004) The shape of non-Gaussianities, *Journal of Cosmology and Astroparticle Physics*, 8, 009.

[9] Balbi, A. (2007), *The Music of the Big Bang*, Springer-Verlag.

[10] Baldi, P., Marinucci, D. (2007). Some characterizations of the spherical harmonics coefficients for isotropic random fields, *Statistics & Probability Letters*, 77(5), 490-496.

[11] Baldi, P., Marinucci, D., Varadarajan, V.S. (2007) On the characterization of isotropic random fields on homogeneous spaces of compact groups, *Electronic Communications in Probability*, 12, 291-302.

[12] Baldi, P., Kerkyacharian, G., Marinucci, D., Picard, D. (2008) High frequency asymptotics for wavelet-based tests for Gaussianity and isotropy on the torus, *Journal of Multivariate Analysis*, 99(4), 606–636.

[13] Baldi, P., Kerkyacharian, G., Marinucci, D., Picard, D. (2009) Asymptotics for Spherical Needlets, *Annals of Statistics*, 37(3), 1150–1171, arxiv:math/0606599.

[14] Baldi, P., Kerkyacharian, G. Marinucci, D., Picard, D. (2009) Subsampling Needlet Coefficients on the Sphere, *Bernoulli*, 15(2), 438-463, arxiv 0706.4169.

[15] Baldi, P., Kerkyacharian, G., Marinucci, D., Picard, D. (2009) Density estimation for directional data using needlets, *Annals of Statistics*, 37(6A), 3362–3395.

[16] Baldi, P., Kerkyacharian, G., Marinucci, D., Picard, D. (2009) Besov spaces for sections of spin fiber bundles on the sphere, preprint.

[17] Balkar, E., Lovesey, S.W. (2009), *Introduction to the Graphical Theory of Angular Momentum*, Springer Tracts on Modern Physics, Springer.

[18] Bartolo, N. , Komatsu, E., Matarrese, S., Riotto, A. (2004). Non-Gaussianity from inflation: theory and observations, *Physical Reports*, 402, 103-266.

[19] Bartolo, N., Matarrese, S., Riotto, A. (2010) Non-Gaussianity and the Cosmic Microwave Background anisotropies, *Advances in Astronomy*, in press, arXiv: 1001.3957.

[20] Bartolo, N., Fasiello, M., Matarrese, S., Riotto, A. (2010) Large non-Gaussianities in the effective field theory approach to single-field inflation: the bispectrum, *Journal of Cosmology and Astroparticle Physics*, 1008:08, arXiv: 1004.0893.

[21] Bennett, C. L., Halpern, M., Hinshaw, G., Jarosik, N., Kogut, A., Limon, M., Meyer, S. S., Page, L., Spergel, D. N., Tucker, G. S., Wollack, E., Wright, E. L., Barnes, C., Greason, M. R., Hill, R. S., Komatsu, E., Nolta, M. R., Odegard, N., Peiris, H. V., Verde, L., Weiland, J. L. (2003) First -Year Wilkinson Microwave Anisotropy Probe (WMAP) observations: preliminary maps and basic results, *Astrophysical Journal Supplement Series*, Volume 148, Issue 1, pp. 1-27.

[22] Bennett, C. L., Hill, S., Hinshaw, G., Larson, D., Smith, K. M., Dunkley, J., Gold, B., Halpern, M., Jarosik, N., Kogut, A., Komatsu, E., Limon, M., Meyer, S. S., Nolta, M. R., Odegard, N., Page, L., Spergel, D. N., Tucker, G. S., Weiland, J. L., Wollack, E., Wright, E. L. (2010) Seven-year Wilkinson Microwave Anisotropy Probe (WMAP) observations: are there Cosmic Microwave Background anomalies?, arXiv: 1001.4758.

[23] Biedenharn, L.C., Louck J.D. (1981) *The Racah-Wigner Algebra in Quantum Theory*, Encyclopedia of Mathematics and its Applications, Volume 9, Addison-Wesley.

[24] Billingsley, P. (1968) *Convergence of Probability Measures*, J. Wiley.

[25] Bishop, R.L., Goldberg, S. (1980) *Tensor Analysis on Manifolds*, Dover.

[26] Blei, R. (2001) *Analysis in Integer and Fractional Dimensions*, Cambridge University Press.

[27] Breuer, P., Major, P. (1983) Central limit theorems for nonlinear functionals of Gaussian fields, *Journal of Multivariate Analysis*, 13, no. 3, 425-441.

[28] Bridles, S. et al. (2009) Handbook for the GREAT08 Challenge: An image analysis competition for cosmological lensing, *Annals of Applied Statistics*, Vol. 3, No. 1, 6-37.

[29] Brillinger, D. W. (1975) *Time series. Data Analysis and Theory*, Holt, Rinehart and Winston.

[30] Brockwell, P.J., Davis, R.A. (1991) *Time Series: Theory and Methods*, Second edition, Springer Series in Statistics, Springer-Verlag.

[31] Brocker, T., tom Dieck, T. (1985) *Representations of Compact Lie Groups*, Graduate Texts in Mathematics, 98, Springer-Verlag.

[32] Bump, D. (2005) *Lie Groups*, Graduate Texts in Mathematics, 225, Springer-Verlag.

[33] Cabella, P., Kamionkowskii, M. (2005) *Theory of Cosmic Microwave Background Polarization*, Lectures given at the 2003 Villa Mondragone School of Gravitation and Cosmology: "The Polarization of the Cosmic Microwave Background," Rome, arxiv: astro.ph/0403392.

[34] Cabella, P., Hansen, F.K., Marinucci, D., Pagano, D., Vittorio, N. (2004) Search for non-Gaussianity in pixel, harmonic, and wavelet space: compared and combined, *Physical Review D*, 69, 063007.

[35] Cabella, P., Hansen, F.K., Liguori, M., Marinucci, D., Matarrese, S., Moscardini, L., Vittorio, N. (2005) Primordial non-Gaussianity: local curvature method and statistical significance of constraints on f_{NL} from WMAP data, *Monthly Notices of the Royal Astronomical Society*, Vol. 358, pp. 684-692.

[36] Cabella, P., Hansen, F.K., Liguori, M., Marinucci, D., Matarrese, S., Moscardini, L., Vittorio, N. (2006) The integrated bispectrum as a test of CMB non-Gaussianity: detection power and limits on f_{NL} with WMAP data, *Monthly Notices of the Royal Astronomical Society*, 369, 819-824, arxiv:astro-ph/0512112.

[37] Cabella, P., Marinucci, D. (2009) Statistical challenges in the analysis of Cosmic Microwave Background radiation, *Annals of Applied Statistics*, 3(1), 61-95.

[38] Chambers, D., Slud, E. (1989) Necessary conditions for nonlinear functionals of Gaussian processes to satisfy central limit theorems, *Stochastic Processes and their Applications*, 32(1), 93–107.

[39] Chambers, D., Slud, E. (1989) Central Limit Theorems for nonlinear functionals of stationary Gaussian processes, *Probability Theory and Related Fields*, 80(3), 323–346.

[40] Cruz, M., Cayon, L., Martinez-Gonzalez, E., Vielva, P., Jin, J., (2007) The non-Gaussian Cold Spot in the 3-year WMAP Data, *Astrophysical Journal*, 655 , 11-20.

[41] Cruz, M., Cayon, L., Martinez-Gonzalez, E., Vielva, P., (2006) The non-Gaussian Cold Spot in WMAP: significance, morphology and foreground contribution, *Monthly Notices of the Royal Astronomical Society*, 369, 57-67.

[42] Dahlke, S., Steidtl, G., Teschke, G. (2007) Frames and coorbit theory on homogeneous spaces with a special guidance on the sphere, *Journal of Fourier Analysis and its Applications*, 13, 387-404.

[43] Davidson, J. (1994), *Stochastic Limit Theory*, Oxford University Press.

[44] De Bernardis, P. et al. (2000) A flat Universe from high-resolution maps of the Cosmic Microwave Background radiation, *Nature*, Vol. 404, Issue 6781, pp. 955-959.

[45] de Gasperis, G., Balbi, A., Cabella, P., Natoli, P., Vittorio, N. (2005) ROMA: A map-making algorithm for polarised CMB data sets, *Astronomy and Astrophysics*, Vol. 436, Issue 3, pp.1159-1165.

[46] Delabrouille, J., Cardoso, J.-F., Le Jeune, M., Betoule, M., Fay, G., Guilloux, F. (2009) A full sky, low foreground, high resolution CMB map from WMAP, *Astronomy and Astrophysics*, Vol. 493, Issue 3, pp.835-857, arXiv:0807.0773.

[47] Dennis, M. (2004), Canonical representation of spherical functions: Sylvester's theorem, Maxwell's multipoles and Majorana's sphere, *Journal of Physics A*, 37, 9487-9500.

[48] Dennis, M. (2005) Correlations between Maxwell's multipoles for Gaussian random functions on the sphere, *Journal of Physics A*, 38, 1653-1658.

[49] Diaconis, P. (1988) *Group Representations in Probability and Statistics*, IMS Lecture Notes – Monograph Series, 11, Hayward.

[50] Diaconis, P., Freedman, D. (1987) A dozen de Finetti-style results in search of a theory, *Annales Institute Henri Poincaré Probabilités et Statistiques*, 23(2), 397–423.

[51] Dodelson, S. (2003) *Modern Cosmology*, Academic Press.

[52] Doré, O., Colombi, S., Bouchet, F.R. (2003) Probing non-Gaussianity using local curvature, *Monthly Notices of the Royal Astronomical Society*, 344, 905-916.

[53] Doroshkevich, A.G., Naselsky, P.D., Verkhodanov, O.V. , Novikov, D.I., Turchaninov, V.I., Novikov, I.D. , Christensen, P.R. , Chiang, L.-Y. (2005) Gauss-Legendre Sky Pixelization (GLESP) for CMB Maps, *International Journal of Modern Physics D*, 14, 275.

[54] Doukhan, P. (1988) Formes de Toeplitz associées à une analyse multi-échelle, (French) [Toeplitz forms associated to a multiscale analysis] *Comptes Rendus de l'Académie des Sciences. Série I. Mathématique*, 306, no.15, 663–666.

[55] Doukhan, P., Leon, J. R. (1990) Formes quadratique d'estimateurs de densité par projections orthogonales. (French) [Quadratic deviation of projection density estimates] *Comptes Rendus de l'Académie des Sciences. Série I. Mathématique*, 310, no. 6, 425–430.

[56] Dudley, R.M. (2002) *Real Analysis and Probability*, revised reprint of the 1989 original, Cambridge Studies in Advanced Mathematics, 74, Cambridge University Press.

[57] Duistermaat, J.J., Kolk, J.A.C. (1997) *Lie Groups*, Springer-Verlag.

[58] Duffin, R.J., Schaeffer, A.C. (1952) A class of nonharmonic Fourier series, *Transactions of the American Mathematical Society*, 72, 341-366.

[59] Durastanti, C., Geller, D., Marinucci, D. (2010) *Nonparametric Regression on Spin fiber Bundles*, under revision for the *Journal of Multivariate Analysis*, arXiv preprint 1009.4345.

[60] Durrer, R. (2008) *The Cosmic Microwave Background*, Cambridge University Press.

[61] Efstathiou, G. (2004) Myths and truths concerning estimation of power spectra: the case for a hybrid estimator, *Monthly Notices of the Royal Astronomical Society*, 349, Issue 2, pp. 603-626.

[62] Eriksen, H.K., Hansen, F.K., Banday, A.J., Gorski, K.M., Lilje, P.B. (2004) Asymmetries in the CMB anisotropy field, *Astrophysical Journal*, 605, 14-20.

[63] Faraut, J. (2006) *Analyse sur le Groupes de Lie*, Calvage et Mounet.

[64] Faÿ, G., Guilloux, F., Betoule, M., Cardoso, J.-F., Delabrouille, J., Le Jeune, M. (2008) CMB power spectrum estimation using wavelets *Physical Review D*, 78:083013.

[65] Faÿ, G., Guilloux, F. (2008) Consistency of a Needlet Spectral Estimator on the Sphere, arXiv:0807.2162.

[66] Feller, W. (1970) *An Introduction to Probability Theory and its Applications*, Volume II, 2nd Edition, J. Wiley.

[67] Fergusson, J.R., Liguori, M., Shellard, E.P.S. (2009) General CMB and Primordial Bispectrum Estimation I: Mode Expansion, Map-Making and Measures of f_{NL}, arXiv: 0912.5516.

[68] Fergusson, J.R., Liguori, M., Shellard, E.P.S. (2010) The CMB Bispectrum, arXiv: 1006.1642.

[69] Foulds, L.R. (1992) *Graph Theory and Applications*, Universitext, Springer-Verlag.

[70] Freeden, W., Schreiner, M. (1998) Orthogonal and nonorthogonal multiresolution analysis, scale discrete and exact fully discrete wavelet transform on the sphere. *Constructive Approximations*, 14, 4, 493–515.

[71] Gangolli, R. (1967) Positive definite kernels on homogeneous spaces and certain stochastic processes related to Lévy's Brownian motion of several parameters. *Annales de l'Institut H. Poincaré Sect. B*, Vol. 3, 121-226.

[72] Geller, D., Hansen, F.K., Marinucci, D., Kerkyacharian, G., Picard, D. (2008), Spin needlets for Cosmic Microwave Background Polarization data analysis, *Physical Review D*, D78:123533, arXiv:0811.2881.

[73] Geller, D., Lan, X., Marinucci, D. (2009) Spin needlets spectral estimation, *Electronic Journal of Statistics*, Vol. 3, 1497-1530, arXiv:0907.3369.

[74] Geller, D., Marinucci, D. (2008) Spin wavelets on the sphere, *Journal of Fourier Analysis and its Applications*, Vol. 16, Issue 6, pages 840-884, arXiv: 0811.2835.

[75] Geller, D., Marinucci, D. (2011) Mixed needlets, *Journal of Mathematical Analysis and Applications*, Vol. 375, n.2, pp.610-630, arXiv: 1006.3835.

[76] Geller, D., Mayeli, A. (2009) Continuous wavelets on manifolds, *Mathematische Zeitschrift*, Vol. 262, pp. 895-927, arXiv: math/0602201.

[77] Geller, D., Mayeli, A. (2009) Nearly Tight frames and space-frequency analysis on compact manifolds, *Mathematische Zeitschrift*, Vol. 263 (2009), pp. 235-264, arXiv: 0706.3642.

[78] Geller, D., Mayeli, A. (2009) Besov spaces and frames on compact manifolds, *Indiana University Mathematics Journal*, Vol. 58, pp. 2003-2042, arXiv:0709.2452.

[79] Geller, D., Mayeli, A. (2009) Nearly tight frames of spin wavelets on the sphere, *Sampling Theory in Signal and Image Processing*, in press, arXiv:0907.3164.

[80] Geller, D., Pesenson, I . (2010), Band-limited localized Parseval frames and Besov spaces on compact homogeneous manifolds, *Journal of Geometric Analysis*, in press, arXiv:1002.3841.

[81] Genovese, C.R., Perone-Pacifico, M., Verdinelli, I., Wasserman, L. (2009) On the path density of a gradient field. *Annals of Statistics*, 37(6A), 3236–3271.

[82] Genovese, C.R., Perone-Pacifico, M., Verdinelli, I., Wasserman, L. (2010) Nonparametric filament estimation, arXiv:1003.5536.

[83] Ghosh, T., Delabrouille, J., Remazeilles, M., Cardoso, J.-F., Souradeep, T. (2010) Foreground maps in WMAP frequency bands, arxiv: 1006.0916.

[84] Goldberg, J.N., Newman, E.T., (1967) Spin-s Spherical Harmonics and ð, *Journal of Mathematical Physics*, 8(11), 2155-2166.

[85] Gorski, K. M. , Hivon, E., Banday, A. J., Wandelt, B. D., Hansen, F. K. , Reinecke, M. , Bartelman, M., (2005) HEALPix – A framework for high resolution discretization, and fast analysis of data distributed on the sphere, *Astrophysical Journal*, 622, 759-771.

[86] Gradshteyn, I. S., Ryzhik, I. M. (1980) *Table of Integrals, Series, and Products*, Academic Press.

[87] Guilloux, F., Fay, G., Cardoso, J.-F. (2008) Practical wavelet design on the sphere, *Applied and Computational Harmonic Analysis*, 26, no. 2, 143–160.

[88] Guionnet, A. (2009) Large random matrices: lectures on macroscopic asymptotics. Lecture Notes in Mathematics, Vol. 1957, Springer-Verlag.

[89] Guivarc'h, Y. , Keane, M. and Roynette, B. (1977) *Marches Aléatoires sul les Groupes de Lie*, Lecture Notes in Mathematics, Vol. 624, Springer-Verlag.

[90] Hamann, Jan, Wong, Yvonne Y. Y. (2008) The effects of Cosmic Microwave Background (CMB) temperature uncertainties on cosmological parameter estimation , *Journal of Cosmology and Astroparticle Physics*, Issue 03, pp. 025.

[91] Hanany, S., Ade, P., Balbi, A., Bock, J., Borrill, J., Boscaleri, A., de Bernardis, P., Ferreira, P. G., Hristov, V. V., Jaffe, A. H., Lange, A. E., Lee, A. T., Mauskopf, P. D., Netterfield, C. B., Oh, S., Pascale, E., Rabii, B., Richards, P. L., Smoot, G. F., Stompor, R., Winant, C. D., Wu, J. H. P. (2000) MAXIMA-1: A measurement of the Cosmic Microwave Background anisotropy on angular scales of 10'-5°, *The Astrophysical Journal*, Vol. 545, Issue 1, L5-L9.

[92] Hannan, E.J. (1970) *Multiple Time Series*. J. Wiley.

[93] Hansen, F.K., Cabella, P., Marinucci, D., Vittorio, N. (2004) Asymmetries in the local curvature of the WMAP data, *Astrophysical Journal Letters*, L67-L70.

[94] Hardle, W., Kerkyacharian, G., Picard, D. and Tsybakov, A. (1998) *Wavelets, Approximation, and Statistical Applications*, Springer Lecture Notes in Statistics, 129.

[95] Hausman, J.A. (1978) Specification tests in econometrics, *Econometrica*, 6, 1251-1271.

[96] Havin, V. and Joricke, B. (1994) *The Uncertainty Principle in Harmonic Analysis*, Springer-Verlag.

[97] Hernandez, E., Weiss, G. (1996) *A First Course on Wavelets*, Studies in Advanced Mathematics, CRC Press.

[98] Hikage, C., Matsubara, T., Coles, P., Liguori, M., Hansen, F.K., Matarrese, S. (2008) Primordial non-Gaussianity from Minkowski functionals of the WMAP temperature anisotropies, *Monthly Notices Royal Astronomical Society*, 389:1439-1446.

[99] Hinshaw, G., Weiland, J. L., Hill, R. S., Odegard, N., Larson, D., Bennett, C. L., Dunkley, J., Gold, B., Greason, M. R., Jarosik, N., Komatsu, E., Nolta, M. R., Page, L., Spergel, D. N., Wollack, E., Halpern, M., Kogut, A., Limon, M., Meyer, S. S., Tucker, G. S., Wright, E. L. (2009) Five-Year Wilkinson Microwave Anisotropy Probe (WMAP) observations: data processing, sky maps, and basic results, *Astrophysical Journal Supplement Series*, 180:225-245.

[100] Hivon, E., Gorski, K.M., Netterfield, C.B., Crill, B.P., Prunet, S., Hansen F.K. (2002) MASTER of the Cosmic Microwave Background anisotropy power spectrum: a fast method for statistical analysis of large and complex Cosmic Microwave Background data sets, *Astrophysical Journal*, Volume 567, Issue 1, pp. 2-17.

[101] Holschneider, M., Iglewska-Nowak., I. (2007) Poisson wavelets on the sphere *Journal of Fourier Analysis and its Applications*, 13, 405 - 420.

[102] Hu, W. (2001) The angular trispectrum of the Cosmic Microwave Background, *Physical Review D*, Volume 64, Issue 8, id.083005.

[103] Hu, Y., Nualart, D. (2005) Renormalized self-intersection local time for fractional Brownian motion, *The Annals of Probability*, 33(3), 948-983.

[104] Ivanov, A.V., Leonenko, N.N. (1989), *Statistical Analysis of Random Fields*, Kluwer.

[105] Jansson, S. (1997) *Gaussian Hilbert Spaces*, Cambridge University Press.

[106] Johnson, N.L., Kotz S.J. (1972) *Distributions in Statistics: Continuous Multivariate Distributions*, J. Wiley.

[107] Kagan, A.M., Linnik, Y.V., Rao, C.R. (1973) *Characterization Problems in Mathematical Statistics*, J. Wiley.

[108] Kamionkowski, M., Kosowski, A., Stebbins, A. (1997) Statistics of Cosmic Microwave Background Polarization, *Physical Review D*, 55, 7368-7388.

[109] Keihänen, E., Kurki-Suonio, H., Poutanen, T. (2005) MADAM- a map-making method for CMB experiments, *Monthly Notices of the Royal Astronomical Society*, Vol. 360, Issue 1, pp. 390-400.

[110] Kerkyacharian, G., Petrushev, P., Picard, D., Willer, T. (2007) Needlet algorithms for estimation in inverse problems, *Electronic Journal of Statistics*, 1, 30-76.

[111] Kerkyacharian, G., Nickl, R., Picard, D. (2010) Concentration inequalities and confidence bands for needlet density estimators on compact homogeneous manifolds, *Probability Theory and Related Fields*, in press, arXiv:1102.2450.

[112] Kerkyacharian, G., Pham Ngoc, T.M., Picard, D. (2009) Localized spherical deconvolution, *Annals of Statistics*, in press, arXiv: 0908.1952.

[113] Kim, P.T., Koo, J.-Y. (2002) Optimal spherical deconvolution, *Journal of Multivariate Analysis*, 80, 21-42.

[114] Kim, P.T., Koo, J.-Y., Luo, Z.-M. (2009) Weyl eigenvalue asymptotics and sharp adaptation on vector bundles, *Journal of Multivariate Analysis*, 100, 1962-1978.

[115] Kitching, T. et al. (2010) Gravitational lensing accuracy testing 2010 (GREAT10) challenge handbook, preprint, arXiv: 1009.0779.

[116] Kolb, E., Turner, M. (1994), *The Early Universe*, Cambridge University Press.

[117] Komatsu, E., Spergel, D.N. (2001) Acoustic signatures in the primary Microwave Background bispectrum, *Physycal Review D* , 63, 063002.

[118] Komatsu, E. Wandelt, B.D., Spergel, D.N., Banday, A.J., Gorski, K.M. (2002), Measurement of the Cosmic Microwave Backgroun bispectrum on the COBE DMR sky maps, *Astrophysical Journal*, 566, 19-29.

[119] Komatsu E., Yadav, A., Wandelt, B. (2007) Fast estimator of primordial non-Gaussianity from temperature and polarization anisotropies in the Cosmic Microwave Background, *Astrophysical Journal*, 664:680-686.

[120] Komatsu E., Yadav, A., Wandelt, B., Liguori, M., Hansen, F.K., Matarrese, S. (2008) Fast estimator of primordial non-Gaussianity from temperature and polarization anisotropies in the Cosmic Microwave Background II: partial sky coverage and inhomogeneous noise, *Astrophysical Journal* 678:578.

[121] Komatsu et al. (2009) Five-Year Wilkinson Microwave Anisotropy Probe observations: cosmological interpretation, *Astrophysical Journal Supplement Series*, 180, 2, 330-376.

[122] Koornwinder, T.H. (2008), Representations of $SU(2)$ and Jacobi polynomials, preprint, available online http://staff.science.uva.nl/ thk/edu/orthopoly.pdf.

[123] Lan, X., Marinucci, D. (2008) The needlets bispectrum, *Electronic Journal of Statistics*, 2, 332-367.

[124] Lan, X., Marinucci, D. (2009) On the dependence structure of wavelet coefficients for spherical random fields, *Stochastic Processes and their Applications*, 119, 3749-3766.

[125] Leonenko, N. (1999) *Limit Theorems for Random Fields with Singular Spectrum*, Kluwer.

[126] Leonenko, N., Sakhno L. (2009) On spectral representations of tensor random fields on the sphere, arXiv:0912.3389.

[127] Liboff, R.L. (1999) *Introductory Quantum Mechanics*, Addison-Wesley.

[128] Magnus, J.R., Neudecker, H. (1988) *Matrix Differential Calculus with Applications to Statistics and Econometrics*, J. Wiley.

[129] Malyarenko, A. (2009) Invariant random fields in vector bundles and application to cosmology, preprint arXiv: 0907.4620.

[130] Marinucci, D. (2004) Testing for non-Gaussianity on Cosmic Microwave Background radiation: a review, *Statistical Science*, 19, 294-307.

[131] Marinucci, D. (2006) High-resolution asymptotics for the angular bispectrum of spherical random fields, *Annals of Statistics*, 34, 1-41.

[132] Marinucci, D. (2008) A central limit theorem and higher order results for the angular bispectrum, *Probability Theory and Related Fields*, 141(3-4), 389-409.

[133] Marinucci, D., Piccioni, M. (2004) The empirical process on Gaussian spherical harmonics, *Annals of Statistics*, 32, 1261-1288.

[134] Marinucci, D., Peccati, G. (2008) High-frequency asymptotics for subordinated stationary fields on an Abelian compact group, *Stochastic Processes and their Applications*, 118 (4), 585-613.

[135] Marinucci, D., Peccati, G. (2010) Group representations and high-resolution Central Limit Theorems for subordinated spherical random fields, *Bernoulli*, 16, 798-824.

[136] Marinucci, D.; Peccati, G. (2010) Representations of *SO*(3) and angular polyspectra, *Journal of Multivariate Analysis*, 101, 77-100.

[137] Marinucci, D.; Peccati, G. (2010) Ergodicity and Gaussianity for spherical random fields, *Journal of Mathematical Physics*, 51, n. 4, 043301, 23 pp.

[138] Marinucci, D., Wigman, I. (2010) On the excursion sets of spherical Gaussian eigenfunctions, preprint, arXiv:1009.4367.

[139] Marinucci, D., Pietrobon, D., Balbi, A., Baldi, P., Cabella, P., Kerkyacharian, G., Natoli, P., Picard, D., Vittorio, N. (2008) Spherical needlets for CMB data analysis, *Monthly Notices of the Royal Astronomical Society*, Vol. 383, 539-545, arXiv: 0707.0844.

[140] Mayeli, A. (2010) Asymptotic uncorrelation for Mexican needlets, *Journal of Mathematical Analysis and Applications*, Vol. 363, Issue 1, pp. 336-344, arXiv: 0806.3009.

[141] McEwen, J.D., Vielva, P., Wiaux, Y., Barreiro, R.B., Cayon, L., Hobson, M.P., Lasenby, A.N., Martinez-Gonzalez, E., Sanz, J. (2007) Cosmological applications of a wavelet analysis on the sphere, *Journal of Fourier Analysis and its Applications*, 13, 495-510.

[142] Miller, W. *Topics in Harmonic Analysis with Applications to Radar and Sonar*, preprint, available online http://www.ima.umn.edu/ miller/radarla.pdf .

[143] Narcowich, F.J., Petrushev, P., Ward, J.D. (2006) Localized tight frames on spheres, *SIAM Journal of Mathematical Analysis* , 38, 2, 574–594.

[144] Narcowich, F.J., Petrushev, P., Ward, J.D. (2006) Decomposition of Besov and Triebel-Lizorkin spaces on the sphere, *Journal of Functional Analysis*, 238, 2, 530–564.

[145] Natoli, P., Degasperis, G., Marinucci, D., Vittorio, N. (2002) Non-iterative methods to estimate the in-flight noise properties of CMB detectors, *Astronomy and Astrophysics*, 383, pp. 1100-1112.

[146] Newman, E. T., Penrose, R. (1966) Note on the Bondi-Metzner-Sachs group, *Journal of Mathematical Physics*, 7, 863–870.

[147] Nourdin, I., Peccati, G. and Reinert, G. (2010) Invariance principles for homogeneous sums: universality of the Gaussian Wiener chaos, *Annals of Probability*, 38(5), 1947-1985.

[148] Nourdin, I., Peccati, G. (2009). Stein's method on Wiener chaos, *Probability Theory and Related Fields*, 145(1), 75-118.

[149] Nourdin, I., Peccati, G. (2009) Stein's method meets Malliavin calculus: a short survey with new estimates. In the volume: *Recent Advances in Stochastic Dynamics and Stochastic Analysis*, World Scientific.

[150] Nourdin, I., Peccati, G., Réveillac, A. (2008). Multivariate normal approximation using Stein's method and Malliavin calculus, *Annales de l'Institut H. Poincaré (B)*, 46(1), 45-58.

[151] Nualart, D. (2006) *The Malliavin Calculus and Related Topics. Second edition*, Springer-Verlag.

[152] Nualart, D., Peccati, G. (2005) Central limit theorems for sequences of multiple stochastic integrals, *Annals of Probability*, 33, 177-193.

[153] Patanchon, G., Delabrouille, J., Cardoso, J.-F., Vielva, P. (2005) CMB and foreground in WMAP first-year data, *Monthly Notices of the Royal Astronomical Society*, 364, pp. 1185-1194.

[154] Peacock, J.A. (1999) *Cosmological Physics*, Cambridge University Press.

[155] Peccati, G. (2001) On the convergence of multiple random integrals, *Studia Sc. Math. Hungarica*, 37, 429-470.

[156] Peccati, G. (2007) Gaussian approximations of multiple integrals, *Electronic Communications in Probability* 12, 350-364.

[157] Peccati, G., Pycke, J.-R. (2010) Decompositions of stochastic processes based on irreducible group representations, *Theory of Probability and Applications*, 54(2), 217-245.

[158] Peccati, G., Taqqu, M.S. (2008) Stable convergence of multiple Wiener-Itô integrals, *Journal of Theoretical Probability*, 21(3), 527-570.

[159] Peccati, G., M.S. Taqqu (2010) *Wiener Chaos: Moments, Cumulants and Diagrams. A Survey with Computer Implementation*, Springer-Verlag.

[160] Peccati, G., Tudor, C.A. (2005) Gaussian limits for vector-valued multiple stochastic integrals. In: *Séminaire de Probabilités XXXVIII*, 247-262, Springer Verlag.

[161] Peebles, J. (1993), *Principles of Cosmology*, Princeton University Press.

[162] Pietrobon, D., Balbi, A., Marinucci, D. (2006) Integrated Sachs-Wolfe effect from the cross correlation of WMAP 3-Year and the NRAO VLA Sky Survey Data: new results and constraints on dark energy, *Physical Review D*, 74, 043524.

[163] Pietrobon, D. Amblard, A., Balbi, A., Cabella, P., Cooray, A., Marinucci, D. (2008) Needlet detection of features in the WMAP CMB sky and the impact on anisotropies and hemispherical asymmetries, *Physical Review D*, vol. 78, Issue 10, id. 103504.

[164] Pietrobon, D. Amblard, A., Balbi, A., Cabella, P., Cooray, A., Vittorio, N. (2009) Constraints on primordial non-Gaussianity from a needlet analysis of the WMAP-5 data, *Monthly Notices of the Royal Astronomical Society*, Volume 396, Issue 3, pp. 1682-1688.

[165] Pietrobon, D. Amblard, A., Balbi, A., Cabella, P., Cooray, A., Vittorio, N. (2009) Needlet bispectrum asymmetries in the WMAP 5-year Data, *Monthly Notices of the Royal Astronomical Society*, L367, arXiv: 0905.3702.

[166] Pietrobon, D., Balbi, A., Cabella, P. Gorski, K. M. (2010) Needatool: A Needlet Analysis Tool for Cosmological Data Processing, *Astrophysical Journal*, 723, 1.

[167] Polenta, G., Marinucci, D., Balbi, A., De Bernardis, P., Hivon, E., Masi, S., Natoli, P., Vittorio, N. (2005) Unbiased estimation of angular power spectra, *Journal of Cosmology and Astroparticle Physics*, Issue 11, n.1, pp.1-17.

[168] Pycke, J.-R. (2007) A decomposition for invariant tests of uniformity on the sphere, *Proceedings of the American Mathematical Society*, 135, 2983-2993.

[169] Revuz, D., Yor, M. (1999) *Continuous Martingales and Brownian motion*, Third edition, Grundlehren der Mathematischen Wissenschaften [Fundamental Principles of Mathematical Sciences], 293, Springer-Verlag.

[170] Robinson, P.M. (1995) Log-periodogram regression of time series with long range dependence, *Annals of Statistics*, 23, 1048-1072.

[171] Robinson, P.M. (1995) Gaussian semiparametric estimation of long range dependence, *Annals of Statistics*, 23, 1630-1661.

[172] Rosca, D. (2007) Wavelet bases on the sphere obtained by radial projection, *Journal of Fourier Analysis and its Applications*, 13, 421-434.

[173] Rudin, W. (1962) *Fourier Analysis on Groups*, Wiley Classics Library, Wiley.

[174] Rudin, W. (1975) *Real and Complex Analysis*, McGraw-Hill.

[175] Rudjord, O., Hansen, F.K. Lan, X., Liguori, M., Marinucci, D., Matarrese, S. (2009) An estimate of the primordial non-Gaussianity parameter f_{NL} using the needlet bispectrum from WMAP, *The Astrophysical Journal*, 701:369-376, arXiv:0901.3154.

[176] Rudjord, O., Hansen, F.K. Lan, X., Liguori, M., Marinucci, D., Matarrese, S. (2010), Directional variations of the non-Gaussianity parameter f_{NL}, *Astrophysical Journal*, Vol. 708, 2, 1321-1325.

[177] Schreiber, M. (1969) Fermeture en probabilité de certains sous-espaces d'un espace L^2, *Zeitschrift für Wahrscheinlichkeitstheorie und Verwandte Gebiete* 14, 36-48.

[178] Schwartzman, A., Mascarenhas, W.F. and Taylor, J.E.T. (2008) Inference for eigenvalues and eigenvectors of Gaussian symmetric matrices, *Annals of Statistics*, 36, no. 6, 2886–2919.

[179] Scodeller S., Rudjord, O., Hansen, F.K., Marinucci, D., Geller, D., Mayeli, A. (2010) Introducing Mexican needlets for CMB analysis: Issues for practical applications and comparison with standard needlets, *Astrophysical Journal*, in press, arXiv: 1004.5576.

[180] Simon, B. (1996) *Representations of Finite and Compact Groups*, Graduate Studies in Mathematics, 10, American Mathematical Society.

[181] Seljak, U., Zaldarriaga, M. (1996) Line-of-Sight integration approach to Cosmic Microwave Background anisotropies, *Astrophysical Journal*, Vol.469, p.437.

[182] Senatore, L., Smith, K.M., Zaldarriaga, M. (2010) Non-Gaussianities in single field inflation and their optimal limits from the WMAP 5-year data, *Journal of Cosmology and Astroparticle Physics*, 1001:028.

[183] Serre, J.P. (1977) *Linear Representation of Finite Groups*, Springer-Verlag.

[184] Regan, D.M., Shellard, E.P.S. (2009), Cosmic string power spectrum, bispectrum and trispectrum, arXiv:0911.2491.

[185] Shigekawa, I. (1986) De Rham–Hodge–Kodaira's decomposition on an abstract Wiener space, *Journal of Mathematics of the Kyoto University*, 26, 191-202.

[186] Shyraev, A.N. (1984) *Probability*, Springer-Verlag.

[187] Smoot, G. F., Bennett, C. L., Kogut, A., Wright, E. L., Aymon, J., Boggess, N. W., Cheng, E. S., de Amici, G., Gulkis, S., Hauser, M. G., Hinshaw, G., Jackson, P. D., Janssen, M., Kaita, E., Kelsall, T., Keegstra, P., Lineweaver, C., Loewenstein, K., Lubin, P., Mather, J., Meyer, S. S., Moseley, S. H., Murdock, T., Rokke, L., Silverberg, R. F., Tenorio, L., Weiss, R., Wilkinson, D. T. (1992) Structure in the COBE differential microwave radiometer first-year maps, *Astrophysical Journal, Part 2 - Letters*, Vol. 396, no. 1, pp. L1-L5.

[188] Spergel, D.N. et al. (2003) First-Year Wilkinson Microwave Anisotropy Probe (WMAP) observations: determination of cosmological parameters, *Astrophysical Journal Supplement Series*, 148, 1, pp. 175-194.

[189] Spergel, D.N. et al. (2007) Three-Year Wilkinson Microwave Anisotropy Probe (WMAP) observations: implications for cosmology, *Astrophysical Journal Supplement Series*, 170, 2, 377-408.

[190] Stein, E.M., Weiss, G. (1971) *Introduction to Fourier Analysis on Euclidean Spaces*, Princeton University Press.

[191] Sternberg, S. *Group Theory and Physics*, Cambridge University Press.

[192] Surgailis, D. (2003) CLTs for polynomials of linear sequences: Diagram formula with illustrations. In *Theory and Applications of Long Range Dependence*, 111-128, Birkhäuser.

[193] Szego, G. (1975) *Orthogonal Polynomials*, American Mathematical Society Colloquium Publications, Volume 23 Reprinted version of the 1939 original.

[194] Varadarajan, V.S. (1999) *An Introduction to Harmonic Analysis on Semisimple Lie Groups*, Corrected reprint of the 1989 original, Cambridge University Press.

[195] Varshalovich, D.A., Moskalev, A.N., Khersonskii, V.K. (1988). *Quantum Theory of Angular Momentum*, World Scientific Press.

[196] Vielva, P., Martínez-González, E., Gallegos, J. E., Toffolatti, L., Sanz, J. L. (2003) Point source detection using the spherical Mexican hat wavelet on simulated all-sky *Planck* maps, *Monthly Notice of the Royal Astronomical Society*, Vol. 344, Issue 1, 89-104.

[197] Vilenkin, N.Ja. and Klimyk, A.U. (1991) *Representation of Lie Groups and Special Functions*, Kluwer.

[198] Yadav, A.P.S., Komatsu, E., Wandelt, B.D. (2007) Fast estimator of primordial non-Gaussianity from temperature and polarization anisotropies in the Cosmic Microwave Background, *Astrophysical Journal*, 664:680-686.

[199] Yadav, A.P.S. and Wandelt, B.D. (2008) Evidence of primordial non-Gaussianity (f_{NL}) in the Wilkinson Microwave Anisotropy Probe 3-Year Data at 2.8 sigma, *Physical Review Letters*, vol. 100, Issue 18, id. 181301.

[200] Yadav, A.P.S. and Wandelt, B.D. (2010) Primordial non-Gaussianity in the Cosmic Microwave Background, Advances in Astronomy, in press, arXiv: 1006.0275.

[201] Yadrenko, M.I. (1983) *Spectral Theory of Random Fields*, Translated from the Russian, Translation Series in Mathematics and Engineering, Optimization Software, Inc., Publications Division.

[202] Wiaux, Y., McEwen, J.D., Vielva, P., (2007) Complex data processing: fast wavelet analysis on the sphere, *Journal of Fourier Analysis and its Applications*, 13, 477-494.

[203] Wiaux, Y., Jacques, L., Vandergheynst, P. (2005) Correspondence principle between spherical and Euclidean wavelets, *The Astrophysical Journal*, Vol. 632, Issue 1, pp. 15-28.

[204] Wiaux, Y., Jacques, L., Vandergheynst, P. (2007) Fast spin +-2 spherical harmonics transforms and application in cosmology, *Journal of Computational Physics*, 226:2359-2371.

[205] Wiener, N. (1938), The homogeneous chaos, *American Journal of Mathematics*, 60, 879-936.

[206] Wigman, I. (2009) On the distribution of the nodal sets of random spherical harmonics, *Journal of Mathematical Physics*, 50, no. 1, 013521, 44 pp.

[207] Wigman, I. (2010) Fluctuations of the nodal length of random spherical harmonics, *Communications in Mathematical Physics,* , Vol. 298, n. 3, pp. 787-831, arXiv: 0907.1648.

[208] Zaldarriaga, M., Seljak, U. (2000) CMBFAST for spatially closed Universes, *Astrophysical Journal Supplements Series*, 129, 431-434.

Index

Printed in the United States
By Bookmasters